Symmetry in Many-Body Physics

Symmetry in Many-Body Physics

Editors

V.I. Yukalov
V. S. Bagnato
Rashid G. Nazmitdinov

MDPI • Basel • Beijing • Wuhan • Barcelona • Belgrade • Manchester • Tokyo • Cluj • Tianjin

Editors

V.I. Yukalov
Bogoliubov Laboratory of Theoretical Physics
Joint Institute for Nuclear Research
Dubna, Moscow region
Russia

V. S. Bagnato
Instituto de Fisica de São Carlos
Universidade de São Paulo
São Carlos
Brazil

Rashid G. Nazmitdinov
Bogoliubov Laboratory of Theoretical Physics
Joint Institute for Nuclear Research
Dubna, Moscow region
Russia

Editorial Office
MDPI
St. Alban-Anlage 66
4052 Basel, Switzerland

This is a reprint of articles from the Special Issue published online in the open access journal *Symmetry* (ISSN 2073-8994) (available at: www.mdpi.com/journal/symmetry/special_issues/Symmetry_Many-Body_Physics).

For citation purposes, cite each article independently as indicated on the article page online and as indicated below:

LastName, A.A.; LastName, B.B.; LastName, C.C. Article Title. *Journal Name* **Year**, *Volume Number*, Page Range.

ISBN 978-3-0365-6729-7 (Hbk)
ISBN 978-3-0365-6728-0 (PDF)

Contents

Preface to "Symmetry in Many-Body Physics"

Symmetry plays a crucial role in many-body physics. However, finding an exact solution to the problem of interacting particles very often presents a fundamental problem. Nevertheless, the knowledge of basic symmetries plays an important role in finding the approximate solutions and determines the accuracy of the used approximations. The purpose of this issue is to demonstrate the principal role of exact and approximate symmetries in solving various problems of many-particle physics, as well as in finding approximate solutions for the systems typical of condensed matter, trapped Fermi and Bose gases, nuclear matter, and field theory.

V.I. Yukalov, V. S. Bagnato, and Rashid G. Nazmitdinov
Editors

symmetry MDPI

Editorial

Symmetry in Many-Body Physics

Vanderlei S. Bagnato [1], **Rashid G. Nazmitdinov** [2] **and Vyacheslav I. Yukalov** [1,2,*]

1 Instituto de Fisica de São Carlos, Universidade de São Paulo, São Carlos 13560-970, Brazil
2 Bogolubov Laboratory of Theoretical Physics, Joint Institute for Nuclear Research, 141980 Dubna, Russia
* Correspondence: yukalov@theor.jinr.ru

The harmony of nature is expressed through the implementation of symmetry providing optimal structures for complex systems from snowflakes to graphene lattices. However, sometimes nature breaks symmetry, for example: in sugar molecules, the predominance of people with the heart on the left side, or in various phase transitions. Symmetry plays a crucial role in many-body physics. For instance, chiral symmetry is important in the unusual properties of graphene and the theory of strong interactions. Symmetry breaking and restoration constantly occurs in the world around us. Usually, finding exact solution to the problem of interacting particles presents a fundamental challenge. Therefore, we have to restrict ourselves to approximate solutions that reflect the essential features of the entire problem as a whole and contain an indication of the range of applicability of these solutions. An important role in finding the approximate solutions is played by the knowledge of basic symmetries that determine the accuracy of the used approximations. The purpose of this issue is to demonstrate the principal role of exact and approximate symmetries in solving various problems in many-particle physics, as well as finding approximate solutions for the systems typical of condensed matter, trapped Fermi and Bose gases, nuclear matter, and field theory [1–14].

The review article "Non-Thermal Fixed Points in Bose Gas Experiments", by L. Madeira and V.S. Bagnato, considers one of the most challenging tasks in physics for understanding the route of an out-of-equilibrium system to its thermalized state. This problem can be particularly overwhelming when one considers a many-body quantum system. Several recent theoretical and experimental studies have indicated that some far-from-equilibrium systems display universal dynamics close to the so-called non-thermal fixed point (NTFP), following a rescaling of both space and time. This opens up the possibility of a general framework for studying and categorizing out-of-equilibrium phenomena into well-defined universality classes. This paper reviews the recent advances in observing NTFPs in experiments involving Bose gases. A brief introduction is provided to the theory behind this universal scaling, focusing on the experimental observations of NTFPs. The benefits of the NTFP universality classes are presented, using the analogy with renormalization group theory in equilibrium critical phenomena.

The review article "A Review of Many-Body Interactions in Linear and Nonlinear Plasmonic Nanohybrids", by M.R. Singh, discusses the many-body interactions in plasmonic nanohybrids made from an ensemble of quantum emitters and metallic nanoparticles. A theory of the linear and nonlinear optical emission intensity is developed by using the many-body quantum-mechanical density matrix method. The ensemble of quantum emitters and metallic nanoparticles interact with each other via the dipole–dipole interaction. Surface plasmon polaritons are located near to the surface of the metallic nanoparticles. It is shown that the nonlinear Kerr intensity enhances due to the weak dipole–dipole coupling limits. On the other hand, in the strong dipole–dipole coupling limit, the single peak in the Kerr intensity splits into two peaks. The splitting of the Kerr spectrum is due to the creation of dressed states in the plasmonic nanohybrids within the strong dipole–dipole interaction. Further, it is found that the Kerr nonlinearity is enhanced due to the interaction between

Citation: Bagnato, V.S.; Nazmitdinov, R.G.; Yukalov, V.I. Symmetry in Many-Body Physics. *Symmetry* **2023**, *15*, 72. https://doi.org/10.3390/sym15010072

Received: 20 December 2022
Accepted: 22 December 2022
Published: 27 December 2022

the surface plasmon polaritons and excitons of the quantum emitters. Next, the effect of the spontaneous decay-rate enhancement is predicted due to dipole–dipole coupling. The enhancement of the Kerr intensity due to the surface plasmon polaritons can be used to fabricate nanosensors. The splitting of one peak into two peaks can be used for the fabrication of nanoswitches for nanotechnology and nanomedical applications.

In the paper "Spin Interference Effects in a Ring with Rashba Spin-Orbit Interaction Subject to Strong Light-Matter Coupling in Magnetic Field", by M. Pudlak and R. Nazmitdinov, the authors studied electron transport through a one-dimensional quantum ring, subject to Rashba spin–orbit interaction and connected with two external leads in the presence of external fields. They include optical radiation, produced by an off-resonant high-frequency electric field, and a perpendicular magnetic field. By means of the Floquet theory of periodically driven quantum systems the interference effects under these fields are described in detail. Specific analytic conditions are found for reaching the spin-filtering effect, caused by the interplay of the external fields and Rashba spin–orbit interactions.

The paper "Long-Time Bit Storage and Retrieval without Cold Atom Technology", by R. Friedberg and J.T. Manassah, reports computer investigations showing how the duration of memory for storage and retrieval of a classical bit can be increased to 100 times the decay time of an isolated atom, with no use of high-tech cold-atom preparations recently developed in the light-matter field. This low-tech procedure can greatly enlarge the number of experimenters able to enter this field. The role of symmetry in this procedure arises in a careful interplay of incoherent and coherent excitations of a large collection of two-level atoms, the level separation being matched by the dominant frequency of the electromagnetic fields (short pulses and continuing field) applied to the system.

In the paper "Formation of Matter-Wave Droplet Lattices in Multi-Color Periodic Confinements", by M.R. Pathak and A. Nath, a new model is introduced that addresses the generation of quantum droplets (QDs) in binary Bose–Einstein condensate (BEC) mixtures with mutually symmetric spinor components, loaded in multi-color optical lattices (MOLs) of commensurate wavelengths and tunable intensities. The considered MOL confinement is the combination of the four-color optical lattice with an exponential periodic trap, which includes the complete set of the Fourier harmonics. Employing the one-dimensional (1D) extended nonlinear Schrodinger equation (eGPE), the exact analytical form of the wavefunction, mean field/beyond mean field nonlinearities, and MOL trap parameters are described. Utilizing the exact solutions, the formation of supersolid-like spatially periodic matter-wave droplet lattices and superlattices are illustrated under the space-periodic nonlinearity management. The precise positioning of the density maxima/minima of the droplet patterns at the center of the trap and tunable Anderson-like localization are observed by tuning the symmetry and amplitude of the considered MOL trap. The stability of the obtained solution is confirmed using the Vakhitov–Kolokolov criterion.

The paper "Josephson-like Oscillations in Toroidal Spinor Bose-Einstein Condensates: A Prospective Symmetry Probe", by M.H. Figlioli Donato and S.R. Muniz, studies Josephson junctions that are essential ingredients in the superconducting circuits used in many existing quantum technologies. Additionally, ultra-cold atomic quantum gases have also become essential platforms to study superfluidity. The analogy between superconductivity and superfluidity is discussed concentrating on an intriguing effect caused by a thin finite barrier in a quasi-one-dimensional toroidal spinor Bose–Einstein condensate (BEC). In this system, the atomic current density flowing through the edges of the barrier oscillates similar to the electrical current through a Josephson junction in a superconductor, but in that case there is no current circulation through the barrier. It is also shown how the nontrivial broken-symmetry states of spinor BECs change the structure of this Josephson-like current, creating the possibility to probe the spinor symmetry, solely using measurements of this superfluid current.

In the paper "Spontaneous Symmetry Breaking: The Case of Crazy Clock and Beyond", by M.C. Pagnacco, J.P. Maksimovic′, M. Dakovic′, B. Bokic, S.R. Mouchet, T. Verbiest, Y. Caudano and Branko Kolaric, the authors describe the crazy-clock phenomenon involving the state I (low iodide and iodine concentration) to state II (high iodide and iodine concentration with new iodine phase) transition after a Briggs–Rauscher (BR) oscillatory process. While the BR crazy-clock phenomenon is known, this is the first time that crazy-clock behavior has been linked and explained with the symmetry-breaking phenomenon, highlighting the entire process in a novel way. The presented phenomenon has been thoroughly investigated by running more than 60 experiments, and evaluated by using statistical cluster K-means analysis. The mixing rate, as well as the magnetic bar shape and dimensions, have a strong influence on transition appearance. Although the transition for both mixing and non-mixing conditions take place completely randomly, by using statistical cluster analysis the authors obtain a different number of clusters (showing the time-domains where the transition is most likely to occur). In the case of stirring, clusters are more compact and separated, revealing new hidden details regarding the chemical dynamics of nonlinear processes. The significance of the presented results is beyond oscillatory reaction kinetics since the described example belongs to the small class of chemical systems that show intrinsic randomness in their response and it may be considered a real example of a classical liquid random number generation.

Interesting results are reported in the paper "Exact Solutions for Solitary Waves in a Bose-Einstein Condensate under the Action of a Four-Color Optical Lattice", by B. Halder, S. Ghosh, P. Basu, J. Bera, B. Malomed and U. Roy who address the dynamics of Bose–Einstein condensates (BECs) loaded into a one-dimensional four-color optical lattice (FOL) potential with commensurate wavelengths and tunable intensities. This configuration lends system-specific symmetry properties. The analysis identifies specific multi-parameter forms of the FOL potential which admits exact solitary-wave solutions. This newly found class of potentials includes more particular species, such as frustrated double-well superlattices and bichromatic and three-color lattices, which are subject to respective symmetry constraints. The exact solutions provide options for the controllable positioning of density maxima of the localized patterns, and tunable Anderson-like localization in the frustrated potential. A numerical analysis is performed to establish dynamic and structural stability of the obtained solutions, making them relevant for experimental realization. The newly found solutions offer applications to the design of schemes for quantum simulations and processing quantum information.

In the paper "Zeroth-Order Nucleation Transition under Nanoscale Phase Separation", by V.I. Yukalov and E.P. Yukalova, materials with nanoscale phase separation are considered. A system representing a heterophase mixture of ferromagnetic and paramagnetic phases was studied. After averaging the over phase configurations, a renormalized Hamiltonian is derived describing the coexisting phases. The system is characterized by direct and exchange interactions and an external magnetic field. The properties of the system are studied numerically. The stability conditions define the stable state of the system. At a temperature of zero, the system is in a pure ferromagnetic state. However, at a finite temperature, for some interaction parameters, the system exhibits a zeroth-order nucleation transition between the pure ferromagnetic phase and the mixed state with coexisting ferromagnetic and paramagnetic phases. At the nucleation transition, the finite concentration of the paramagnetic phase appears via a jump.

The paper "Small-Angle Scattering from Fractional Brownian Surfaces", by E.M. Anitas, investigates materials with fractal-like geometry. Recent developments in nanotechnology have allowed the fabrication of a new generation of advanced materials with various fractal-like geometries. Fractional Brownian surfaces (fBs) are often used as models to simulate and characterize these complex geometries, such as the surfaces of particles in dilute particulate systems (e.g., colloids) or the interfaces in non-particulate two-phase systems (e.g., semicrystalline polymers with crystalline and amorphous phases). However, for such systems, a realistic simulation involves parameters averaged over a macroscopic volume. The author proposes a method, based on small-angle scattering technique, to extract the main structural parameters of surfaces/interfaces from experimental data. The method involves the analysis of scattering intensities and the corresponding pair distance distribution functions. This allows the extraction of information with respect to the overall size, fractal dimension, Hurst and spectral exponents. The method is applied to several classes of fBs, and it is demonstrated that the obtained numerical values of the structural parameters are in good agreement with the theoretical ones.

The paper "Characteristic Length Scale during the Time Evolution of a Turbulent Bose-Einstein Condensate", by L. Madeira, A.D. Garci'a-Orozco, M.A. Moreno-Armijos, F.E. Alves dos Santos and V.S. Bagnato, addresses the topic of quantum turbulence that is currently highly studied. Quantum turbulence is characterized by many degrees of freedom interacting non-linearly to produce disordered states, both in space and time. In this work, the decaying regime of quantum turbulence in a trapped Bose–Einstein condensate was investigated. An alternative way of exploring this phenomenon is presented, by defining and computing a characteristic length scale, which possesses the relevant characteristics to study the establishment of the quantum turbulent regime. To reconstruct the three-dimensional momentum distributions, the inverse Abel transform was employed, as has been successfully done in other works. The analysis is presented for both two- and three-dimensional momentum distributions, discussing their similarities and differences. It is argued that the characteristic length allows us to intuitively visualize the time evolution of the turbulent state.

In the paper "Probing Many-Body Systems near Spectral Degeneracies", K. Ziegler discusses how the employment of the diagonal elements of the time correlation matrix can be used to probe closed quantum systems that are measured at random times. This enables one to extract two distinct parts of the quantum evolution, a recurrent part and an exponentially decaying part. This separation is strongly affected when spectral degeneracies occur, for instance, in the presence of spontaneous symmetry breaking. Moreover, the slowest decay rate is determined by the smallest energy level spacing, and this decay rate diverges at the spectral degeneracies. Probing of the quantum evolution with the diagonal elements of the time correlation matrix is discussed as a general concept and tested in the case of a bosonic Josephson junction. This reveals for the latter, characteristic properties at the transition to Hilbert-space localization.

The paper "Morphology of an Interacting Three-Dimensional Trapped Bose-Einstein Condensate from Many-Particle Variance Anisotropy", by O.E. Alon, analyzes the characteristic properties of a Bose–Einstein condensate (BEC) through the variances of observable quantities. The variance of the position operator is associated with how wide or narrow a wave-packet is, the momentum variance is similarly correlated with the size of a wave-packet in momentum space, and the angular-momentum variance quantifies to what extent a wave-packet is non-spherically symmetrical. The interacting three-dimensional trapped BEC in the limit of an infinite number of particles was investigated, and its position, momentum, and angular-momentum anisotropies are described. Computing the variances of the three Cartesian components of the position, momentum, and angular-momentum operators, it is shown that there exist scenarios where the anisotropy of a BEC is different at the many-body and mean-field levels of theory, despite having the same many-body and mean-field densities per particle. This suggests a way to classify correlations via the morphology of 100%-condensed bosons in a three-dimensional trap in the limit of an infinite number of particles. Implications are briefly discussed.

The paper "Acoustic Plasmons in Graphene Sandwiched between Two Metallic Slabs", by L. Salasnich, studied the effect of two metallic slabs on the collective dynamics of electrons in graphene positioned between the two slabs. It is shown that, if the slabs are perfect conductors, the plasmons of graphene display a linear dispersion relationship. The velocity of these acoustic plasmons crucially depends on the distance between the two metal gates and the graphene sheet. In the case of generic slabs, the dispersion relationship of the graphene plasmons is much more complicated, but it was found that acoustic plasmons can still be obtained under specific conditions.

We hope that the papers of this Issue hold useful information and new ideas for the readers.

Author Contributions: All authors contributed to this editorial equally. All authors have read and agreed to the published version of the manuscript.

Funding: This research received no external funding.

Data Availability Statement: This study only used theoretical results and their applications are published in the cited references.

Conflicts of Interest: The authors declare no conflict of interest.

References

1. Madeira, L.; Bagnato, V. Non-Thermal Fixed Points in Bose Gas Experiments. *Symmetry* **2022**, *14*, 678. [CrossRef]
2. Singh, M.R. A Review of Many-Body Interactions in Linear and Nonlinear Plasmonic Nanohybrids. *Symmetry* **2021**, *13*, 445. [CrossRef]
3. Pudlak, M.; Nazmitdinov, R. Spin Interference Effects in a Ring with Rashba Spin-Orbit Interaction Subject to Strong Light–Matter Coupling in Magnetic Field. *Symmetry* **2022**, *14*, 1194. [CrossRef]
4. Friedberg, R.; Manassah, J.T. Long-Time Bit Storage and Retrieval without Cold Atom Technology. *Symmetry* **2022**, *14*, 1505. [CrossRef]
5. Pathak, M.R.; Nath, A. Formation of Matter-Wave Droplet Lattices in Multi-Color Periodic Confinements. *Symmetry* **2022**, *14*, 963. [CrossRef]
6. Donato, M.H.F.; Muniz, S.R. Muniz. Josephson-like Oscillations in Toroidal Spinor Bose–Einstein Condensates: A Prospective Symmetry Probe. *Symmetry* **2022**, *14*, 867. [CrossRef]
7. Pagnacco, M.C.; Maksimović, J.P.; Daković, M.; Bokic, B.; Mouchet, S.R.; Verbiest, T.; Caudano, Y.; Kolaric, B. Spontaneous Symmetry Breaking: The Case of Crazy Clock and Beyond. *Symmetry* **2022**, *14*, 413. [CrossRef]
8. Halder, B.; Ghosh, S.; Basu, P.; Bera, J.; Malomed, B.; Roy, U. Exact Solutions for Solitary Waves in a Bose-Einstein Condensate under the Action of a Four-Color Optical Lattice. *Symmetry* **2022**, *14*, 49. [CrossRef]
9. Yukalov, V.I.; Yukalova, E.P. Zeroth-Order Nucleation Transition under Nanoscale Phase Separation. *Symmetry* **2021**, *13*, 2379. [CrossRef]
10. Anitas, E.M. Small-Angle Scattering from Fractional Brownian Surfaces. *Symmetry* **2021**, *13*, 2042. [CrossRef]
11. Madeira, L.; García-Orozco, A.D.; Moreno-Armijos, M.A.; dos Santos, F.E.A.; Bagnato, V.S. Characteristic Length Scale during the Time Evolution of a Turbulent Bose-Einstein Condensate. *Symmetry* **2021**, *13*, 1865. [CrossRef]
12. Ziegler, K. Probing Many-Body Systems near Spectral Degeneracies. *Symmetry* **2021**, *13*, 1796. [CrossRef]
13. Alon, O.E. Morphology of an Interacting Three-Dimensional Trapped Bose-Einstein Condensate from Many-Particle Variance Anisotropy. *Symmetry* **2021**, *13*, 1237. [CrossRef]
14. Salasnich, L. Acoustic Plasmons in Graphene Sandwiched between Two Metallic Slabs. *Symmetry* **2021**, *13*, 684. [CrossRef]

symmetry

MDPI

Article

Long-Time Bit Storage and Retrieval without Cold Atom Technology

Richard Friedberg [1,*] and Jamal T. Manassah [2,*]

1 Department of Physics and Astronomy, Barnard College, Columbia University, New York, NY 10027, USA
2 Department of Electrical Engineering, City College of New York, New York, NY 10031, USA
* Correspondence: rfriedberg49@gmail.com (R.F.); jmanassah@ccny.cuny.edu (J.T.M.)

Abstract: We report computer studies showing how the duration of memory for storage and retrieval of a classical bit can be increased to 100 times the decay time of an isolated atom, with no use of high-tech cold-atom preparations recently developed in the light-matter field. We suggest that our low-tech procedure can greatly enlarge the number of experimenters able to enter this field. The role of symmetry in this procedure arises in a careful interplay of incoherent and coherent excitations of a large collection of "two-level" atoms, the level separation being matched by the dominant frequency of the electromagnetic fields (short pulses and continuing field) applied to the system.

Keywords: bit retrieval; long-time memory; Nyxion; continuous pump; coalescence; Aurora

Citation: Friedberg, R.; Manassah, J.T. Long-Time Bit Storage and Retrieval without Cold Atom Technology. *Symmetry* **2022**, *14*, 1505. https://doi.org/10.3390/sym14081505

Academic Editors: V.I. Yukalov, V. S. Bagnato and Rashid G. Nazmitdinov

Received: 27 May 2022
Accepted: 20 July 2022
Published: 22 July 2022

1. Introduction

The subradiant regime of an ensemble of two-level atoms has been studied in [1–5], and the use of this system as an optical memory device has been explored inter alia in [6–10]. We study in this paper the state of this ensemble after excitation by a short pulse up to the edge of the subradiant regime; we call this dark state a "Nyxion" (Nyx was the goddess of the night in Greek mythology).

We emphasize that we are not trying to compete with the startling achievements in this field made possible by the controlled positioning of individual atoms in a cold-atom trap. In particular, our system does not aim at storage and retrieval of *quantum* information. We deal only with *classical* bit storage and retrieval. The value of our calculations depends on their applicability to a system under quite ordinary conditions. We hope that the results will stimulate significant work by experimenters who do not have access to the high-tech methods of cold-atom trapping and manipulation.

In a system under ordinary conditions, without cold-atom technology, a single Nyxion is stable only for a duration comparable to the longitudinal decay time Γ_1^{-1} of an isolated atom. If a second Nyxion is produced by a new pulse within this duration, it can coalesce with the first to produce a flash of light; without such coalescence the system remains dark. This system can be used as an optical memory device in which the first Nyxion stores a (classical) bit and the second Nyxion reads it. However, the device as described above yields a memory lasting only for a time $O(\Gamma_1^{-1})$.

We find, however, that the lifetime of the memory can be increased 100-fold by supplementing the Nyxion producing pulses with a pump of well-chosen constant strength (a "dc pump").

The role of symmetry in this procedure arises in a careful interplay of incoherent and coherent excitations of a large collection of "two-level" atoms, the level separation being matched by the dominant frequency of the electromagnetic fields (short pulses and continuing field) applied to the system.

This is a theoretical paper. Our findings are based on quasinumerical simulations. Our calculations use the complete basis formed by the eigenfunctions of the Liénard–Wiechert one-dimensional Green function, details of which can be found in [11–13]. These eigenfunctions are found analytically; the sums over many eigenstates constitute the numerical calculation. The numerical part is by far the more computer intensive, as it must be done independently for each value of the time T during a simulation.

2. Theoretical Model (Methods)

Consider a sample of two-level atoms arranged in the so-called slab geometry, interacting with a classical electromagnetic field (see [11]). The atoms are taken as uniformly distributed in the "long" dimension (thickness of the slab) from $Z = -1$ to $Z = 1$. The evolution in time T is given [11] by

$$\frac{\partial\chi(Z,T)}{\partial T} = -[\Gamma_2 - i\Omega_L n(Z,T)]\chi(Z,T) + \frac{i}{2}n(Z,T)\psi(Z,T) \tag{1}$$

$$\begin{aligned}\frac{\partial n(Z,T)}{\partial T} &= -i[\chi^*(Z,T)\psi(Z,T) - \chi(Z,T)\psi^*(Z,T)] + \Gamma_1(1-n(Z,T)) \\ &- \frac{R(T)}{2}(1+n(Z,T)),\end{aligned} \tag{2}$$

$$\psi(Z,T) = iu_0\int_{-1}^{1} dZ'\chi(Z',T)\exp(iu_0|Z-Z'|), \tag{3}$$

where χ (complex) and n (real) are, respectively, the active medium polarization and the degree of excitation of the matter ($n = 1$ if the atoms are all in the ground state and $n = -1$ if all in the excited state), and ψ represents the normalized complex electric field.

2.1. Notation

All symbols in Equations (1)–(3) and their origin are detailed in [11]. However, in order to make this exposition as self-contained as possible, we summarize as follows the relations between the "normalized" (dimensionless) variables appearing in (1)–(3) and the physical quantities they represent:

The physical thickness of the slab is $L = 2z_0$, and the position variable z has the range $-z_0 \le z \le z_0$. The dimensionless variable corresponding to z is $Z = z/z_0$, which has the range $-Z_0 \le Z \le Z_0$. (Similarly, z', Z'.)

The physical resonant wavelength is λ_0, and the corresponding wavenumber is $k_0 = 2\pi c/\lambda_0$ where c is the speed of light in vacuo. The dimensionless partner of k_0 is $u_0 = k_0 z_0$, so that $\exp(iu_0 Z) = \exp(ik_0 z)$.

Physical time variables t, t_0, etc., are converted to dimensionless times T, T_0, etc., by multiplication with the inverse time $C = 4\pi\mathcal{P}^2\rho/\hbar$ where $\mathcal{P}$ is the reduced density matrix of the transition and ρ is the number of atoms per volume, assumed to be constant throughout the slab.

Physical angular frequencies ω, ω_0 have dimensionless partners $\Omega = \omega/C$, $\Omega_0 = \omega_0/C$, so that $\exp(-i\Omega T) = \exp(-i\omega t)$, etc. Likewise, the longitudinal and transverse isolated-atom decay rates γ_1, γ_2 have dimensionless partners $\Gamma_1 = \gamma_1/C$, $\Gamma_2 = \gamma_2/C$.

The Lorentz shift in our units is $\omega_L = (4\pi/3)\mathcal{P}^2\rho/\hbar = C/3$ so that its dimensionless partner is $\Omega_L = \omega_L/C = 1/3$.

2.2. Skeleton Formulas

A "skeleton" version of (1) and (2) is

$$\begin{aligned}\frac{\partial\chi(Z,T)}{\partial T} &= +\frac{i}{2}n(Z,T)\psi(Z,T) \\ \frac{\partial n(Z,T)}{\partial T} &= -i[\chi^*(Z,T)\psi(Z,T) - \chi(Z,T)\psi^*(Z,T)]\end{aligned} \tag{4}$$

which resembles the familiar magnetic precession equation $\dot{\vec{B}} = \vec{H} \times \vec{B}$ if we regard $\Re\chi, \Im\chi, n/2$ as the x, y, z components of the Bloch vector [14] and $\Re\psi, \Im\psi$ as the x, y components of a transverse applied magnetic field.

The skeletal version of (3) would be the same as the original:

$$\psi(Z,T) = iu_0 \int_{-1}^{1} dZ' \chi(Z',T) \exp(iu_0|Z-Z'|). \tag{5}$$

in which $\psi(Z,T)$ is displayed as an *instantaneous* function of χ at other positions, neglecting retardation so that we write $\chi(Z',T)$ instead of $\chi(Z',T')$. This can be justified only if the propagation time L/c is too short to allow any significant evolution to take place in accordance with (1) and (2) while the light signal is traveling through the thickness of the slab. All our calculations throughout the paper are consequently contingent on this assumption.

It should be understood that the quantities χ, ψ in (1)–(3) have the rapid temporal oscillation $\exp(-i\Omega_0 T) = \exp(-i\omega_0 t)$ factored out. (This is equivalent to writing Bloch's Equation [14] in the rotating coordinate system.) On the other hand, the rapid spatial oscillation is retained and appears in (3) and (5) through the prefactor $u_0 = k_0 z_0$ and the exponential factor $\exp(iu_0|Z-Z'|) = \exp(ik_0|z-z'|)$.

2.3. Limits on n

The reader may wonder how it is certain, from (1) and (2), that n always remains within the interval from -1 to 1. This can be reasoned out as follows.

Resolve $\chi = \chi_1 + i\chi_2$ and $\psi = \psi_1 + i\psi_2$. Define the "Bloch vector" $\vec{B}$ to be $(\chi_1, \chi_2, n/2)$. Consider the skeletal forms (4). We wish to interpret them as $\dot{\vec{B}} = \vec{H} \times \vec{B}$. Multiplying out the expression for dn/dT, we find

$$(d/dT)\frac{n}{2} = \chi_1\psi_2 - \chi_2\psi_1 \tag{6}$$

which is exactly in the desired form, with $\vec{H} = (\psi_1, \psi_2)$ (the fact that $\vec{H}$ changes with time makes no difference to the argument). Turning to the skeletal expression for $d\chi/dT$, we have

$$\begin{aligned} (d/dT)\chi_1 &= -\frac{n}{2}\psi_2 \\ (d/dT)\chi_2 &= +\frac{n}{2}\psi_1 \end{aligned} \tag{7}$$

which is good as far as it goes, but the full analogue to Bloch's equation would be

$$\begin{aligned} (d/dT)\chi_1 &= -\frac{n}{2}\psi_2 + \chi_2\psi_3 \\ (d/dT)\chi_2 &= +\frac{n}{2}\psi_1 - \chi_1\psi_3 \end{aligned} \tag{8}$$

and so far, we have introduced no quantity ψ_3. What should we do?

The missing component ψ_3 is in fact the large frequency Ω that has not appeared because $\exp(-i\Omega T)$ has been factored out of χ, and hence out of ψ_1, ψ_2, as remarked above. If we restore it, we obtain exactly the desired form (8). This shows that $\dot{\vec{B}} = \vec{H} \times \vec{B}$ is instantaneously satisfied and the skeletal forms (4) will not cause (χ, n) to grow in magnitude. Since n starts out between -1 and 1, it will remain so. This conclusion about n remains true if the fast precession of χ and of ψ_1, ψ_2 is factored out as in (7).

The third component of H in NMR is usually a very strong unchanging magnetic field which creates an energy difference between the "up" and "down" states of the dipole. The corresponding quantity $\psi_3 = \Omega$ is likewise (apart from the factor $\hbar$) the energy difference between the two resonant levels. (Physically, it is the incorporation of the frequency Ω into all of the exciting pulses that selects the resonant levels 1 and 2 and renders all other atomic levels insignificant [15,16] except for the rôle of level 3 in realizing the incoherent pump, see Section 2.4 below).

Now, we must consider the effect of the "nonskeletal" terms. In (1), we have the terms in Γ_2 and Ω_L. The first of these simply causes the real decay of χ_1 and χ_2 independently. The second is just a correction to the frequency Ω that was factored out. Being imaginary, it does not change the magnitude of χ.

Finally, we look at the terms in Γ_1 and R appearing in (2). The term in Γ_1 drives n toward the ground state $n = 1$. However, the ground state is never quite reached under the influence of this term, because Γ_1 is multiplied by the factor $1 - n$, which vanishes at that point. Hence, this term can cause only exponential decay *toward* $n = 1$, not beyond it.

Likewise, we study the term in R. R is never negative because it always consists of a sum of positive quantities α (see below). Hence, it can only drive n toward the totally inverted state $n = -1$. However, this state is not reached because R is multiplied by $1 + n$. Hence, at most, n can only approach the inverted state exponentially. (Actually, such an approach would trigger superradiance, which also would drive n away from total inversion; however, the foregoing argument shows independently that the inverted limit cannot be reached).

2.4. Pump

The pump (by which we mean the two pulses as well as the continuous (dc) pump) is described by the last term in (2), in which R is a real function of T to be chosen at the start of each simulation. This term creates only an incoherent excitation of the two-level system, since n carries no phase. The incoherent pump may be achieved by adding a third level, higher than the first two, which is driven by a resonant interaction from the first (ground) level, but which decays with an extremely fast time constant to the second (middle) level, much faster than any other time parameter in the problem. This rapid decay destroys the initial 1–3 coherence, leaving the system in an incoherent mixture of levels 1 and 2. The mixture is parametrized by the population difference n ($n = 1$ for all atoms in level 1, $n = -1$ for all in level 2).

If the incoherence is perfect, the system is left in a state with $\chi = 0$, from which (1)–(3) provide no escape. This standstill, however, is unstable, and it requires only a minuscule extra pulse at the 1–2 resonance frequency to activate a positive-feedback loop between the polarization and the internal field, which stabilizes at a magnitude determined by n. We do not give the stabilized value of χ analytically, as its calculation requires the nonlinear methods of [11]. Thus, the only effect of level 3 is to produce the incoherent pumping effect modulated by $R(T)$; the Equations (1)–(3) are written only in terms of levels 1 and 2 (however, see the issues raised in our Conclusion).

In Equation (1), the transverse decay rate Γ_2 (called Γ_T in [11]) has been calculated [12,13,17] to be $2.33/4 + \Gamma_1/2$, where the first (large) term is the normalized resonance width in a gas; this is its value for the $(J = 1) \to (J = 0)$ transition in Helium-like atoms, in the statistical approximation [13]. This value must be adjusted by the individual experimenter depending on the multiplicities in the atomic transition under study. For the much smaller longitudinal decay rate Γ_1 appearing in Γ_2 and in Equation (2), we take a convenient multiple .002 of Γ_2 that lies in the range often studied by experimenters on gas lasers at ordinary conditions. Thus, $\Gamma_2 = 1.001 \times 2.33/4$. In this paper, we fix the physical slab thickness to be $L = 2.25\lambda_0$ where λ_0 is the resonance wavelength between levels 1 and 2.

The function $R(T)$ is parametrized by six quantities $\alpha_0, \alpha_1, \alpha_2, \beta, R_0$ and δ. The α's are each either 1 or 0 to indicate whether the corresponding input is to be applied or not. The parameter β determines the delay between the time $T_0 = 5$ when the dc pump, if $\alpha_0 = 1$, and the first Nyxion, if $\alpha_1 = 1$, are turned on, and the time βT_0 when the second Nyxion, if $\alpha_2 = 1$, is turned on. (This means that the actual delay is only $(\beta - 1)T_0$, but in this paper, we shall always have $\beta >> 1$.) The reason for making $T_0 > 0$ is only to show the rapid excitation from the ground state in our graphs, for $\alpha_1 = 1$. The parameter δ controls the strength of the dc pump, when $\alpha_0 = 1$. (The purpose of the tanh function multiplying δ is explained in Section 3 below).

We take for the complete form of the R-function

$$R(T) = R_0[\alpha_1 \operatorname{sech}^2((T - T_0)/T_s) + \alpha_2 \operatorname{sech}^2((T - \beta T_0)/T_s) + \alpha_0 \delta \tanh(2T/T_0)]. \quad (9)$$

The values $T_0 = 5$ and $T_s = 2$ are fixed throughout the whole paper. Besides these, there are four parameters that remain fixed during a simulation as T increases from T_0 to its value at the end: the three α's and the overall coefficient R_0 controlling the pump strength $R(T)$. To this we may add a fifth constant β, which controls the delay of the second Nyxion pulse if $\alpha_2 = 1$, and a sixth, δ, which gives the *relative* strength of the dc pump if $\alpha_0 = 1$.

For our slab thickness $2.25\lambda_0$, we find a pump strength $R_0 = 3.6\Gamma_2$ to be suitable. It is small enough so that the system remains dark after a single Nyxion generating pulse ($\alpha_1 = 1$, $\alpha_2 = \alpha_0 = 0$) but large enough so that the noise coming from machine rounding for a 32-bit word can be disregarded.

For our main results, we set the relative pump strength at $\delta = 0.001686$. This choice will be explained in Section 3.

3. Physics (Preliminary Results)

Suppose ($\alpha_1 = 1$, $\alpha_2 = \alpha_0 = 0$) that a single pulse is applied at $T = T_0$ with no continuous (dc) pump. With $R_0 = 3.6$, the system is rapidly excited (Figure 1a) from the ground state $n = 1$ to the superradiant threshold $n = -0.2$. (The reason that the threshold is not at $n = 0$ is that when n becomes negative the transverse decay rate Γ_2 must be overcome to keep the system on the Bloch sphere; the laser amplification for a thin slab such as we are studying is not strong enough to do this easily.) Afterward, the system relaxes to ground (Figure 1b) at the rate Γ_1. (Figure 1c shows the tiny emission coming from the dark state shortly after reaching threshold).

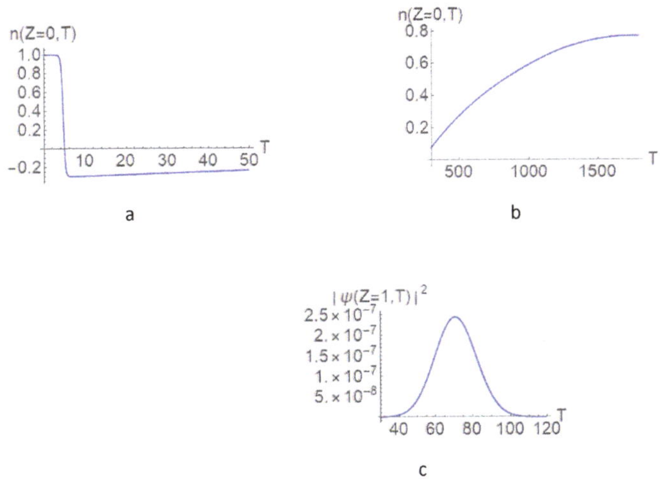

Figure 1. $\alpha_1 = 1, \alpha_2 = \alpha_0 = 0$. Creation (**a**) and later decay (**b**) at rate Γ_1 of single Nyxion with no continuous (dc) pump; early subradiant emission (**c**) from "dark" state.

By adding a second Nyxion pulse, also of subradiant strength in itself (Figure 2a,b) within this time ($\alpha_1 = \alpha_2 = 1$, $\alpha_0 = 0, 5\beta < 1000$), we can cause the second pulse to "read" the bit that has been "stored" by the first pulse. The reading produces a superradiant flash shown in Figure 2c.

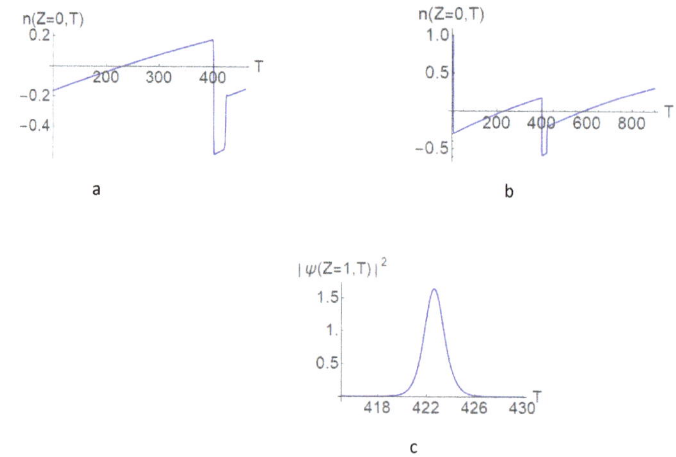

Figure 2. $\alpha_1 = \alpha_2 = 1, \alpha_0 = 0$ and $5\beta = 400$. A bit "stored" at $T = 5$ (creation not shown here, similar to Figure 1a) is "read" by a second pulse at $T = 400$ (**a**,**b**), and a superradiant flash (**c**) is produced.

However, suppose that instead of the first pulse, we apply the dc pump at $T = T_0$ ($\alpha_0 = 1$, $\alpha_2 = \alpha_1 = 0$). The dc pump is much weaker ($\delta << 1$) than the pulse would have been, but it continues to operate. (The tanh function multiplying δ in (9) serves only the purpose of softening the discontinuity which might generate irregularities in the calculations. For practical purposes, we may think of it as a step function.) Under this continuous influence, n decreases exponentially (Figure 3a) at the rate Γ_1 toward the threshold level -0.2, at which the dc pump is just strong enough to prevent the decay toward ground. By maintaining the dc pump, one can prolong the excited state indefinitely in the ideal slab environment.

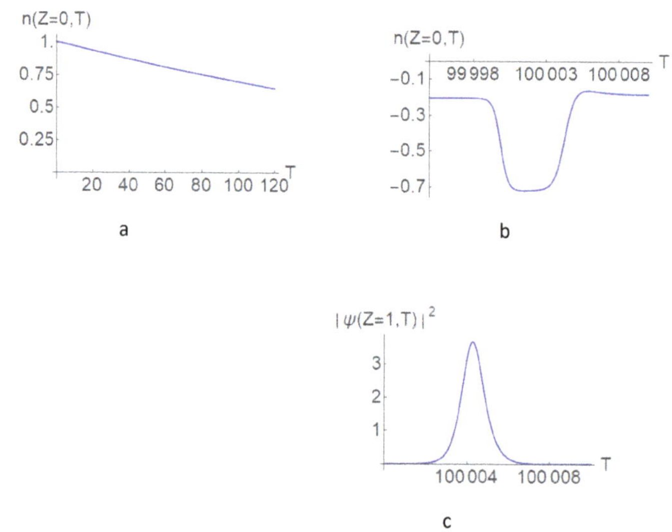

Figure 3. $\alpha_1 = 0, \alpha_2 = \alpha_0 = 1$, 5β = 100,000 and $\delta = 0.001687$. The dc pump is turned on at $T = 5$, without an accompanying pulse, but with the relative continuous (dc) pump strength slightly greater than δ_{crit}. Excitation grows (**a**) with time constant Γ_1 toward steady-state value $n = -0.2$. At T =100,000 = $100\Gamma_1^{-1}$, a "reading pulse" is applied. An "Aurora" results, shown as a temporary superexcitation of n (**b**) and an immediately following superradiant emission (**c**) in the forward direction.

The parameter δ has a peculiar influence on what happens if, after using the dc pump alone as in the preceding paragraph to reach the state with $n = -0.2$, we apply a "second" pulse at 5β, which now may be be taken as $O(100\Gamma_1^{-1})$ ($\alpha_0 = \alpha_2 = 1, \alpha_1 = 0, 5\beta = 100{,}000$). If δ is sufficiently high, the second pulse may produce a flash (the "Aurora", Figure 3b,c) although there was no input pulse at $T = T_0$.

We have found by changing δ in steps of 10^{-7} that the Aurora never takes place when $\delta < \delta_{crit} = 0.0016866$. Therefore, we have set $\delta = 0.0016860$ in our main calculations, so as not to provoke an Aurora when $\alpha_1 = 0$. (See Figure 4: $\alpha_0 = \alpha_2 = 1, \alpha_1 = 0$, $5\beta = 100{,}000$ and $\delta = 0.0016860$).

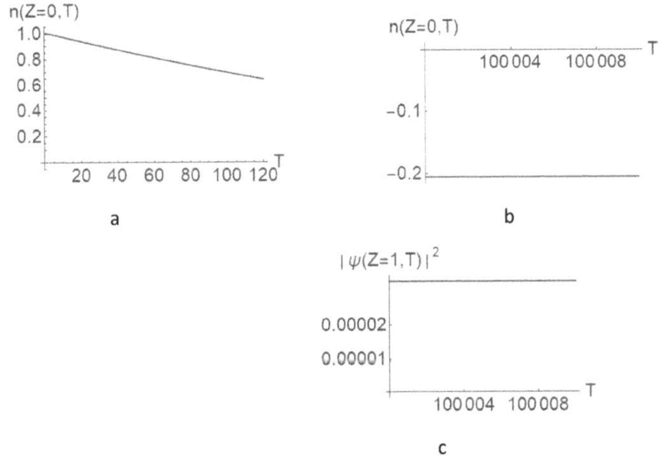

Figure 4. (**a–c**) $\alpha_1 = 0, \alpha_2 = \alpha_0 = 1$, $5\beta = 100{,}000$ and $\delta = 0.001686$. Same parameters as in Figure 3 except that δ has been set slightly *lower* than δ_{crit}. The Aurora following the "reading pulse" does not take place (**b**,**c**).

4. Delayed Bit Retrieval (Final Results and Discussion)

For our main calculation, we set all three $\alpha_0 = \alpha_1 = \alpha_2 = 1$ and $\beta = 20{,}000$, $\delta = 0.0016860$. The first input pulse at $T = T_0$ rapidly excites the system (Figure 5a) to $n = -0.2$. Meanwhile, the continuous (dc) pump has started to operate and *maintains* the excitation at this level, which is its equilibrium point. At $T = 5\beta = 100{,}000$, a second Nyxion pulse is applied (Figure 5b), and a flash is observed (Figure 5c), qualitatively resembling the Aurora seen (Figure 3) without an input pulse when δ is set too high.

During the long delay while the dc is maintaining the excitation at $n = -0.2$, there is no sign that the first pulse has had any effect. Everything looks just as though we had set $\alpha_1 = 0$ as in Figure 4. However, there is a hidden effect, revealed (Figure 5) when the second pulse is applied. Thus, we may consider the first pulse as the "input" of a bit which is "read" by the second pulse. In the presence of the dc pump, the delay between storing and reading is 100 times the isolated atom decay time Γ_1^{-1}, and the output signal (Figure 5c) is 4–5 orders of magnitude greater than that produced (too small to see in Figure 4c but shown by the underlying data from the simulation) in the absence of an input.

Thus, the simulations indicate that by including a dc pump in the function $R(T)$, a bit of information (pulse or no pulse at $T = T_0$) can be read with essentially no error after a delay as long as 100 times the isolated atom decay rate. What is more, the multiple 100 has been chosen arbitrarily; there is no reason, within the framework of our calculations, that it could not be made arbitrarily large. Perhaps the practical limit will have to do with the possibility of experimentally achieving the ideal slab picture, or with the assumption ("pressure broadening") that Γ_2 is so large that all broadening can be regarded as homogeneous.

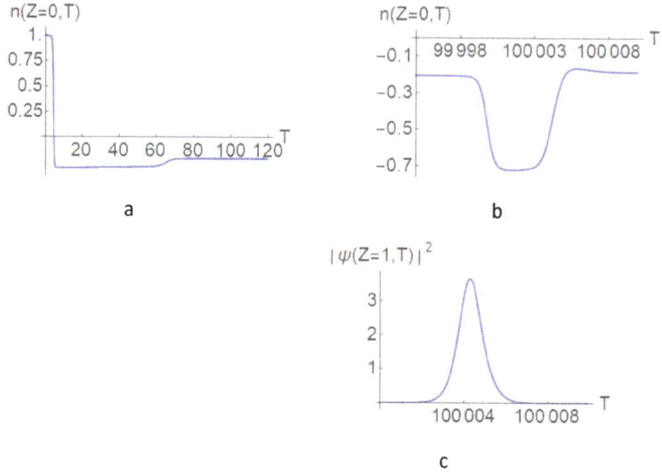

Figure 5. $\alpha_1 = \alpha_2 = \alpha_0 = 1$, 5β = 100,000 and δ = 0.0016860. Both continuous (dc) pump and "storage" pulse are turned on together at $T = 5$. The initial reaction (**a**) is dominated by the initial pulse. (Slight overshoot of $n = -0.2$ is corrected at $T = 70$.) After reaching $n = -0.2$, the system does *not* decay as shown in Figure 1a,b, but is *maintained* by the dc pump at $n = -0.2$ until it is discharged (**b**,**c**) by the "reading" pulse. The relative dc pump strength has been set *below* δ_{crit} so that the discharge would not have taken place in the absence of the initial storage pulse (compare with Figure 4b,c).

5. Conclusions

We repeat that these theoretical predictions can in principle be tested experimentally *without* cold-atom technology, and do not bear on the storage of quantum bits. We hope that the present paper will encourage research in classical bit storage and retrieval on the part of experimental scientists who seek to achieve a significant result with low-tech methods.

At the same time, we warn that the realization of an incoherent pump as described in Section 2 may pose challenging problems in practice. Although first observed [4] in 1985, the production of subradiant states has always gained far less attention than that of superradiant states.

Perhaps the most troubling possibility is the production of local pockets of quasi-superradiance inherited from the initial $1 \to 3$ excitation. If such pockets are distributed unevenly through the lateral extent of the slab, they will destroy the crucial assumption of Section 2, that the value of χ at any time depends spatially only on Z. Perhaps this could be prevented by selecting a substance for which the $3 \to 2$ transition is strongly inhomogeneously broadened (in contrast to the $2 \to 1$ transition which is assumed essentially homogeneous).

On the other hand, a slight deviation from total incoherence in noninteracting atoms is probably harmless—in fact, it may obviate the requirement of a $1 \to 2$ micropulse to kick χ out of the null state.

It must be admitted that our scheme appears quite naive in comparison with some recent studies of incoherent pumping [18,19]. On the other hand, we aim for less. We are not trying to use subradiant pumping to target a particular sublevel of level 2. On the contrary, we assume that Γ_2 is large enough so that any such sublevels will merge effectively into a single spectral line.

One can also hope that experimenters interested in our proposal will think of improvements that would never have occurred to the present authors. We can only claim that if the configuration described in (1)–(3) can be produced, our calculations indicate that some interesting long-life advances in classical bit storage and retrieval will be possible.

Author Contributions: Both authors contributed equally to this work. All authors have read and agreed to the published version of the manuscript.

Funding: This research did not receive any specific grant from funding agencies in the public, commercial, or not-for-profit sectors.

Institutional Review Board Statement: Not applicable.

Informed Consent Statement: Not applicable.

Data Availability Statement: Not applicable.

Acknowledgments: R.F. thanks Ana Asenjo-Garcia of the Columbia Physics Department for a very illuminating discussion of current advances in cold-atom technology. We are also grateful for the input of anonymous reviewers, which has contributed measurably to the quality of our manuscript.

Conflicts of Interest: The authors declare that they have no known competing financial interest or personal relationship that could have appeared to influence the work reported in this paper.

References

1. Dicke, R.H. Coherence in spontaneous radiation processes. *Phys. Rev.* **1954**, *93*, 99. [CrossRef]
2. Stroud, C.S.; Eberly, J.; Lama, W.L.; Mandel, L. Superradiant effects in systems of two-level atoms. *Phys. Rev. A* **1972**, *5*, 1094. [CrossRef]
3. Friedberg, R.; Hartmann, S.R. Temporal evolution of superradiance in a small sphere. *Opt. Commun.* **1974**, *10*, 298. [CrossRef]
4. Pavolini, D.; Crubellier, A.; Pillet, P.; Cabaret, L.; Liberman, S. Experimental evidence for subradiance. *Phys. Rev. Lett.* **1985**, *54*, 1917. [CrossRef] [PubMed]
5. Asenjo-Garcia, A.; Moreno-Cardoner, M.; Albright, A.; Kimble, H.J.; Chang, D.E. Exponential improvement in photon storage fidelities using subradiance and selective radiance in atomic arrays. *Phys. Rev. X* **2017**, *7*, 031024. [CrossRef]
6. Sinha, K.; Meystre, P.; Goldschmidt, E.A.; Fatemi, F.K.; Rolston, S.A.; Solano, P. Non-markovian collective emission from macroscopically separated emitters. *Phys. Rev. Lett.* **2020**, *124*, 043603. [CrossRef] [PubMed]
7. Ferioli, G.; Glicenstein, A.; Henriet, L.; Ferrier-Barbut, I.; Browaeys, A. Storage and Release of subradiant excitation in a dense atomic cloud. *Phys. Rev. X* **2021**, *11*, 021031. [CrossRef]
8. Cipris, A.; Bachelard, R.; Kaiser, R.; Guerin, W. Van der Waals dephasing for Dicke subradiance in cold atomic clouds. *Phys. Rev. A* **2021**, *163*, 033714. [CrossRef]
9. Sokolov, I.M. Subradiance of cold and diluted atomic ensembles excited by resonant pulsed radiation. *J. Exp. Theor. Phys.* **2021**, *132*, 68. [CrossRef]
10. Terhal, B.M. Quantum error correction for quantum memories. *Rev. Mod. Phys.* **2015**, *87*, 307. [CrossRef]
11. Friedberg, R.; Manassah, J.T. Solutions of the Maxwell-Bloch equations in (1,1)-d in the presence of a pulsed pump. *Opt. Comm.* **2021**, *501*, 127389. [CrossRef]
12. Friedberg, R.; Manassah, J.T. Eigenfunctions and eigenvalues in superradiance with x-y translational symmetry. *Phys. Lett. A* **2008**, *372*, 2787. [CrossRef]
13. Manassah, J.T. Quantum electrodynamics of two-level atoms in 1D configurations. *Adv. At. Opt. Phys.* **2014**, *63*, 359.
14. Bloch, F. Nuclear induction. *Phys. Rev.* **1946**, *70*, 460. [CrossRef]
15. Allen, L.; Eberly, J.H. *Optical Resonance and Two-Level Atoms*; Wiley: Hoboken, NJ, USA, 1975.
16. Berman, P.R.; Malinovsky, V.S. *Principles of Laser Spectroscopy and Quantum Optics*; Princeton University Press: Princeton, NJ, USA, 2011.
17. Friedberg, R.; Hartmann, S.R.; Manassah, J.T. Frequency shifts in emission and absorption by resonant systems of two-level atoms. *Phys. Rep.* **1973**, *7*, 104. [CrossRef]
18. Cipris, A.; Moreira, N.A.; Santo, T.S.d.E.; Weiss, P.; Villas-Boas, C.J.; Kaiser, R.; Guerin, W.; Bachelard, R. Subradiance with saturated atoms: Population enhancement of the long-lived state. *Phys. Rev. Lett.* **2021**, *126*, 103604. [CrossRef] [PubMed]
19. Glicenstein, A.; Feriola, G.; Browaeys, A.; Ferrier-Barbut, I. From superradiance to subradiance: Exploring the many-body Dicke ladder. *Opt. Lett.* **2022**, *47*, 1541. [CrossRef] [PubMed]

symmetry MDPI

Article

Spin Interference Effects in a Ring with Rashba Spin-Orbit Interaction Subject to Strong Light–Matter Coupling in Magnetic Field

Michal Pudlak [1] and R. Nazmitdinov [2,3,*]

1 Institute of Experimental Physics, Slovak Academy of Sciences, 040 01 Kosice, Slovakia; pudlak@saske.sk
2 Bogoliubov Laboratory of Theoretical Physics, Joint Institute for Nuclear Research, 141980 Dubna, Moscow Region, Russia
3 Faculty of Natural and Engineering Science, Dubna State University, 141982 Dubna, Russia
* Correspondence: rashid@theor.jinr.ru

Abstract: Electron transport through a one-dimensional quantum ring, subjected to Rashba spin–orbit interaction and connected with two external leads, is studied in the presence of external fields. They include the optical radiation, produced by an off-resonant high-frequency electric field, and a perpendicular magnetic field. By means of the Floquet theory of periodically driven quantum systems the interference effects under these fields are described in detail. It is found analytically the specific conditions to reach the spin-filtering effect, caused by the interplay of the external fields and Rashba spin-orbit interaction.

Keywords: Rashba spin–orbit interaction; semiconductor quantum ring; Floquet theory; light–matter coupling; magnetic field

Citation: Pudlak, M.; Nazmitdinov, R. Spin Interference Effects in a Ring with Rashba Spin-Orbit Interaction Subject to Strong Light–Matter Coupling in Magnetic Field. *Symmetry* **2022**, *14*, 1194. https://doi.org/10.3390/sym14061194

Academic Editor: Charalampos Moustakidis

Received: 24 May 2022
Accepted: 6 June 2022
Published: 9 June 2022

1. Introduction

Progress in nanotechnology raised a tremendous activity in the field of quantum electronics. In particular, a special attention is paid to the possibility to use spin–orbit interaction (SOI) for the design of nanoelectronic devices, based on control of electron spin without application of the magnetic field. In semiconductors there are two mechanisms of SOI: the Dresselhaus SOI [1], caused by the inversion asymmetry of the crystal lattice; the Rashba SOI [2], produced by the inversion asymmetry of a heterostructure. It is important to note that the strength of the Dresselhaus SOI is determined exclusively by the material, while the strength of the Rashba SOI can be by altered externally, for example, by means of a gate voltage. Therefore, the vast majority of literature, devoted to spin-dependent transport in nanostructures, is focused on materials with spin–orbit interaction of Rashba type (e.g., Refs. [3,4]).

It was recently proposed to use a strong off-resonant optical field to manipulate spin-orbit coupling [5]. In this case there is no real absorption of the wave. This is so-called regime of strong light–matter interaction, when quantum nature of light can drastically modify the properties of the matter itself. In fact, recent progress in laser physics provides the possibility to use optical high-frequency fields to control various atomic and condensed-mater structures, based on the Floquet theory of periodically driven quantum systems (e.g., Refs. [6–8]). The concept of radiation-dressed states in atom [9] is the fundamental background for this consideration. In this case, the hybrid electron-field object (dressed electron) represents an elementary quasiparticle, which physical properties can differ sufficiently from the "bare" electron.

Thanks to new generation of high-efficient lasers, this phenomenon may render possible its wide application in semiconductor physics. In particular, physical properties of dressed electrons have been studied in quantum wells [10], quantum rings [11,12], and topological insulators [13,14]. Evidently, this phenomenon becomes quite attractive for spintronics as well, since the spin of individual carriers can be controlled by optical means with or without application of the magnetic field, indeed. From this point of view semiconductor quantum rings with the Rashba SOI represent a fertile ground in the regime of strong light–matter coupling for applied physics, as well as for study of effects of different geometric phases [15]. The control of electron spin by means of the optical method and by a weak external magnetic field, and its consequences for transport properties in the above system have been escaped in previous studies (see, e.g., Refs. [3,16–20] and references therein). The main goal of this paper is to fill this gap in the case of the ring with the Rashba SOI for dressed electrons in magnetic field.

2. Model

2.1. The Hamiltonian

To analyse the regime of strong light–matter interaction, we consider the two-dimensional (2D) Hamiltonian describing ballistic electrons of charge $-e(e > 0)$ and the effective mass m, in the presence of the Rashba SOI, a magnetic field and a high-frequency electric field:

$$\hat{H} = \frac{1}{2m}\boldsymbol{\Pi}^2 + \alpha(\boldsymbol{\sigma} \times \boldsymbol{\Pi})_z + g\mu\boldsymbol{\sigma}\boldsymbol{B} + V(\boldsymbol{r})\,. \tag{1}$$

Here $\boldsymbol{\sigma}$ is the vector of the Pauli spin matrices, $\boldsymbol{\Pi} = \boldsymbol{p} + e\boldsymbol{A}$, α is the strength of the Rashba SOI. The vector potential of a linear polarised electromagnetic wave $\boldsymbol{A} = ([E_0/\omega]\cos(\omega t) - By/2, Bx/2, 0)$ includes the magnetic field $\boldsymbol{B}$, pointing in the z direction (perpendicular to the plane). The electric field is characterised by the amplitude E_0 and by the wave frequency ω. We consider a narrow ring, characterised by a steep confining potential $V(\boldsymbol{r})$. If the field is time-independent and $E_0 = 0$, in such a ring geometry the electron energy spectrum is determined by the 1D Hamiltonian in polar coordinates (see also [17,21]):

$$\hat{H}_R^{(0)} = \frac{\hbar^2}{2mR^2}\left(-\mathrm{i}\frac{\partial}{\partial\varphi} + \frac{\Phi}{\Phi_0}\right)^2 + \frac{\hbar\omega_B}{2}\sigma_z + \hbar\omega_R\sigma_x(\varphi)\left(-\mathrm{i}\frac{\partial}{\partial\varphi} + \frac{\Phi}{\Phi_0}\right) - \mathrm{i}\frac{\hbar\omega_R}{2}\sigma_y(\varphi)\,. \tag{2}$$

Here, $\omega_B = 2\mu B/\hbar$, $\omega_R = \alpha/R$, $\sigma_x(\varphi) = \cos\varphi\sigma_x + \sin\varphi\sigma_y$, $\sigma_y(\varphi) = \cos\varphi\sigma_y - \sin\varphi\sigma_x$, φ is the polar angle of the electron on the ring, $\Phi = \pi BR^2$ is the magnetic flux through the ring, and the magnetic flux quantum $\Phi_0 = h/e$. Once we add a time-dependent electric field the Hamiltonian of an irradiated ring takes the following form

$$\hat{H}_{1D} = \hat{H}_R + \left[\sum_{n=1}^{2}\hat{V}_n e^{in\omega t} + H.c\right], \tag{3}$$

where the stationary term is complemented by a field-induced constant energy shift

$$\hat{H}_R = \hat{H}_R^{(0)} + E_{shift}^{(0)}, \quad E_{shift}^{(0)} = \frac{e^2E_0^2}{4m\omega^2}\,. \tag{4}$$

The periodic term consists of two harmonics, raised by the irradiation,

$$\hat{V}_1 = -\frac{eE_0}{2mR\omega}\left(\sin\varphi\hat{l}_{z,\Phi} - \mathrm{i}\hbar\frac{\cos\varphi}{2}\right] - \frac{\alpha eE_0}{2\omega}\sigma_y\,, \tag{5}$$

$$\hat{V}_2 = \frac{e^2E_0^2}{8m\omega^2}\,, \tag{6}$$

and we introduce the notation $\hat{l}_{z,\Phi} = -i\hbar\partial_\varphi + \hbar\Phi/\Phi_0$.

In the following we employ the high-frequency approximation for a periodically driven quantum system (for a review see, e.g., Ref. [22]). Such the approach provides a systematic high-frequency expansion for the effective Hamiltonian. In our analysis, we consider only the leading terms in the high-frequency limit. As a result, we can reduce the time-dependent Hamiltonian (3) to the effective time-independent one (see also [19]):

$$\hat{\mathcal{H}} = \hat{H}_R + \sum_{n=1}^{2} \frac{[\hat{V}_n, \hat{V}_n^\dagger]}{n\hbar\omega} + \sum_{n=1}^{2} \frac{[\hat{V}_n, \hat{H}_R], \hat{V}_n^\dagger] + H.c.}{2(n\hbar\omega)^2} = \hat{\mathcal{H}}_0 + \hat{\mathcal{V}}, \tag{7}$$

where

$$\begin{aligned} \hat{\mathcal{H}}_0 &= \frac{\hat{l}^2_{z,\Phi}}{2m^* R^2} + \omega_R \left[\sigma_x(\varphi)\hat{l}_{z,\Phi} - \mathrm{i}\hbar\frac{\sigma_y(\varphi)}{2} \right] - \left(\frac{eE_0\alpha}{R\omega^2}\right)^2 \frac{\hat{l}_{z,\Phi}}{m\hbar}\sigma_z \\ &+ \frac{\hbar\omega_B}{2}\sigma_z + E^{(0)}_{shift} + E^{(1)}_{shift}, \quad E^{(1)}_{shift} = \frac{1}{2m}\left(\frac{\hbar e E_0}{4mR^2\omega^2}\right)^2 \end{aligned} \tag{8}$$

$$\hat{\mathcal{V}} = \frac{\hbar^2}{2mR^2}\left[\hat{V}_a + \hat{V}_b + \left(\frac{\gamma_1 \hat{l}_{z,\Phi}}{2\hbar}\right)^2 \cos 2\varphi\right], \tag{9}$$

$$\hat{V}_a = \left[\frac{3}{16}\gamma_1^2 \cos 2\varphi - \mathrm{i}\gamma_1^2\gamma_2\left(\gamma_2^2 - \frac{1}{4}\right)\sigma_x \sin\varphi\right], \tag{10}$$

$$\hat{V}_b = \left[\frac{\mathrm{i}}{2}\gamma_1^2 \sin 2\varphi - 2\gamma_1^2\gamma_2\left(\gamma_2^2 - \frac{1}{4}\right)\sigma_x \cos\varphi\right]\frac{\hat{l}_{z,\Phi}}{\hbar}. \tag{11}$$

Here, we introduce the following notations:

$$\gamma_1 = eE_0/(mR\omega^2), \quad \gamma_2 = mR\alpha/\hbar, \quad m^* = \frac{m}{1 + 3(\gamma_1/2)^2}. \tag{12}$$

The irradiation leads to the mass renormalization, i.e., $m \to m^*$. It yields as well the energy shift to the zero energy $\Delta E = E^{(0)}_{shift} + E^{(1)}_{shift}$, and introduces the coupling between the strength of the Rashba SOI and the ring radius by means of the parameter γ_2.

2.2. The Eigenvalue Problem

The Hamiltonian (7) possess the azimuthal symmetry. The operator $\hat{J}_z = I \otimes (-\mathrm{i}\hbar\partial_\varphi) + \hbar\sigma_z/2$, defined in the laboratory frame, is an integral of motion $[\mathcal{H}, \hat{J}_z] = 0$. Let us analyse, first, the eigenvalue problem for the Hamiltonian (8). It is convenient to transform this Hamiltonian in the rotating frame by means of the unitary transformation $R = \exp[\mathrm{i}(\sigma_z/2)\varphi]$. As a result, we obtain

$$\hat{\hat{H}} = R\mathcal{H}_0 R^+ = \frac{\hat{X}^2_{z,\Phi}}{2m^* R^2} + \frac{\alpha}{R}\left[\sigma_x \hat{X}_{z,\Phi} - \mathrm{i}\hbar\frac{\sigma_y}{2}\right] - \left(\frac{eE_0\alpha}{R\omega^2}\right)^2 \frac{\hat{X}_{z,\Phi}}{m\hbar}\sigma_z + \frac{\hbar\omega_B}{2}\sigma_z + \Delta E. \tag{13}$$

$$\hat{X}_{z,\Phi} = -\mathrm{i}\hbar\partial_\varphi + \hbar\Phi/\Phi_0 - \hbar\frac{\sigma_z}{2}. \tag{14}$$

In the rotating frame the operator $\hat{J}_z$ takes the following form

$$\hat{\hat{J}}_z = R\hat{J}_z R^+ = I \otimes (-\mathrm{i}\hbar\partial_\varphi). \tag{15}$$

Consequently, we search the eigenfunctions of the Hamiltonian (13) in a general form

$$\Phi^s_j(\varphi) = e^{\mathrm{i}j\varphi}\chi^s_j, \quad \chi^s_j = \begin{pmatrix} \chi^s_1 \\ \chi^s_2 \end{pmatrix}. \tag{16}$$

Evidently, the eigenvalues of the operator $\hat{J}$ are expected to be half-integers in the laboratory frame, that should be hold in the rotating frame as well ($RR^{-1} = 1$)

$$\hat{\hat{J}}_z \Phi^s_j(\varphi) = \hbar j \Phi^s_j(\varphi), \quad j = \lambda n + 1/2, \quad n = 1, 2, 3, \ldots. \tag{17}$$

Here, the orbital quantum number n corresponds to the electron rotation either in the counterclockwise direction $\lambda = +1$, or in the clockwise one $\lambda = -1$. The solution of the Schrödinger equation by means of the probe functions (16) yields the eigenenergies

$$E_j^s = \hbar\omega_0 \left[\left(j + \frac{\Phi}{\Phi_0} \right)^2 + \frac{1}{4} + s\sqrt{\Omega^2 + \left(\frac{m^*}{m}\right)^2 Q_R^2 \left(j + \frac{\Phi}{\Phi_0} \right)^2} \right] + \Delta E; \quad s = \pm 1, \quad (18)$$

where

$$\Omega = \frac{\omega_B}{2\omega_0} - \left[1 + 2Q_E^2 \frac{m^*}{m}\right] \left(j + \frac{\Phi}{\Phi_0} \right). \quad (19)$$

Here, we introduce the following definitions: $\omega_0 = \hbar/(2m^* R^2)$, $Q_E = eE_0\alpha/(\hbar\omega^2)$, and $Q_R = 2m\alpha R/\hbar$. For the eigenfunctions we obtain two sets

$$\chi_j^{(s=1)} = \begin{pmatrix} \sin\frac{\gamma}{2} \\ \cos\frac{\gamma}{2} \end{pmatrix}, \quad \chi_j^{(s=-1)} = \begin{pmatrix} \cos\frac{\gamma}{2} \\ -\sin\frac{\gamma}{2} \end{pmatrix}, \quad (20)$$

where

$$\tan\gamma = \frac{\alpha\hbar/R\left(j + \frac{\Phi}{\Phi_0}\right)}{\Omega} = \frac{Q_R m^*/m}{1 + 2Q_E^2 m^*/m - \omega_B/[2\omega_0(j + \Phi/\Phi_0)]}, \quad (21)$$

γ is the angle between the local spin quantization axis and the z-axis. The high-frequency (dressing) field decreases this angle relative to its value $\tan\gamma = Q_R$, obtained in Ref. [20] without the external electric field ($E_0 = 0$) and Zeeman interaction. Evidently, a proper choice of the Rashba coupling and parameters of the external high-frequency electric field may lead to new features of the considered system (see below).

The eigenstates of the Hamiltonian (8) are defined in the laboratory frame as

$$\Psi_j^s(\varphi) = e^{-i\frac{\sigma_z}{2}\varphi}\Phi_j^s(\varphi) = e^{ij\varphi}e^{-i\frac{\sigma_z}{2}\varphi}\chi_j^s. \quad (22)$$

Before proceeding further, there are a few comments required. The total effective Hamiltonian (7) consists of the discussed Hamiltonian (8) and the term $\hat{V}$. The terms (9)–(11) are of order $\sim\gamma_1^2$. In the high-frequency approximation, considered in our paper, $\gamma_1 \ll 1$. Consequently, we neglect the contribution of the above terms, and will analyse the transport properties of semiconductor quantum rings with the reduced Hamiltonian (8) (see also the discussion in Ref. [19]). For a typical semiconductor (for example, GaAs) the magnetic orbital effect is much enhanced in comparison with the magnetic spin effect (see, e.g., Ref. [23]). Moreover, we will consider the effect of the weak magnetic field (see below). Consequently, without loss of generality, we can neglect the Zeeman effect in Equations (19) and (21) and obtain

$$\begin{aligned} E_{\lambda,n}^s &= \hbar\omega_0 \left[\left(\lambda n + \frac{1}{2} + \frac{\Phi}{\Phi_0} \right)^2 + \frac{1}{4} + s\left| \lambda n + \frac{1}{2} + \frac{\Phi}{\Phi_0} \right| \times \right. \\ &\left. \times \sqrt{\left[1 + 2Q_E^2 \frac{m^*}{m}\right]^2 + \left(\frac{m^*}{m}\right)^2 Q_R^2} \right], \end{aligned} \quad (23)$$

where the energy shift ΔE is omitted, since it is the same constant for the electron dressed levels. In the case: (i) $E_0 = 0$; (ii) $\boldsymbol{B} = 0$, the reduced energies Equation (23) are equivalent to the energies obtained in Ref. [17]. These energies are maximised for the up spin states $|\uparrow\rangle$, i.e., for $s = 1$; and the factor Q_R plays the important role in transport properties. The external high-frequency electric field gives rise to the additional factor Q_E. As we will see below, its interplay with the factor Q_R leads to novel phenomena in transport properties of the semiconductor rings.

3. Transport Properties

In this section, we investigate the effect of two factors, Q_R and Q_E, on the conductance and the polarization of the ballistic current in our 1D model. To model realistic transport, it is desirable to take into account the effects of disorder. Since we consider the high-frequency limit (a semiclassical regime), there are different classical paths connecting the entrance and exit attached leads. It seems reasonable to assume that only pairs with the same length could contribute essentially to the conductance. In our consideration we assume a perfect coupling between leads and ring, neglecting the backscattering effects. In other words, in our model the interference effects arising from counterclockwise and clockwise waves.

Thus, the wave function of an incoming electron from the left lead, attached to the ring, is split at the ring entrance into four partial waves. In particular, we consider that electrons with Fermi energy E_F move from the entrance to the exit with four different wave numbers $n \Rightarrow \lambda n_\lambda^s$. In this case the eigenstates (22) can be written as (see also the discussion in Ref. [17])

$$\Psi_{+n}^{1}(\varphi) = e^{in_+^{+1}\varphi}\begin{pmatrix} \sin\frac{\gamma}{2} \\ \cos\frac{\gamma}{2}e^{i\varphi} \end{pmatrix}, \tag{24}$$

$$\Psi_{+n}^{-1}(\varphi) = e^{in_+^{-1}\varphi}\begin{pmatrix} \cos\frac{\gamma}{2} \\ -\sin\frac{\gamma}{2}e^{i\varphi} \end{pmatrix}, \tag{25}$$

$$\Psi_{-n}^{1}(\varphi) = e^{-in_-^{+1}\varphi}\begin{pmatrix} \cos\frac{\gamma}{2} \\ -\sin\frac{\gamma}{2}e^{i\varphi} \end{pmatrix}, \tag{26}$$

$$\Psi_{-n}^{-1}(\varphi) = e^{-in_-^{-1}\varphi}\begin{pmatrix} \sin\frac{\gamma}{2} \\ \cos\frac{\gamma}{2}e^{i\varphi} \end{pmatrix}. \tag{27}$$

They meet at the exit (a right attached lead), exactly opposite to the entrance. The wave, propagating clockwise, travels the angle $-\pi$ from the entrance to the exit. The wave, propagating counterclockwise, travels the angle π from the entrance to the exit.

3.1. Conductance

In order to analyse the conductance we use the Landauer–Büttiker formalism. In this case the conductance at zero temperature has the following form (see, e.g., Ref. [24])

$$G = \frac{e^2}{h}\sum_{s,s'} T_{s,s'}\,. \tag{28}$$

Here, $T_{s,s'} = |t_{s,s'}|^2$ is the quantum probability of transmission between incoming state with spin s' and outgoing state with spin s; $t_{s,s'}$ is the corresponding transmission amplitude. Using the results from Appendix A, we arrive to the expression

$$G = \frac{e^2}{h}\left[1 + \frac{1}{2}[\cos\pi(n_-^{-1} - n_+^{+1}) + \cos\pi(n_-^{+1} - n_+^{-1})]\right]. \tag{29}$$

The quantities n_λ^s can be obtained from the solution of Equation (18) at the condition $E_{\lambda,n}^s = E_F$ for different spin orientation $s = \pm 1$.

$$n_-^{-1} - n_+^{+1} = 1 + 2\frac{\Phi}{\Phi_0} + \sqrt{\left[1 + 2Q_E^2\frac{m^*}{m}\right]^2 + \left(\frac{m^*}{m}\right)^2 Q_R^2}\,, \tag{30}$$

$$n_-^{+1} - n_+^{-1} = 1 + 2\frac{\Phi}{\Phi_0} - \sqrt{\left[1 + 2Q_E^2\frac{m^*}{m}\right]^2 + \left(\frac{m^*}{m}\right)^2 Q_R^2}\,. \tag{31}$$

Taking into account Equations (30) and (31), we arrive to the final expression for the conductance

$$G = \frac{e^2}{h}\left[1 - \cos\left(2\pi\frac{\Phi}{\Phi_0}\right)\cos\pi\sqrt{\left[1+2Q_E^2\frac{m^*}{m}\right]^2 + \left(\frac{m^*}{m}\right)^2 Q_R^2}\right]. \quad (32)$$

Now we are ready to trace the conductance behaviour as a function of the following variables: (i) the strength of the Rashba interaction α; (ii) the electric field E_0; and (iii) the magnetic field. As a typical example, we consider InGaAs-based quantum rings with the following parameters: the effective mass $m = 0.045m_e$, radius $R \approx 200$ nm, and the strength of the Rashba SOI $\alpha \approx 10^4$ m/s. The effective mass and the radius determine the energy scale $\hbar\omega_0 \equiv \hbar^2/(2mR^2) \approx 2.16 \times 10^{-5}$ eV at $m^* = m$. Assuming the maximal flux $\Phi = 0.5\Phi_0$ through the ring in our consideration, we obtain

$$\mu B = \frac{e\hbar}{2m_e} \times \frac{\hbar}{e}\frac{1}{R^2} \ll \hbar\omega_0. \quad (33)$$

Our approximation (neglecting the Zeeman term) is quite satisfactory, indeed.

At zero magnetic and electric fields the conductance is modulated by the strength Q_R alone [17]. Taking into account that the amplitude

$$E_0 = \sqrt{\frac{2I}{\varepsilon_0 c}}, \quad (34)$$

where I is the irradiance intensity, ε_0 is the vacuum permittivity, we can enrich the interference of the conductance from the destructive to constructive and vice versa by altering the intensity at a fixed value of the Rashba SOI α (see Figure 1). Once we switch on the magnetic field, the conductance modulations are reversed (see Figure 2). Moreover, the oscillations are removed with a proper choice of of the Rashba SOI strength.

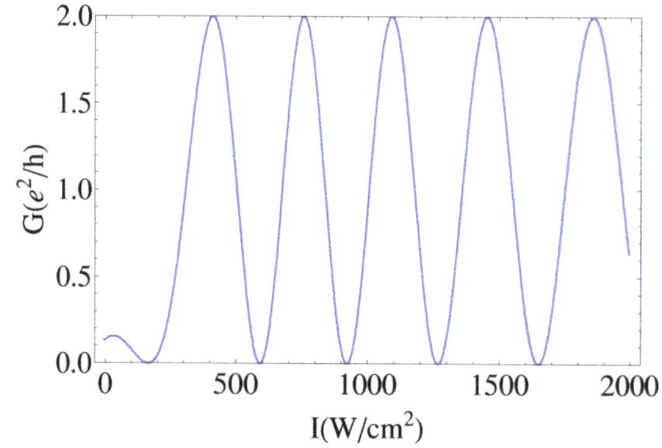

Figure 1. Conductance versus irradiation intensity I. Electron effective mass $m = 0.045m_e$, the Rashba coupling constant $\alpha = 5 \times 10^4$ ms^{-1}, and the ring radius is $R = 200$ nm. The dressing field has the frequency $\omega = 1.6 \times 10^{12}$ s^{-1}, the magnetic flux $\Phi = 0$.

Thus, the magnetic field provides the additional key element of possible ring-shaped spintronic devices operated by light.

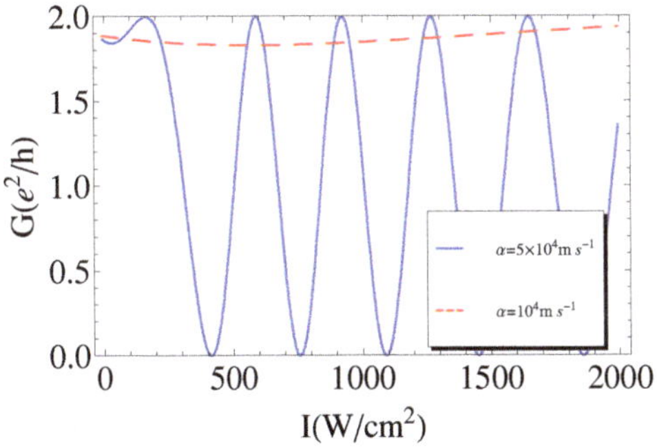

Figure 2. Conductance versus irradiation intensity I for different Rashba coupling constant α. Electron effective mass $m = 0.045m_e$ and the ring radius is $R = 200$ nm. The dressing field has the frequency $\omega = 1.6 \times 10^{12}\ \mathrm{s}^{-1}$, the magnetic flux $\Phi = 0.5\Phi_0$.

To get deeper inside let us consider the minimum of the conductance at $m^* = m$. At zero magnetic field, we have the condition

$$\cos \pi \sqrt{\left[1 + 2Q_E^2 \frac{m^*}{m}\right]^2 + \left(\frac{m^*}{m}\right)^2 Q_R^2} = 1, \tag{35}$$

which is subject to the equation

$$\sqrt{[1 + 2Q_E^2]^2 + Q_R^2} = 2n, \quad n = 1, 2, \ldots. \tag{36}$$

By introducing the variables

$$\sin \gamma = \frac{Q_R}{2n}, \quad \cos \gamma = \frac{1 + 2Q_E^2}{2n}, \tag{37}$$

we arrive to Equation (21) at $\omega_B = 0$. Thus, the single-valuedness of the eigenfunctions (20) determines the character of the conduction modulations (35).

3.2. Spin-Filtering Effect

The question we address in this section is could we control the polarization of the electron beam by means of our quantum ring with the aid of the intensity of the external electric field and with the strength of the vertical magnetic field?

The spin polarization P, determined as

$$P = \frac{T_{\uparrow\uparrow} + T_{\uparrow\downarrow} - T_{\downarrow\uparrow} - T_{\downarrow\downarrow}}{T_{\uparrow\uparrow} + T_{\downarrow\uparrow} + T_{\uparrow\downarrow} + T_{\downarrow\downarrow}}, \tag{38}$$

in virtue of the results for transmission probabilities $T_{\sigma\sigma'}$ (see Appendix A), yields

$$P = \cos 2\gamma \frac{\sin\left(2\pi \frac{\Phi}{\Phi_0}\right) \sin \pi \sqrt{[1 + 2Q_E^2 \frac{m^*}{m}]^2 + (\frac{m^*}{m})^2 Q_R^2}}{1 - \cos\left(2\pi \frac{\Phi}{\Phi_0}\right) \cos \pi \sqrt{[1 + 2Q_E^2 \frac{m^*}{m}]^2 + (\frac{m^*}{m})^2 Q_R^2}}. \tag{39}$$

From Equation (39) it follows that the presence of the magnetic field is the basic condition for the polarization process, since $P = 0$ at $\Phi = n\Phi_0/2, n = 0, 1, 2, \ldots$. On the other hand, once the magnetic field takes, for example, the value $\Phi/\Phi_0 = 1/4$, we can require that

$$\sin \pi \sqrt{\left[1 + 2Q_E^2 \frac{m^*}{m}\right]^2 + \left(\frac{m^*}{m}\right)^2 Q_R^2} = 1\,. \tag{40}$$

Without loss of generality we consider the case $m^* = m$ and obtain

$$\sin\gamma = \frac{Q_R}{k}, \quad \cos\gamma = \frac{(1+2Q_E^2)}{k}, \quad k = 2n + 1/2, \quad n = 0, 1, 2, \ldots \tag{41}$$

In this case the polarization is defined as

$$P = \cos 2\gamma = [(1+2Q_E^2)^2 - Q_R^2]/k^2\,. \tag{42}$$

At a fixed value of the strength of the Rashba SOI, we can define the value of the intensity of the electric field that could provide the maximal polarization $P = 1$ (see Figure 3).

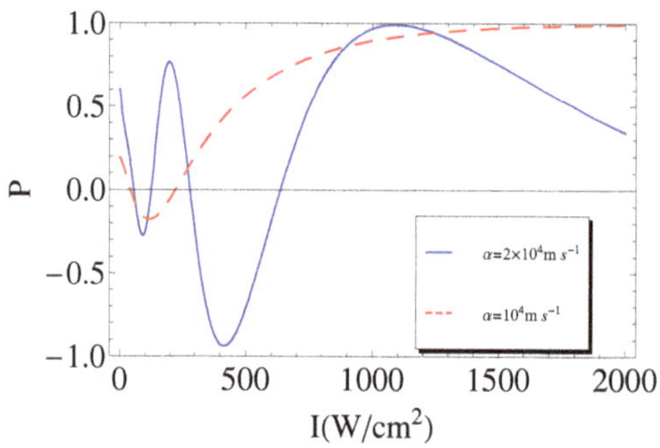

Figure 3. Spin polarization P versus the irradiation intensity I. The calculations are performed at the magnetic flux $\Phi = 0.25\Phi_0$; the dressing field frequency is $\omega = 0.8 \times 10^{12}$ s^{-1}. The solid (blue) line corresponds to the strength $\alpha = 2 \times 10^4$ m/s, while the dashed (red) line corresponds to $\alpha = 10^4$ m/s.

For example, taking into account the definition of $Q_E = eE_0\alpha/(\hbar\omega^2)$ and the intensity (34), we have at $k = 1/2$

$$2Q_E^2 = \sqrt{Q_R^2 + \frac{1}{4}} - 1 = \frac{4I}{\varepsilon_0 c}\left(\frac{\hbar\omega^2}{e\alpha}\right)^2. \tag{43}$$

From Equation (43) it follows evidently that $Q_E > 0$ if the following relation takes place (taking into account the definition $Q_R = (2mR/\hbar)\alpha$)

$$\alpha > \frac{\sqrt{3}}{2} R\omega_0\,. \tag{44}$$

Once this condition is fulfilled, the minimal value of the intensity is

$$I = \frac{\varepsilon_0 c}{4}\left(\frac{\hbar\omega^2}{e\alpha}\right)^{-2}\left[\sqrt{Q_R^2 + \frac{1}{4}} - 1\right], \tag{45}$$

that allows to observe the spin-filtering effect in our system. Altering the value of the magnetic field, we can decrease, as well, the dynamic threshold intensity field at a fixed value of the strength of the Rashba SOI (see Figure 4).

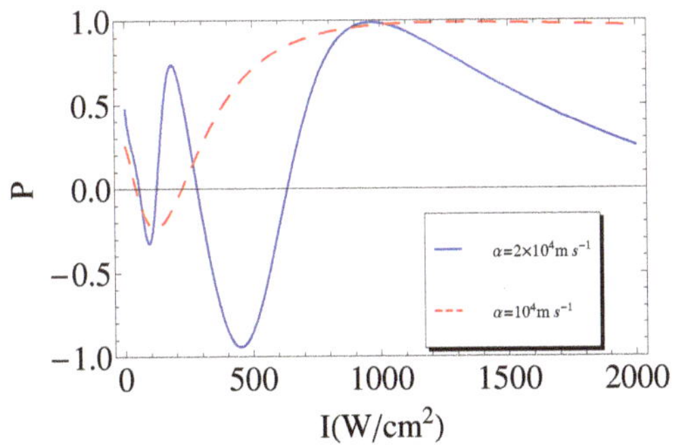

Figure 4. Similar to Figure 3 at the magnetic flux $\Phi = 0.2\Phi_0$.

4. Conclusions

The effect of a high-frequency optical field and an external magnetic field on quantum transport through the one-dimensional quantum ring subject to Rashba SOI is manifested in a rich variety of phenomena. To carry on our analysis of the external fields, we employed the Floquet theory of periodically driven quantum system. In our consideration we assumed a perfect coupling between leads and ring, neglecting the backscattering effects. In this limit, several mechanisms, responsible for quantum interference effects have been proposed. In particular, it shown that the conductance oscillations, produced by the ring irradiated by the dressing field, can be reversed by the application of the weak magnetic field (compare Figures 1 and 2). In other words, our system behaves like a diode, operating at a certain intensity of the dressing field, that allows the current flow only at a certain value of the external magnetic field. Finally, we formulated analytically the requirements to reach the spin- filtering effects under the external fields (see Section 3.2). Our findings may provide new capabilities for spintronics devices, exploiting the combined effect of optical and magnetic fields.

Author Contributions: All authors contributed equally to this work. All authors have read and agreed to the published version of the manuscript.

Funding: This research received no external funding.

Institutional Review Board Statement: Not applicable.

Informed Consent Statement: Not applicable.

Data Availability Statement: Not applicable.

Conflicts of Interest: The authors declare no conflict of interest.

Abbreviations

The following abbreviations are used in this manuscript:

MDPI	Multidisciplinary Digital Publishing Institute
DOAJ	Directory of open access journals
TLA	Three letter acronym

Appendix A. Transmission Probabilities

Let us consider the case of the incoming electron with spin $\uparrow$ $(s = +1)$ [see Equation (20)] entering the ring at $\varphi = 0$

$$|\uparrow\rangle = \begin{pmatrix} \sin\frac{\gamma}{2} \\ \cos\frac{\gamma}{2} \end{pmatrix}. \tag{A1}$$

Evidently, states of a particular spin split equally into the clockwise path ($\lambda = -1$) and the counterclockwise path ($\lambda = 1$) in the ring

$$|\Psi(\uparrow,\varphi)\rangle = \frac{1}{2}\exp[in_{+}^{+1}\varphi]\begin{pmatrix} \sin\frac{\gamma}{2} \\ \cos\frac{\gamma}{2}e^{i\varphi} \end{pmatrix} + \frac{1}{2}\exp[-in_{-}^{-1}\varphi]\begin{pmatrix} \sin\frac{\gamma}{2} \\ \cos\frac{\gamma}{2}e^{i\varphi} \end{pmatrix} \tag{A2}$$

Additionally, in particular, we have at the exit of the ring

$$|\Psi(\uparrow,\pi)\rangle = \frac{1}{2}\exp[in_{+}^{+1}\pi]\begin{pmatrix} \sin\frac{\gamma}{2} \\ \cos\frac{\gamma}{2}e^{i\pi} \end{pmatrix} + \frac{1}{2}\exp[-in_{-}^{-1}(-\pi)]\begin{pmatrix} \sin\frac{\gamma}{2} \\ \cos\frac{\gamma}{2}e^{-i\pi} \end{pmatrix} \tag{A3}$$

Consequently, the probability amplitude without the spin flip for the incoming electron with spin $\uparrow$ is

$$t_{\uparrow\uparrow} = \langle \uparrow | \Psi(\uparrow,\pi)\rangle\,, \tag{A4}$$

which determines the corresponding transmission probability as

$$T_{\uparrow\uparrow} = |t_{\uparrow\uparrow}|^2 = \frac{1}{2}\cos^2\gamma\Big[1 + \cos\pi(n_{-}^{-1} - n_{+}^{+1})\Big]\,. \tag{A5}$$

The amplitude of probability that the incoming electron with spin $\uparrow$ is outgoing with the spin $\downarrow$ [see Equation (20)] is

$$t_{\downarrow\uparrow} = \langle \downarrow | \Psi(\uparrow,\pi)\rangle\,, \quad |\downarrow\rangle = \begin{pmatrix} \cos\frac{\gamma}{2} \\ -\sin\frac{\gamma}{2} \end{pmatrix}, \tag{A6}$$

which yields the following result

$$T_{\downarrow\uparrow} = |t_{\downarrow\uparrow}|^2 = \frac{1}{2}\sin^2\gamma\Big[1 + \cos\pi(n_{-}^{-1} - n_{+}^{+1})\Big]\,. \tag{A7}$$

Let us consider the case of the incoming electron with spin $\downarrow$ ($s = -1$) [see Equation (20)] entering the ring at $\varphi = 0$. In this case electron traverses the ring, and its wave function [see Equations (25) and (26)] is

$$|\Psi(\downarrow,\varphi)\rangle = \frac{1}{2}\exp[in_{+}^{-1}\varphi]\begin{pmatrix} \cos\frac{\gamma}{2} \\ -\sin\frac{\gamma}{2}e^{i\varphi} \end{pmatrix} + \frac{1}{2}\exp[-in_{-}^{+1}\varphi]\begin{pmatrix} \cos\frac{\gamma}{2} \\ -\sin\frac{\gamma}{2}e^{i\varphi} \end{pmatrix}. \tag{A8}$$

It takes the following form at the exit of the ring

$$|\Psi(\downarrow,\varphi)\rangle = \frac{1}{2}\exp[in_{+}^{-1}\pi]\begin{pmatrix} \cos\frac{\gamma}{2} \\ -\sin\frac{\gamma}{2}e^{i\pi} \end{pmatrix} + \frac{1}{2}\exp[-in_{-}^{+1}(-\pi)]\begin{pmatrix} \cos\frac{\gamma}{2} \\ -\sin\frac{\gamma}{2}e^{i(-\pi)} \end{pmatrix}. \tag{A9}$$

For the amplitude of probability that the incoming electron with spin $\downarrow$ is outgoing with spin $\downarrow$ we have

$$t_{\downarrow\downarrow} = \langle \downarrow | \Psi(\downarrow,\pi)\rangle\,. \tag{A10}$$

Consequently, the corresponding transmission probability is

$$T_{\downarrow\downarrow} = |t_{\downarrow\downarrow}|^2 = \frac{1}{2}\cos^2\gamma\Big[1 + \cos\pi(n_{-}^{+1} - n_{+}^{-1})\Big]\,. \tag{A11}$$

The amplitude of probability that the incoming electron with spin $\downarrow$ is outgoing with spin $\uparrow$ is

$$t_{\uparrow\downarrow} = \langle \uparrow | \Psi(\downarrow,\pi)\rangle\,. \tag{A12}$$

It results in the following transmission probability

$$T_{\uparrow\downarrow} = |t_{\uparrow\downarrow}|^2 = \frac{1}{2}\sin^2\gamma\Big[1 + \cos\pi(n_{-}^{+1} - n_{+}^{-1})\Big]\,. \tag{A13}$$

References

1. Dresselhaus, G. Spin-orbit coupling effects in zinc blende structures. *Phys. Rev.* **1955**, *100*, 580. [CrossRef]
2. Bychkov, Y.A.; Rashba, E.I. Properties of a 2D electron gas with lifted spectral degeneracy. *JETP Lett.* **1984**, *39*, 78.
3. Bercioux, D.; Lucignano, P. Quantum transport in Rashba spin-orbit materials: A review. *Rep. Prog. Phys.* **2015**, *78*, 106001. [CrossRef] [PubMed]
4. Pichugin, K.; Puente, A.; Nazmitdinov, R. Kramers degeneracy and spin inversion in a lateral quantum dot. *Symmetry* **2020**, *12*, 2043. [CrossRef]
5. Sheremet, A.S.; Kibis, O.V.; Kavokin, A.V.; Shelykh, I.A. Datta-and-Das spin transistor controlled by a high-frequency electromagnetic field. *Phys. Rev. B* **2016**, *93*, 165307. [CrossRef]
6. Goldman, N.; Dalibard, J. Periodically Driven Quantum Systems: Effective Hamiltonians and Engineered Gauge Fields. *Phys. Rev. X* **2014**, *4*, 031027. [CrossRef]
7. Holthaus, M. Floquet engineering with quasienergy bands of periodically driven optical lattices. *J. Phys. B* **2016**, *49*, 013001. [CrossRef]
8. Meinert, F.; Mark, M.J.; Lauber, K.; Daley, A.J.; Nägerl, H.-C. Floquet Engineering of Correlated Tunneling in the Bose-Hubbard Model with Ultracold Atoms. *Phys. Rev. Lett.* **2016**, *116*, 205301. [CrossRef]
9. Cohen-Tannoudji, C.; Dupont-Roc, J.; Grynberg, G. *Atom-Photon Interactions: Basic Processes and Applications*; Wiley–VCH: Hoboken, NJ, USA, 2004.
10. Teich, M.; Wagner, M.; Schneider, H.; Helm, M. Semiconductor quantum well excitons in strong, narrowband terahertz fields. *New J. Phys.* **2013**, *15*, 065007. [CrossRef]
11. Joibari, F.K.; Blanter, Y.M.; Bauer, G.E.W. Light-induced spin polarizations in quantum rings. *Phys. Rev. B* **2014**, *90*, 155301. [CrossRef]
12. Koshelev, K.L.; Kachorovskii, V.Y.; Titov, M. Resonant inverse Faraday effect in nanorings. *Phys. Rev. B* **2015**, *92*, 235426. [CrossRef]
13. Foa Torres, L.E.F.; Perez-Piskunow, P.M.; Balseiro, C.A.; Usaj, G. Multiterminal Conductance of a Floquet Topological Insulator. *Phys. Rev. Lett.* **2014**, *113*, 266801. [CrossRef] [PubMed]
14. Mikami, T.; Kitamura, S.; Yasuda, K.; Tsuji, N.; Oka, T.; Aoki, H. Brillouin-Wigner theory for high-frequency expansion in periodically driven systems: Application to Floquet topological insulators. *Phys. Rev. B* **2016**, *93*, 144307. [CrossRef]
15. Nagasawa, F.; Frustaglia, D.; Saarikoski, H.; Richter, K.; Nitta, J. Control of the spin geometric phase in semiconductor quantum rings. *Nat. Commun.* **2013**, *4*, 2526. [CrossRef]
16. Molnár, B.; Peeters, F.M.; Vasilopoulos, P. Spin-dependent magnetotransport through a ring due to spin-orbit interaction. *Phys. Rev. B* **2004**, *69*, 155335. [CrossRef]
17. Frustaglia, D.; Richter, K. Spin interference effects in ring conductors subject to Rashba coupling. *Phys. Rev. B* **2004**, *69*, 235310. [CrossRef]
18. Citro, R.; Romeo, F.; Marinaro, M. Zero-conductance resonances and spin filtering effects in ring conductors subject to Rashba coupling. *Phys. Rev. B* **2006**, *74*, 115329. [CrossRef]
19. Kozin, V.K.; Iorsh, I.V.; Kibis, O.V.; Shelykh, I.A. Quantum ring with the Rashba spin-orbit interaction in the regime of strong light-matter coupling. *Phys. Rev. B* **2018**, *97*, 155434. [CrossRef]
20. Frustaglia, D.; Nitta, J. Geometric spin phases in Aharonov-Casher interference. *Sol. State Comm.* **2020**, *311*, 113864. [CrossRef]
21. Meijer, F.E.; Morpurgo, A.F.; Klapwijk, T.M. One-dimensional ring in the presence of Rashba spin-orbit interaction: Derivation of the correct Hamiltonian. *Phys. Rev. B* **2002**, *66*, 033107. [CrossRef]
22. Eckardt, A.; Anisimovas, E. High-frequency approximation for periodically driven quantum systems from a Floquet-space perspective. *New J. Phys.* **2015**, *17*, 093039. [CrossRef]
23. Heiss, W.D.; Nazmitdinov, R.G. Orbital magnetism in small quantum dots with closed shells. *JETP Lett.* **1998**, *68*, 915. [CrossRef]
24. Ihn, T. *Semiconductor Nanostructures*; Oxford University Press: New York, NY, USA, 2010.

symmetry

MDPI

Article

Formation of Matter-Wave Droplet Lattices in Multi-Color Periodic Confinements

Maitri R. Pathak and Ajay Nath *

Indian Institute of Information Technology Vadodara Gujarat India, Gandhinagar 382028, India; 201773002@iiitvadodara.ac.in
* Correspondence: ajay.nath@iiitvadodara.ac.in

Abstract: In the paper, we introduce a new model that addresses the generation of quantum droplets (QDs) in the binary Bose–Einstein condensate (BEC) mixture with mutually symmetric spinor components loaded in multi-color optical lattices (MOLs) of commensurate wavelengths and tunable intensities. The considered MOL confinement is the combination of the four-color optical lattice with an exponential periodic trap, which includes the complete set of the Fourier harmonics. Employing the one-dimensional (1D) extended Gross–Pitäevskii equation (eGPE), we calculate the exact analytical form of the wavefunction, MF/BMF nonlinearities, and MOL trap parameters. Utilizing the exact solutions, the formation of supersolid-like spatially periodic matter-wave droplet lattices and superlattices is illustrated under the space-periodic nonlinearity management. The precise positioning of the density maxima/minima of the droplet patterns at the center of the trap and tunable Anderson-like localization are observed by tuning the symmetry and amplitude of the considered MOL trap. The stability of the obtained solution is confirmed using the Vakhitov–Kolokolov (VK) criterion.

Keywords: quantum droplets; multi-color periodic confinement; Bose–Einstein condensate

Citation: Pathak, M.R.; Nath, A. Formation of Matter-Wave Droplet Lattices in Multi-Color Periodic Confinements. *Symmetry* **2022**, *14*, 963. https://doi.org/10.3390/sym14050963

Academic Editors: V.I. Yukalov, V. S. Bagnato and Rashid G. Nazmitdinov

Received: 12 April 2022
Accepted: 5 May 2022
Published: 9 May 2022

1. Introduction

Bose–Einstein condensates (BECs) and ultracold atoms confined in an optical lattice (OL) trap constitute an ideal experimental platform for the quantum simulations of emerging quantum many-body phenomena [1–5]. The experimental observation of the zero-temperature quantum phase transition in a strongly interacting Mott insulator [6] and weakly interacting Bose gas [7] utilizing the OL confinement provides a significant example for the same. Further, the OL is widely used to investigate the fundamental physics problems: Anderson localization [8], negative temperature [9–11], supersolid phase [12], etc., and it also provides the basis for the development of quantum technologies: quantum memory [13], registers [14], optical lattice clocks [15], and entanglement [16]. Experimentally, the OL trap is formed by the superposition of two counter propagating laser beams, resulting in the generation of an artificial crystal of light with spatially periodic polarization patterns, which is tunable through the power and period of the overlapping lasers [1,11]. Different engineered forms of OL geometries such as bi-periodic, kagome, hexagonal, double-well superlattices, etc., are regularly experimentally realized by interfering different sets of laser beams [17]. A number of interesting physical phenomena have been reported in the presence of the above-mentioned geometrically frustrated OLs, including frustrated quantum magnetism at negative absolute temperature [18], many-body localization [19], the exploration of the ionic Hubbard model with ultracold fermions [20], and Hund's metal in multicomponent Fermi systems [21].

Currently, a new class of quantum liquids, ultradilute quantum droplets (QDs), has aroused a great deal of attention in the field of BECs [22–24]. Usually, the BEC dynamics is studied in the presence of an external trap, since it is commonly known to exist in a gaseous phase, and in the absence of a container (i.e., external trap), it expands. However, Petrov, in a pioneering theoretical proposal, pointed out that liquid-like QDs can be stabilized in a weakly interacting binary BEC mixture in free space by realizing a subtle balance between the attractive cubic mean-field (MF) interaction and quadratic repulsive beyond mean-field (BMF) interaction generated due to the quantum fluctuations [25]. Like solitons, these dilute QDs are a self-bound many-body state; however, bright solitons collapse under the influence of the attractive cubic MF interaction. Different from that, QDs can be stabilized in 3D by counter balancing MF and BMF interactions. Based on this stabilization mechanism, QDs are observed in Bose–Bose mixtures [26,27] and dipolar gases [28,29]. The idea of droplet formation without any external trapping is not new, and this is studied in classical liquid or liquid helium systems [30]. However, in comparison to classical liquids in which droplets are generated due to a balance of repulsive interactions (generated due to high density) with attractive van der Waals interactions [30], the formation of QDs is a quantum phenomena with its formation dependent on the balance of the Lee–Huang–Yang (LHY) interaction (due to quantum fluctuations), i.e., the BMF interaction with two-body MF interactions. Further, the observed equilibrium density of QDs is 10^8 orders smaller than liquid helium, and due to the realization in ultracold atoms, this provides versatile control of the tunability of MF/BMF interactions and the geometry of the system.

In this work, we address the dynamics of QDs in a two-component binary BEC mixture in the presence of external multi-color optical lattice (MOL) confinement through the spatial periodic management of MF and BMF nonlinearities in 1D geometry. The QD dynamics is extensively explored in the absence of any confinement, i.e., free space [31,32], and some of us have recently reported the exact theoretical model for QDs in harmonic confinement [33]. However, the study of QDs in the presence of the OL or MOL has received less attention in the current literature. In condensed matter physics, the periodic lattice is considered as one of the fundamental problems, and even in the ultracold atoms domain, a variety of solitonic structures have been investigated in OL traps, both analytically and numerically. The study of QDs' behavior in the presence of OL or MOL traps acts as a quantum test bed for exploring advanced solid-state physics concepts, such as topological quantum states, discrete systems, etc. Recently, Morera et al. illustrated the generation of QDs, dimerized QDs, and a variety of phases in OL confinement [34,35], whereas the supersolid-like crystallization of QDs was investigated in 1D [36] and on a periodic lattice in a quasi-2D trapped dipolar BEC [37,38]. Further, the stability of QDs is also studied in OL, and the existence of stable dipole QDs has been proposed [39]. The motivation for studying a two-component BEC in the presence of the MOL is twofold: (i) constructing a family of OL traps: the precise control of the intensity and period of overlapping laser beams results in the formation of the OL, bi-color OL (BOL), frustrated bi-color double-well superlattice, tri-color OL (TOL), and four-color OL (FOL); and (ii) a test bed for quantum simulation: optical superlattice confinements comprise a clean controllable many-body test bed, and a variety of physical phenomena [8,9,16] is observed in this trap. In principle, the multi-color beams can be used to design a variety of optical superlattice potentials necessary for supporting the existence of non-trivial QD patterns [40,41]. Therefore, theoretically, it would be important to investigate the behavior of QDs in the presence of the MOL. In this paper, we solve the 1D eGPE for the considered confinement and calculate the non-trivial exact analytical form of the wavefunction, phase, MF/BMF nonlinearities, and trap parameters. This reveals the specific form of the MOL and its correlation with the MF/BMF nonlinearities, which provide tunability for the generation of various QDs density profiles. As an illustration, we show the generation of supersolid-like density schemes in QDs: periodic lattice, bi-periodic superlattice, and bi-periodic double-well superlattice. For each of these patterns, we write the analytical solutions and identify the specific form of the multi-color OL and its parameter domain. The controllable positioning of density maxima at the center of the trap, compression, fragmentation, and Anderson-like localization of the droplet patterns are also observed by tuning the symmetry of the considered trap.

In the following section, we present the analytical framework for solving the 1D eGPE for a weakly interacting Bose–Bose mixture with equal masses and an equal number of atoms in the components under the MOL. The model for the calculated system variables is explained by finding the MOL potential parameters, i.e., the amplitudes and periodicity, the wavefunction, and the form of MF/BMF interactions. It is shown that with a suitable choice of the laser intensity, one can construct the following experimentally relevant forms of confinements: single-color OL, BOL, TOL, double-well superlattice BOL, and FOL. Next, we study the characteristics of QDs under the influence of the above-mentioned traps and illustrate the interesting supersolid-like periodic lattice and superlattice density patterns in QDs. Finally, the stability of the obtained solutions is confirmed using the Vakhitov–Kolokolov (VK) criterion.

2. The Model and Analytical Framework

We start by considering the two-component mass-balanced binary BEC under the influence of the BMF (LHY corrections for quantum fluctuations) in the presence of spatially varying MOL confinement. The choice of equal masses and an equal number of atoms in the BEC mixture makes the result analysis clearer and easier. In the 1D configuration, the QDs are observed under the subtle balance of the slightly repulsive MF interaction with the attractive BMF. In this geometry, the MF and BMF interactions' contribution to the energy per particle is proportional to n and $\sqrt{n}$, where n is the density of the gas [31], and the system is described by the following equations, the 1D eGPE [25,39]:

$$i\hbar\frac{\partial\psi_1}{\partial t} = -\frac{\hbar^2}{2m}\frac{\partial^2\psi_1}{\partial x^2} + (\Lambda_s(x)|\psi_1|^2 + \Lambda_c(x)|\psi_2|^2)\psi_1 - \Gamma(x)(|\psi_1|^2 + |\psi_2|^2)^{1/2}\psi_1 + v(x)\psi_1, \tag{1}$$

$$i\hbar\frac{\partial\psi_2}{\partial t} = -\frac{\hbar^2}{2m}\frac{\partial^2\psi_2}{\partial x^2} + (\Lambda_c(x)|\psi_1|^2 + \Lambda_s(x)|\psi_2|^2)\psi_2 - \Gamma(x)(|\psi_1|^2 + |\psi_2|^2)^{1/2}\psi_2 + v(x)\psi_2. \tag{2}$$

Here, $v(x)$ is the external MOL confinement present in the considered system with ψ_1 (ψ_2) representing the wavefunctions of binary mixture components. In Equations (1) and (2), we take the interaction strengths controlling the repulsion between the atoms in each components to be equal: $g_{11} = g_{22} \equiv g = 2\hbar^2 a_s(x)/(ma_\perp^2)$ and $g_c = g_{12}$. Here, $\Lambda_s(x) = (g_c + 3g)/2$ represents the self-interaction coefficients, whereas $\Lambda_c(x) = (g_c - g)/2$ is the cross interaction coefficients along with $\Gamma(x) = \sqrt{m}g^{3/2}/(\pi\hbar)$ [39]. $a_s(x)$ represent the space-dependent inter- and intra-components' atomic scattering lengths, which are tunable through the Feshbach resonance technique [42]. Thus, the sign and strength of both (inter- and intra-) components' atomic scattering length, i.e., MF and BMF interactions, can be experimentally modulated. Here, m is the mass of the BEC atoms and $\hbar$ is the scaled Planck's constant.

Next, we reduce the dynamics of the considered system to the space-dependent dimensionless single eGPE by assuming $\psi_1 = \psi_2 = \psi_0\psi$, i.e., mutually symmetric spinor components in the binary mixture [25,31]:

$$i\frac{\partial\psi}{\partial t} = -\frac{1}{2}\frac{\partial^2\psi}{\partial x^2} - g_1(x)|\psi|\psi + g_2(x)|\psi|^2\psi + V(x)\psi. \tag{3}$$

Equation (3) is the extended form of the Gross–Piteäevskii equation (GPE) with the external confinement added. For the case, $V(x) = 0$, it becomes the 1D eGPE in free space, which is extensively explored to investigate the QDs' dynamics [23,24]. In the equation, $\psi(x,t)$ is the condensate wave function of the QDs having mass m, and $g_1(x) = \Gamma(x)$, $g_2(x) = \Lambda_s(x) + \Lambda_c(x)$ are the magnitude coupling strengths of the two-component Bose–Bose mixture representing BMF and MF interactions, respectively. Here, $\frac{\hbar g_1(x)}{\Gamma(x)}\sqrt{\frac{\Lambda_s+\Lambda_c}{2mg_2(x)}}$, $\frac{\hbar(\Lambda_s+\Lambda_c)g_1^2(x)}{2g_2(x)\Gamma^2(x)}$, $\frac{\sqrt{2}g_2(x)\Gamma(x)}{(\Lambda_s+\Lambda_c)g_1(x)}$ are the magnitudes of the scaling parameters x_0, t_0, ψ_0, respectively [31]. To investigate the structure and dynamics of QDs in the presence of MOL, we consider the form of the external trap:

$$V(x) = \sum_{j=1}^{4} V_j \cos^2(jkx) + V_5\, exp[2(p_1 \cos^2(kx) + p_2 \cos^2(2kx))], \tag{4}$$

which is a combination of the FOL with commensurate lattice periods and an exponential periodic trap. The combination results in the generation of an MOL potential trap, which becomes an FOL for $V_5 \to 0$. The choice of the exponential periodic trap ensures the presence of the complete set of the Fourier harmonics in the considered trap combination. Here, p_1 and p_2 are real constants. In Equation (4), $k = 2\pi a_\perp/\lambda$ is the scaled lattice wave vector, which is commensurate for the four-color laser beams $(k, 2k, 3k, 4k)$ with $a_\perp = \sqrt{\hbar/m\omega_\perp}$, and $\omega_\perp$ is the transverse oscillator frequency. Here, V_j ($j = 1,2,3,4,5$) represents the potential depths of each OL and is connected to the recoil energy: $E_R = \frac{2\pi^2\hbar^2}{m\lambda^2}$ for the laser of wavelength (color) λ and mass m of BEC atoms [40]. Recently, ultracold atoms were investigated in the FOL potential to obtain eightfold rotationally symmetric OLs [43] and solitonic solutions [41]. In this paper, we considered the cigar-shaped Li^7 BEC atoms trapped with transverse frequency $\omega_\perp = 2\pi \times 710$ Hz, atomic scattering length $a_s = -0.21$ nm, and a CO_2 laser of wavelength $\lambda = 10.62$ µm [44]. In the experiments, various forms of OLs can be realized by tuning the applied magnetic field, the magnitude of k, and the angle between superimposing laser beams [45].

In order to construct the analytical solution form of ψ for Equation (3), based on the general similarity transformation scheme used for constructing matter-wave solitons in 1D geometry [46,47], we start with following ansatz solution:

$$\psi(x,t) = A(x,t) U[\eta(x,t)] e^{i\phi(x,t)}, \tag{5}$$

where $A(x,t), \phi(x,t)$, and $U[\eta(x,t)]$, being real functions, are the space- and time-modulated amplitude, phase, and similarity variables, respectively. Using the ansatz solution, our goal is to connect Equation (3) to the solvable differential equation:

$$-\frac{\partial^2 U}{\partial \eta^2} - G_1 \mid U(\eta) \mid U + G_2 \mid U(\eta) \mid^2 U = EU, \tag{6}$$

such that we obtain the following consistency conditions on the amplitude and MF and BMF nonlinearities for the chosen potential of Equation (4) (see Appendix A):

$$[A^2(x,t)\eta_x(x,t)]_x = 0, \qquad \eta_t(x,t) + \eta_x(x,t)\phi_x(x,t) = 0, \tag{7}$$

$$G_1\eta_x^2(x,t) - 2A(x,t)g_1(x,t) = 0, \quad G_2\eta_x^2(x,t) - 2A^2(x,t)g_2(x,t) = 0, \tag{8}$$

$$\frac{A_t(x,t)}{A(x,t)} + \frac{1}{2A^2(x,t)}[A^2(x,t)\phi_x(x,t)]_x = 0, \tag{9}$$

$$\frac{A_{xx}(x,t)}{2A(x,t)} - \frac{\phi_x^2(x,t)}{2} - \phi_t(x,t) - \frac{1}{2}E\eta_x^2(x,t) - V(x) = 0. \tag{10}$$

In Equation (6), E is the eigenvalue of the equation, G_1, G_2 denote the constantBMF/MF interactions, which can take a positive or negative magnitude depending on the sign of the inter- and intra-component atomic scattering length. Here, the function with the subscript implies the partial differentiation of the function with respect to the subscripted variable. The above set of consistency conditions are simultaneously solved to obtain the amplitude, phase, and MF/BMF:

$$\eta_x(x,t) = \frac{b(t)}{A^2(x,t)}, \quad \phi_z = -\frac{\eta_t(x,t)}{\eta_x(x,t)}, \quad g_1(x,t) = G_1\frac{\eta_x^2(x,t)}{2A(x,t)}, \quad g_2(x,t) = G_2\frac{\eta_x^2(x,t)}{2A^2(x,t)}, \tag{11}$$

where $b(t)$ is an integration constant. It is evident from Equation (11) that the form of MF/BMF nonlinearities and the phase is directly dependent on the amplitude, which will be determined by solving the consistency Equation (11). For that purpose, we substitute the trap expression from Equation (4) into the set of consistency Equations (7) and (11) and choose $\eta(x,t) = \gamma \int_0^z exp[p_1 \cos^2(kx) + p_2 \cos^2(2kx)]\partial x$, to obtain the exact analytical form of the amplitude, phase, and nonlinearities:

$$A(x,t) = \sqrt{\frac{b(t)}{\gamma \times exp[p_1 \cos^2(kx) + p_2 \cos^2(2kx)]}}, \tag{12}$$

$$\theta(x,t) = [2p_1^2k^2 + 8p_2^2k^2 - 8k^2(p_1 + 4p_2)]t, \tag{13}$$

$$g_1(x,t) = \frac{G_1\gamma^{3/2}}{2b(t)} exp[p_1 \cos^2(kx) + p_2 \cos^2(2kx)]^{\frac{3}{2}}, \tag{14}$$

$$g_2(x,t) = \frac{G_2\gamma^3}{2b(t)} exp[p_1 \cos^2(kx) + p_2 \cos^2(2kx)]^3, \tag{15}$$

with the potential depths of each overlapping OL connected in the following manner:

$$V_1 = p_1k^2\left[1 + \frac{p_2}{2}\right], \; V_2 = k^2\left[\frac{p_1^2}{8} + 4p_2\right] \; V_3 = -\frac{p_1p_2k^2}{2} \; V_4 = -\frac{p_2^2k^2}{2} \; V_5 = \frac{E\gamma^2}{2}. \tag{16}$$

Equation (16) reveals a non-trivial correlation in between trap parameters p_1 and p_2, and this is one of the important results of the article. In principle, by the suitable tuning of these parameters, one can realize various forms of $V(x)$: OL (Figure 1c), BOL (Figure 1d), double-well superlattice (Figure 1e), frustrated double-well optical superlattice (Figure 1f), bi-periodic frustrated double-well optical superlattice (Figure 1g), etc., and study the QDs' profiles with precise knowledge of the MF/BMF interactions. In Equation (12), $b(t) = b$ (constant) $= \gamma^2$.

The solution of Equation (6) can be given as: $U[\eta] = \frac{3(E/G_1)}{1+\sqrt{1-\frac{E}{\mu_0}\frac{G_2}{G_1^2}}\cosh(\sqrt{-E}\eta)}$ with $\mu_0 = -2/9$, $E < 0$, $G_1 < 0$, and $G_2 > 0$ [25,31]. Thus, the complete solution of Equation (3) can be written as:

$$\psi(x,t) = \sqrt{\frac{b}{\gamma \times exp[p_1 \cos^2(kx) + p_2 \cos^2(2kx)]}} \times exp\left[i(2p_1^2k^2 + 8p_2^2k^2 - 8k^2(p_1 + 4p_2)t)\right] \frac{\frac{3E}{G_1}}{1 + \sqrt{1 - \frac{E}{\mu_0}\frac{G_2}{G_1^2}}\cosh(\sqrt{-E}(\gamma \int_0^x exp[p_1 \cos^2(kx) + p_2 \cos^2(2kx)]))}. \tag{17}$$

Further, one can choose the solution of Equation (6) in terms of the Jacobi elliptic function (cn) as: $U(\eta) = B \quad cn[\beta\ \eta,\ q] + D$, with $B = \sqrt{\frac{2}{(2q^2-1)}}D > 0$, $D = \frac{G_1}{3G_2} < 0$, $\beta^2 = -(\frac{6G_2}{(2q^2-1)})\ D^2$, and $q^2 > 1/2$ [48]. For this case, the complete wavefunction form of Equation (3) becomes:

$$\psi(x,t) = \sqrt{\frac{b}{\gamma \times exp[p_1 \cos^2(kx) + p_2 \cos^2(2kx)]}} \times exp\left[i(2p_1^2k^2 + 8p_2^2k^2 - 8k^2(p_1 + 4p_2)t)\right] \left[B\ cn(\beta\ (\gamma \int_0^x exp[p_1 \cos^2(kx) + p_2 \cos^2(2kx)]),\ q) + D\right], \tag{18}$$

where q is the modulus parameter for the Jacobi elliptic function cn. In principle, the cn function can also possess a family of solutions for the range of the modulus parameter, $0 \leq q \leq 1/2$, and at $q = 0$ will signify a periodic QD profile. Thus, it is worth indicating that we constructed a large family of exact analytical solutions of the 1D eGPE for the considered MOL trap configuration.

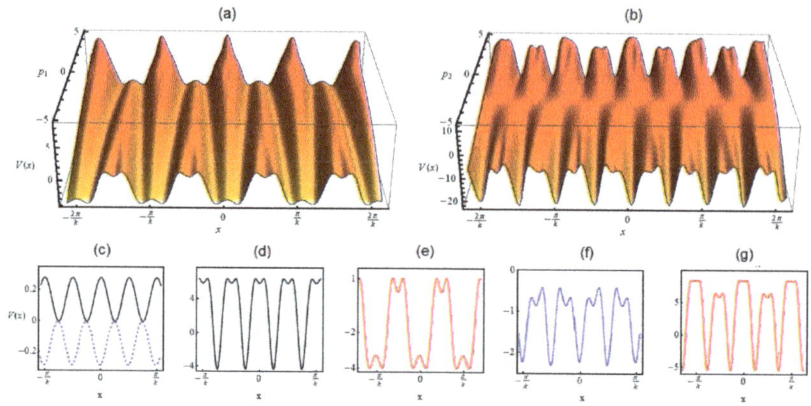

Figure 1. Various potential profiles of the MOL by tuning parameters p_1 and p_2 of Equation (4) for $\gamma = 0.05$, $k = 0.84$, $E = -2/9$ in the interval $[-2\pi/k, 2\pi/k]$: (**a**) for fixed $p_2 = 1$, p_1 varying from -5 to $+5$; (**b**) for fixed $p_1 = 1$, p_2 varying from -5 to $+5$; (**c**) flipping of the OL phase by the half-wavelength: $p_1 = 0$, $p_2 = 0.1$ (black solid line) and $p_1 = 0$, $p_2 = -0.1$ (dashed blue line); (**d**) BOL: $p_1 = 0$, $p_2 = 3.5$ (black line); (**e**) triple-well optical superlattice: $p_1 = -5$, $p_2 = 1$ (red solid line); (**f**) frustrated double-well optical superlattice: $p_1 = 1$, $p_2 = -5$ (blue solid line); (**g**) bi-periodic frustrated double-well optical superlattice: $p_1 = 1$, $p_2 = 4$ (red solid line). The spatial co-ordinate is scaled by the oscillator length.

3. Results

The exact expressions of the distributed confinement parameters are already solved in Equation (16) where the parameters (p_1, p_2) can be widely tuned. We are now in a position to analyze the relevant QDs' profiles along with the corresponding confinement form. In the following, we would like to perform a detailed study of the tuning of the trap under consideration and of the density patterns of the QDs.

3.1. Potential Profiles and Corresponding Trap Parameters

From the constructed model, the general form of the potential can be written as:

$$V(x) = p_1 k^2\left[1 + \frac{p_2}{2}\right]\cos^2(kx) + k^2\left[\frac{p_1^2}{8} + 4p_2\right]\cos^2(2kx) - \frac{p_1 p_2 k^2}{2}\cos^2(3kx) - \frac{p_2^2 k^2}{2}\cos^2(4kx) + \frac{E\gamma^2}{2}\, exp[2(p_1\cos^2(kx) + p_2\cos^2(2kx))], \quad (19)$$

where $-2/9 < E < 0$ and $\gamma > 0$. With suitable tuning of the physical parameters p_1 and p_2, one can construct various potential profiles and obtain the exact analytical expression of the wavefunction. Physically, p_1 and p_2 are connected to the power of overlapping laser intensities. Figure 1 illustrates the various trap patterns by tuning of the trap parameters (p_1, p_2). In Figure 1a,b, the variation of the potential profile is depicted for $\gamma = 0.05$, $k = 0.84$, $E = -2/9$ in the interval $[-2\pi/k, 2\pi/k]$: (a) for fixed $p_2 = 1$, with p_1 varying from -5 to $+5$; and in (b) for fixed $p_1 = 1$, p_2 varying from -5 to $+5$. This clearly indicates the various shapes of the resultant potential by changing the magnitude of p_1 and p_2. For $p_1 = p_2 = 0$, $V(x)$ becomes constant, representing the free space scenario. Further, the trap takes the form of the two-color BOL with periodicity $(k, 2k)$ for $p_1 > 0$ and $p_2 = 0$, whereas if $p_2 \neq 0$, then the MOL becomes a disordered double-well superlattice with frustration present both in the inter- and intra-well separations [41]. Additionally, we observe interesting trap forms for $p_1, p_2 < 0$. For $p_1 < 0$ with p_2 positive constant, this leads to the formation of the triple-well superlattice, which is evident from Figure 1a.

Similarly, we illustrate a half-wavelength shift of this double-well superlattice by tuning $p_2 < 0$ with p_1 positive constant. For better insight into the tunability of the generated MOL trap form, we plot the shape of the trap at some specific points and illustrate the experimentally realizable trap configuration: (a) OL: for $p_1 = 0$, $p_2 = 0.1$ (black solid line) and the flipping of this OL by the half-wavelength for $p_1 = 0$, $p_2 = -0.1$ (dashed blue line) (Figure 1c); (b) BOL: overlapping of two OLs of frequency $3k$ and $4k$ for $p_1 = 0$, $p_2 = 3.5$ (black solid line) (Figure 1d); (c) triple-well optical superlattice: $p_1 = -5$, $p_2 = 1$ (red solid line) (Figure 1e); (d) frustrated double-well optical superlattice: $p_1 = 1$, $p_2 = -5$ (blue solid line) (Figure 1f); (e) bi-periodic frustrated double-well optical superlattice: $p_1 = 1$, $p_2 = 4$ (red line) (Figure 1g). Thus, potential Equation (19) consists of the family of OLs, and in principle, the complete set of the Fourier harmonics can be generated from it.

For $\gamma^2 < 1$ or $p_1, p_2 < 0$, the higher-order terms of the exponential trap tend towards zero, and the trap behaves as an FOL confinement. Utilizing Equation (19), we identify the points in Table 1 at which the shape of the MOL becomes the OL, BOL, or TOL in the (p_1, p_2) space for $\mu = 0$ [41]. On the contrary, the FOL is obtained in the entire space, excluding the points indicated in the table. It needs to be emphasized here that the constructed MOL trap configuration provides a large variety of experimentally useful potential profiles with exact analytical solutions, which may find applications for quantum information processing and simulations [5,16,17]. We further illustrate the results by displaying various QD patterns in the presence of the above-mentioned trap configurations.

Table 1. Various shapes of the MOL potential by tuning the magnitude of the power of the laser beam, i.e., p_1 and p_2.

Multi-Color OL (for $\mu = 0$)		
p_1	p_2	Trap form
0	0	Free space
<1	0	OL (k)
0	≠0	BOL ($2k$, $4k$)
>1	0	BOL (k, $2k$)
8	−2	BOL ($3k$, $4k$)
8	2	BOL ($3k$, $4k$)
≠8	−2	TOL ($2k$, $3k$, $4k$)
≠8	2	TOL ($2k$, $3k$, $4k$)
Other points	Other points	FOL (k, $2k$, $3k$, $4k$)

3.2. Periodic Lattice Density Patterns in QDs

In this section, we correlate the above-mentioned understanding of the confinement engineering with the atomic condensate density for a variety of experimentally relevant forms of the MOL confinement to investigate the droplet characteristics in it. As discussed earlier, a family of OL traps can be constructed from (19), but we begin by investigating the generation of QDs in the presence of the two-color BOL$(k, 2k)$ with competing MF and BMF nonlinearities. The superposition of two-color OLs results in the formation of quasi-periodic optical superlattice confinement, which is used to study various interesting physical phenomena: Anderson localization, frustrated quantum magnetism, negative absolute temperature, etc. [10,18,19]. Motivated by that, we construct the BOL with commensurate period $(k, 2k)$ from potential Equation (20) by taking $p_2 = 0$. Further, the form of the MF and BMF nonlinearities is: $g_2(x) = (G_2\gamma^3/2b)exp[p_1 \cos^2(kx)]^3$ and

$g_1(x) = (G_1\gamma^{3/2}/2b)exp[p_1\cos^2(kx)]^{\frac{3}{2}}$, respectively, with $b(t) = b$ (constant). Thus, the resulting form of the potential from Equation (19) takes the form:

$$V(x) = p_1k^2\cos^2(kx) + \frac{(p_1k)^2}{8}\cos^2(2kx) + \frac{E\gamma^2}{2}exp[2(p_1\cos^2(kx))], \tag{20}$$

for which the wavefunction solution of Equation (3) can be given as:

$$\psi(x) = \sqrt{\frac{b}{\gamma \times exp[p_1\cos^2(kx)]}\frac{\frac{3E}{G_1}}{1+\sqrt{1-\frac{E}{\mu_0}\frac{G_2}{G_1^2}}\cosh(\sqrt{-E}(\gamma\int_0^x exp[p_1\cos^2(kx)]))}} \times exp\left[i(2p_1^2k^2 - 8k^2p_1)t\right]. \tag{21}$$

Using the above wavefunction, we illustrate the density profiles of QDs for $\mp p_1$ for the BOL potential Equation (20) in Figures 2 and 3, respectively. In these figures, each plot (a–d) has two panels: the upper panel shows the density plot, and the lower panel consists of a 2D plot of the density. In Figure 2a, initially, we take $p_1 = 0$, making $V(x) = $ constant, i.e., the free space potential and with the MF interaction ($G_2 = 0.999999999$) and the BMF interaction ($G_1 = -1$). As there is a subtle balance between MF and BMF, thus we observe the flat top density profile, which is the signature feature of QDs.

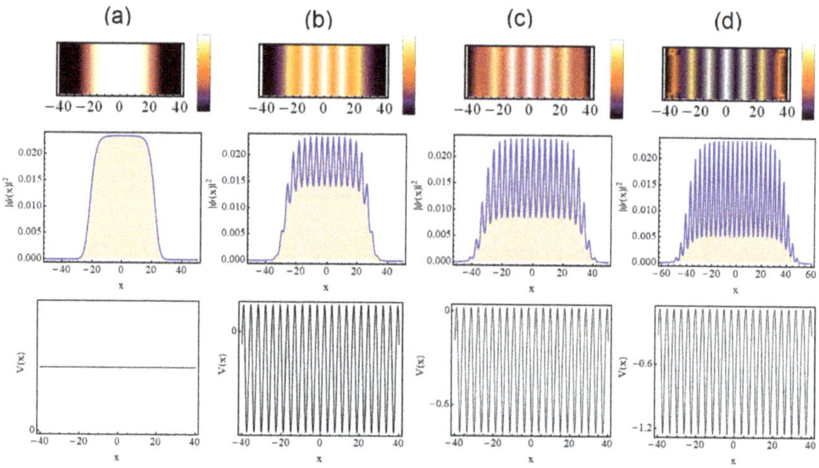

Figure 2. Condensate density patterns for two-color BOL traps with $p_1 < 0$ and $p_2 = 0$: (**a**) $p_1 = 0$ (free space); (**b**) $p_1 = -0.5$; (**c**) $p_1 = -1$; and (**d**) $p_1 = -1.50$. Each plot of (**a–d**) has three panels: the upper panel shows the density plot; the middle panel consists of a 2D plot of the density; the lower panel indicates the corresponding trap profile. Here, the magnitude of the physical parameters: $b = 1$, $\gamma = 1$, $k = 0.84$, $G_1 = -1$, $G_2 = 0.999999999$, $E = -2/9$. The spatial co-ordinate is scaled by the oscillator length.

Similarly, condensate density has previously been illustrated in the free space potential, and the observed pattern is in conformity with the physical situations reported in the literature [32]. Here, the magnitude of other physical parameters: $b = 1$, $\gamma = 1$, $k = 0.84$, $E = -2/9$. Next, we investigate the impact of $p_1 < 0$ on the droplet density profile for the same physical parameters. We took $p_1 = -0.5, -1, -1.5$ and depict their corresponding condensate density patterns in Figure 2b–d. It is evident from the figure that with p_1 tending from $0 \to -1.5$, this leads to the expansion of the QDs, and due to the increase in the lattice depth of the trap, it forms the lattice patterns in the QDs. Importantly, the signature of the potential is superimposed over the flat density profile, and correspondingly, we observe periodic lattice density pattern on the QDs. We observe that for $p_1 < 0$, the density periodic lattice maxima are located at $x = 0$. However, to establish supersolidity in the system, we need to show the spontaneous breaking of the translational symmetry and the spontaneous breaking of the gauge symmetry in the considered system. However, in absence of that, we term this supersolid-like periodic density patterns, as done in previous studies on spinor BECs [38,49]. Further, we note that the decreasing magnitude of p_1 also results in the increase of the width of the QDs, i.e., leading to its expansion. Thus, in principle, the depth and width of these formed periodic lattice density patterns in QDs are connected to the magnitude of p_1, and from the constructed analytical model, we reveal a non-trivial correlation in between them.

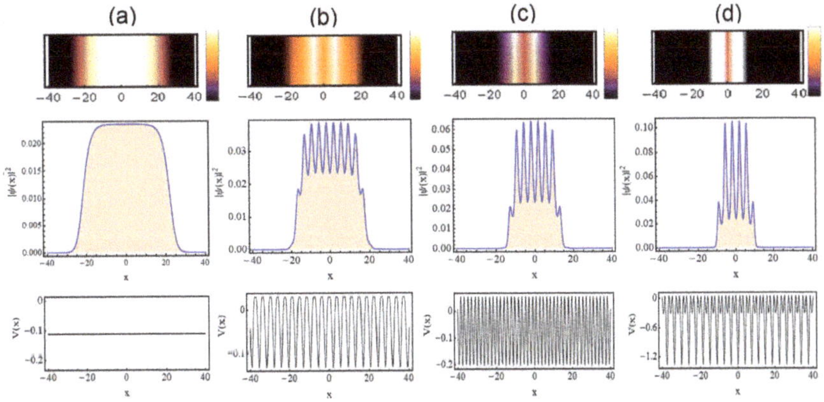

Figure 3. Condensate density patterns for two-color BOL traps with $p_1 > 0$ and $p_2 = 0$ are depicted: (**a**) $p_1 = 0$ (free space); (**b**) $p_1 = 0.5$; (**c**) $p_1 = 1$; and (**d**) $p_1 = 1.5$. Each plot (**a**–**d**) has three panels: the upper panel shows the density plot; the middle panel consists of a 2D plot of the density; the last panel represents the corresponding trap profile. Here, the magnitude of the physical parameters: $b = 1, \gamma = 1, k = 0.84, G_1 = -1, G_2 = 0.999999999, E = -2/9$. The spatial co-ordinate is scaled by the oscillator length.

In Figure 3, we illustrate the density variation of the wavefunction solution (21) for $p_1 > 0$ with its magnitude changing from $0 \to 1.5$ for the identical physical parameter as taken in the case of $p_1 < 0$. Here, we take the parameter values: $b = 1, \gamma = 1, k = 0.84$, $E = -2/9, G_2 = 0.999999999, G_1 = -1$. We begin with $p_1 = 0$ in Figure 3a, i.e., the free space situation. Next, the magnitude of p_1 increased to 0.5, resulting in the formation of multiple peaks in the density profile, which are due to the presence of the OL trap (Figure 3b). However, in comparison to $p_1 = -0.5$ shown in Figure 2b, here, the number of droplet lattice peaks is less and the width is compressed. Subsequently, we observe a reduction in the width of the droplet and the number of density lattice wells with the increase of p_1 (Figure 3c,d). The condensate atomic density illustrates an Anderson-like localization with the increasing magnitude of the laser intensity p_1. This behavior was confirmed by comparing the maximum amplitude of the density for $\pm p_1$. In comparison to $p_1 < 0$, in which the maximum amplitude remains the same (Figure 2), here, its magnitude increases with p_1 tending from 0 to 1.5, showing the localization of condensate atoms. Further, in comparison to $p_1 < 0$, we note here that the density periodic lattice minima are located at $x = 0$ for $p_1 > 0$.

In order to physically understand the reasons for the observation of the QDs' density patterns in Figures 2 and 3, we plot the profile of MF/BMF nonlinearities with respect to the variation of the MOL trap parameters with $p_2 = 0$ and p_1 varying from -1.5 to $+1.5$ in Figure 4a. It is evident from the figure that the magnitude of MF and BMF interactions remains comparable till p_1 changes in the interval $[-1.5, 0]$. This is due to the negative magnitude of p_1, which is present in the exponential term of both interactions, and it ensures the observation of the flat top density profile in Figure 2a–d. Furthermore, the increase in the depth of the QDs' periodic lattice patterns is due to the potential depth of the trap with changing the value of p_1 from 0 to -1.5, which leads to the expansion of the droplet profile. For $p_1 > 0$ and its increase from 0 to $+1.5$, this leads to the amplification of both MF/BMF amplitudes (due to the presence of the exponential factor); however, in this region, the MF term dominates the BMF due to the $exp[p_1 \cos^2(kx)]^3$ term in it. This leads to an imbalance of the MF and BMF interaction strengths, as is evident from Figure 4a, and simultaneously, the system tends towards the soliton region with a dominant repulsive MF. Since the balance of these nonlinearities is essential for the observation of the droplet profile, the increase of p_1 results in the gradual decrease of the flat top density profile (see Figure 3) and the phenomena of the Anderson-like localization observed, which is a characteristic feature of disordered optical lattices [40]. Thus, we illustrate an interesting and non-trivial tunability of QDs' density patterns by changing the sign (i.e., symmetry) and potential depth of the BOL $(k, 2k)$ confinement.

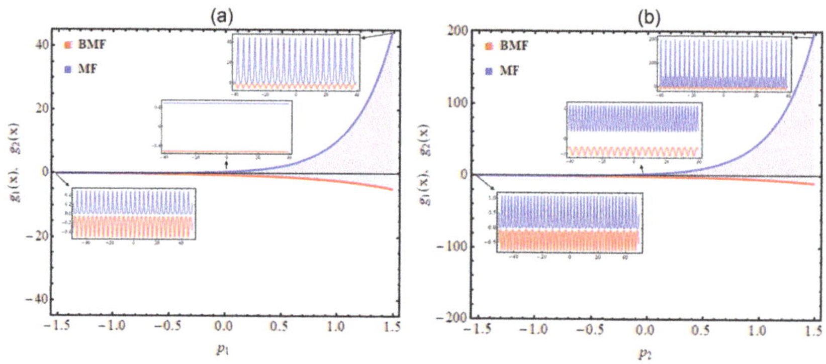

Figure 4. The profile of MF and BMF nonlinearities is plotted with respect to varying (**a**) p_1 in the interval $[-1.5, 1.5]$ at $x = 0$ for $p_2 = 0$ and (**b**) p_2 in the interval $[-1.5, 1.5]$ at $x = 0$ with $p_1 = 0.5$. Each inset plot depicts the variation of $g_1(x)$ (BMF, red line) and $g_2(x)$ (MF, blue line) for the indicated magnitude of p_1 and p_2, respectively. Here, the magnitude of the physical parameters: $b = 1$, $\gamma = 1$, $k = 0.84$, $G_1 = -1$, $G_2 = 0.999999999$, $E = -2/9$.

3.3. *Double-Well Superlattice Density Patterns in QDs*

In this section, we illustrate the formation, expansion, and compression of double-well superlattice density patterns in QDs under the MOL confinement. For that purpose, we take $p_2 \neq 0$ in the potential Equation (20) with $\gamma = 1$ and $k = 0.84$. Utilizing the corresponding wavefunction solution from Equation (20), we plot the atomic condensate density in Figure 5a–d for $p_2 = 0$; $p_2 = -0.5$, $p_2 = -1$, and $p_2 = -1.5$, respectively, for $p_1 = 0.5$, $G_1 = -1$, $G_2 = 0.999999999$, $E = -2/9$. It is evident from Figure 5a that the periodic lattice density patterns are formed for $p_2 = 0$ and $p_1 = 0.5$. However, as p_2 changes from $0 \rightarrow -0.5$, the double-well superlattice density patterns become visible in the QDs' profile (Figure 5b). The decreasing magnitude of p_2 from $-0.5 \rightarrow -1.5$ results in the expansion of the QDs along with the increase in the depths of these double-well superlattice density patterns. Further, in Figure 6, we depict the impact of $p_2 > 0$ and $p_1 = 0.5$ on the QDs' profile for the same physical parameter values. Here, we plot for (a) $p_2 = 0$, (b) $p_2 = 0.5$, (c) $p_2 = 1$, and (d) $p_2 = 1.5$, respectively. In comparison to the double-well superlattice density pattern of QDs for $p_2 = -0.5$, we observe the formation of a bi-periodic density lattice for the case of $p_2 = 0.5$ (Figure 6b). The depth of this bi-periodic density pattern increases, and the width of the droplet decreases with the increasing magnitude of p_2 from $0.5 \rightarrow 1.5$ (Figure 6c,d).

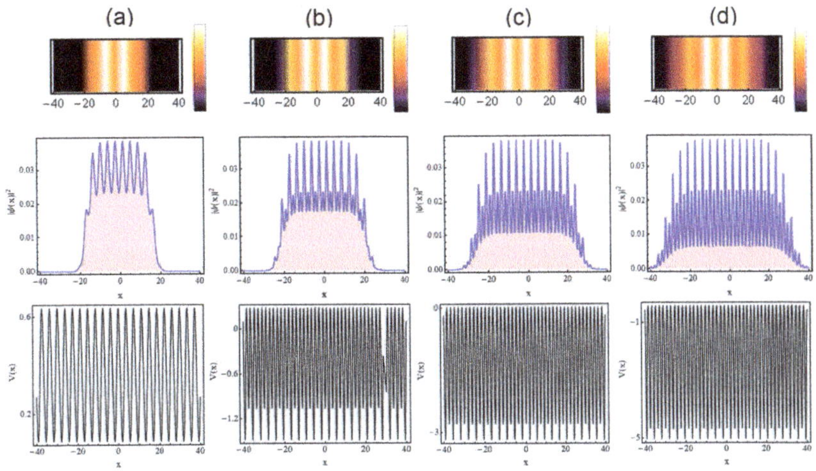

Figure 5. Condensate density patterns for four-color BOL traps with $p_1 = 0.5$ and: (**a**) $p_2 = 0$ (BOL); (**b**) $p_2 = -0.5$; (**c**) $p_2 = -1$; and (**d**) $p_2 = -1.5$. Each plot (**a–d**) has three panels: the upper panel shows the density plot; the middle panel consists of a 2D plot of the density; the lower panel indicates the corresponding trap profile. Here, the magnitude of the physical parameters: $b = 1$, $\gamma = 1$, $k = 0.84$, $G_1 = -1$, $G_2 = 0.999999999$, $E = -2/9$. The spatial co-ordinate is scaled by the oscillator length.

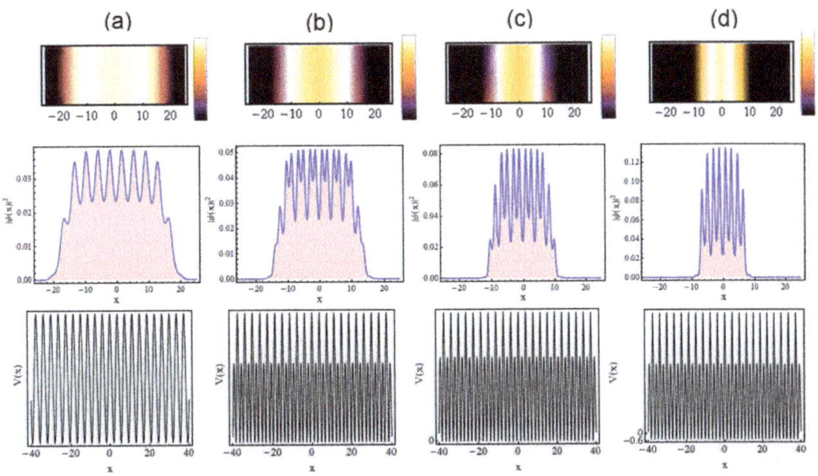

Figure 6. Condensate density patterns for four-color BOL traps with $p_1 = 0.5$ and: (**a**) $p_2 = 0$; (**b**) $p_2 = 0.5$; (**c**) $p_2 = 1$; and (**d**) $p_2 = 1.5$. Each plot (**a–d**) has three panels: the upper panel shows the density plot; the middle panel consists of a 2D plot of the density; the lower panel indicates the corresponding trap profile. Here, the magnitude of the physical parameters: $b = 1$, $\gamma = 1$, $k = 0.84$, $G_1 = -1$, $G_2 = 0.999999999$, $E = -2/9$. The spatial co-ordinate is scaled by the oscillator length.

In order to physically understand the reasons for the observation of the QDs' density superlattice patterns in Figures 5 and 6, we plot the profile of MF/BMF nonlinearities with respect to the variation of the MOL trap parameters with $p_1 = 0.5$ and p_2 varying from -1.5 to $+1.5$ in Figure 4b. In comparison to Figure 4a, in which $p_2 = 0$, here, the double-well superlattice trap is formed due to $p_2 \neq 0$. As discussed earlier, in this regime also, the magnitude of MF and BMF interactions remains comparable till p_2 changes from $[-1.5, 0]$ (shown in Figure 4b). This results in the observation of the flat top density profile as shown in Figure 6a–d. The superlattice QDs' density patterns are due to the superlattice periodicity of MF/BMF interactions and the corresponding resulting trap. For $p_2 > 0$ and tending from 0 to +1.5, then the MF term dominates the BMF due to the $exp[p_1 \cos^2(kx) + p_2 \cos^2(2kx)]^3$ factor. Similar to the previous regime, we observe a decrease in the flat top density profile with the increasing magnitude of p_2. However, different from the previous case, here, the droplet density patterns are bi-periodic. Therefore, we observe that the compression and expansion of the droplet density patterns are due to the spatial periodic management of MF and BMF nonlinearities: $g_1(x,t) = [G_1\gamma^{3/2}/2b(t)]exp[p_1 \cos^2(kx) + p_2 \cos^2(2kx)]^{\frac{3}{2}}$ and $g_2(x,t) = [G_2\gamma^3/2b(t)]exp[p_1 \cos^2(kx) + p_2 \cos^2(2kx)]^3$, respectively, with respect to the changing sign of $\pm p_2$. Thus, we illustrate an interesting transition of the supersolid-like double-well superlattice and bi-periodic density patterns in droplets from the starting periodic density profile by changing the magnitude of p_2 and MF/BMF nonlinearities.

3.4. Stability of QDs in MOL Confinement

In the above sections, we illustrated the generation of a variety of droplet density patterns in the MOL confinement by tuning the magnitude of (p_1, p_2) in Equation (17). In this section, we evaluate the stability of the obtained wavefunction solution (17), and for that purpose, we employed the VK criterion, which is extensively utilized to determine the stability of nonlinear Schrödinger equation solutions [50]. According to the VK criterion, a solution is found to be stable if the slope of the number of atoms with respect to the chemical potential, i.e., $N_E = dN/dE$, is positive. Here, N is the normalization, and E is the chemical potential of the system. For the case $N_E < 0$, the solution is unstable, whereas $N_E = 0$ provides the instability threshold of the obtained solution. Now, using Equation (17) and $N = \int_{-\infty}^{+\infty} |\psi|^2 \partial x$, one can estimate the correlation between normalization N and E as:

$$N = \frac{4}{3}\left[ln\left(\frac{1+\sqrt{\frac{E}{\mu_0}}}{\sqrt{1-\frac{E}{\mu_0}}}\right) - \sqrt{\frac{E}{\mu_0}}\right], \tag{22}$$

where $\frac{G_2}{G_1^2} \approx 1$ and $G_1 = -1$. Equation (22) estimates the magnitude of N in the presence of the MOL and is equal to the N reported for the free space [31,32]. Thus, even in the presence of the MOL confinement, N is conserved, and the considered system shows a continuous symmetry property according to Noether's theorem [51]. In Figure 7, using Equation (22), we plot N_E with respect to E, where $G_1 = -1$. It is evident from Figure 7a that the magnitude of N_E is positive with respect to its variation E, which indicates the stable nature of the obtained solution.

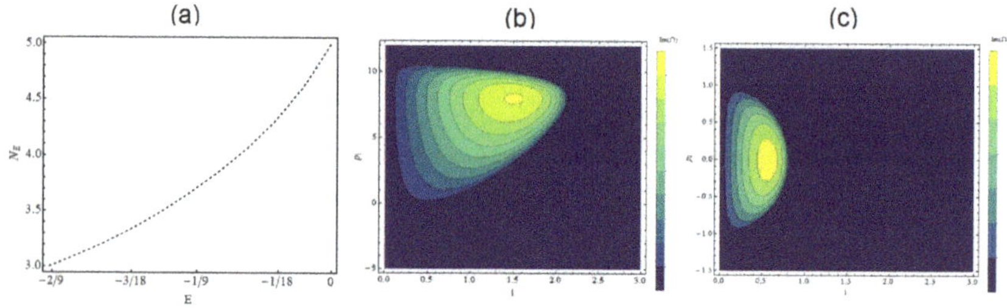

Figure 7. (Color online) (**a**) For the VK stability criterion, the slope of normalization with the chemical potential (N_E) is plotted with respect to a varying chemical potential (E); (**b**) $Im(\Omega)$ is depicted as a function of p_1 and the wavenumber (l) keeping $p_2 = 0$; (**c**) with $p_1 = 0.5$, $Im(\Omega)$ is depicted as a function of p_2 and the wavenumber (l). Here, the magnitude of the physical parameters: $b = 1$, $\gamma = 0.05$, $k = 0.84$, $G_1 = -1$, $G_2 = 0.999999999$, $E = -2/9$ with l varying from 0 to 3.

The small-scale fluctuations in the droplet under the MOL confinement can be estimated by linearizing the 1D eGPE (3) around the ground state given by Equation (17) [32].

We take $\psi(x,t) = \psi_0(x) + \delta\psi(x,t)$, where $\psi_0(x)$ is the stationary solution (17) and $\delta\psi(x,t) = \begin{bmatrix}\delta\psi_R \\ \delta\psi_I\end{bmatrix} << 1$ is the small perturbation with $\delta\psi_R$; $\delta\psi_I$ are the real and imaginary parts of $\delta\psi$. With the substitution of $\delta\psi$ in Equation (3), this leads to the well-known Bogoliubov–de Gennes (BdG) equation:

$$\begin{bmatrix} 0 & 1 \\ -1 & 0 \end{bmatrix} \frac{\partial}{\partial t} \begin{bmatrix}\delta\psi_R \\ \delta\psi_I\end{bmatrix} = \begin{bmatrix} T & 0 \\ 0 & T' \end{bmatrix} \begin{bmatrix}\delta\psi_R \\ \delta\psi_I\end{bmatrix}, \tag{23}$$

with $T = -\frac{1}{2}\frac{\partial^2}{\partial x^2} + 3ng_2(x) - \frac{1}{2}g_1(x)n^{1/2} + v(x)$, $T' = -\frac{1}{2}\frac{\partial^2}{\partial x^2} + g_2(x)n - \frac{1}{2}g_1(x)n^{1/2} + v(x)$, and $n = |\psi_0(x)|^2$. Here, we consider $\delta\psi = exp[i(lx - \omega t)]$, and applying it in Equation (23), then this yields the perturbation eigenmodes, where l denotes the wave number and Ω stands for the frequency. The resulting dispersion relation can be given as:

$$\Omega^2 = \frac{l^4}{4} + l^2\left(2V(x) - 4g_1(x)n^{1/2} + 4ng_2(x)\right), \tag{24}$$

by neglecting the l independent terms. From Equation (24), Ω is imaginary for $4g_1(x)n^{1/2} > 2v(x) + 4ng_2(x) + l^2/4$, and these are the instability region. In Figure 7b, $Im(\Omega)$ is depicted as a function of p_1 and the wavenumber (l), keeping $p_2 = 0$, representing the two-color BOL $((k, 2k))$ case, whereas in Figure 7c, we locate $Im(\Omega)$ with respect to the changing magnitude of p_2 and the wavenumber (l) with $p_1 = 0.5$ for the MOL confinement. The magnitudes of the other physical parameters: $b = 1$, $\gamma = 0.05$, $k = 0.84$, $G_1 = -1$, $G_2 = 0.999999999$, $E = -2/9$.

4. Conclusions

In conclusion, we obtained a family of exact analytical solutions of the 1D eGPE for the generation of QDs in the binary BEC mixture with mutually symmetric spinor components in the presence of the MOL confinement. In the constructed model, the choice of the MOL confinement, which is a combination of the FOL and exponential periodic trap, results in the generation of various experimentally relevant trap profiles: OL, BOL, TOL, FOL, symmetric and asymmetric double-well superlattice, etc., and the corresponding exact wavefunction solution. From the constructed model, we identified the two parameters p_1 and p_2 for tuning the shape of the MOL confinement and revealed interesting potential symmetry with its tuning. Further, by taking the spatial periodic variation of the competing repulsive cubic MF and attractive quadratic BMF interactions, we illustrated the generation of interesting supersolid-like periodic, bi-periodic, and double-well superlattice density patterns in QDs under the BOL and MOL confinements. By tuning the symmetry of the BOL trap, we demonstrated two possible types of periodic lattice density patterns of the droplets: (a) at the central potential site (i.e., at $x = 0$), the maxima of the periodic density lattice for $p_1 < 0$ and $p_2 = 0$ and (b) for $p_1 > 0$ and $p_2 = 0$, its minima $x = 0$. Interestingly, the strength of (p_1, p_2) was identified as a key parameter for the fragmentation, compression, and inter-well transport of droplets. We observed an Anderson-like localization for p_1, $p_2 > 0$, i.e., the compression of the droplet, whereas the expansion of the droplet width was noted for $p_1 > 0$, $p_2 < 0$, forming the MOL, as well as in two-color ($p_2 = 0$) OL confinements. This can be attributed to the change in the subtle balance of the MF and BMF interactions, leading to a decrease in the flat top density profile and the observation of localization. Finally, the stability of the obtained droplet solutions was confirmed using the VK criterion and by the linear stability analysis.

The results of this paper can be tested experimentally by taking binary Bose gases with mass-balanced two-mixture components under the influence of BMF interactions in the presence of spatially varying MOL confinement with the present know-how. The shape of the MOL confinement is dependent on the power of the laser intensity and the wavelength of the laser light, which were connected with (p_1, p_2) in the present work. In particular, for typical values of the physical parameters, one can estimate the potential depths: $E_R = \frac{2\pi^2\hbar^2}{m\lambda^2}$. As the constructed model is based on a 1D geometry, it may be interesting to explore the impact in a 2D/3D confinement setup and droplet crossover to the 2D/3D configuration as future extensions of the present work.

Author Contributions: Conceptualization, formal analysis, investigation, methodology, visualization, writing—original draft: M.R.P.; conceptualization, validation, writing—review and editing, supervision: A.N. All authors have read and agreed to the published version of the manuscript.

Funding: This research received no external funding.

Institutional Review Board Statement: Not applicable.

Informed Consent Statement: Not applicable.

Data Availability Statement: All data generated or analyzed during this study are included in this published article. They can be reproduced by utilizing the form of the wavefunction and the considered trap form.

Conflicts of Interest: The authors declare no conflict of interest.

Appendix A

We begin with the reduced one-component eGPE:

$$i\frac{\partial\psi}{\partial t} = -\frac{\partial^2\psi}{\partial x^2} - g_1(x,t)|\psi|\psi + g_2(x,t)|\psi|^2\psi + V(x)\psi \tag{A1}$$

and our goal is to connect it to the solvable differential equation:

$$-\frac{\partial^2 U}{\partial\eta^2} - G_1 \mid U(\eta) \mid U + G_2 \mid U(\eta) \mid^2 U = EU. \tag{A2}$$

which has the solution form: $U[\eta] = \frac{3(E/G_1)}{1+\sqrt{1-\frac{E}{\mu_0}\frac{G_2}{G_1^2}}\cosh(\sqrt{-E}\eta)}$ with $\mu_0 = -2/9$, $E < 0$, $G_1 < 0$, and $G_2 > 0$ [25,31]. It is a standard mathematical technique to obtain solutions of nonlinear differential equation [47,52].

For that purpose, we substitute an ansatz solution in (A1):

$$\psi(x,t) = A(x,t)U[\eta(x,t)]e^{i\phi(x,t)}. \tag{A3}$$

and separate out the imaginary and real parts of the equation:

$$\frac{\partial A}{\partial t}U + \frac{\partial U}{\partial t}A + 2U\frac{\partial A}{\partial x}\frac{\partial\phi}{\partial x} + 2A\frac{\partial U}{\partial x}\frac{\partial\phi}{\partial x} + AU\frac{\partial^2\phi}{\partial x^2} = 0, \tag{A4}$$

$$-A\frac{\partial\phi}{\partial t}U + \frac{\partial^2 A}{\partial x^2}U + 2\frac{\partial A}{\partial x}\frac{\partial U}{\partial x} + A\frac{\partial^2 U}{\partial x^2} - AU\left[\frac{\partial\phi}{\partial x}\right]^2 + g_1(x,t)A^2|U|U + g_2(x,t)A^3|U|^2U + V(x)UA = 0. \tag{A5}$$

We obtain the following consistency conditions on the amplitude and MF and BMF nonlinearities:

$$[A^2(x,t)\eta_x(x,t)]_x = 0, \quad \eta_t(x,t) + \eta_x(x,t)\phi_x(x,t) = 0, \tag{A6}$$

$$G_1\eta_x^2(x,t) - 2A(x,t)g_1(x,t) = 0, \quad G_2\eta_x^2(x,t) - 2A^2(x,t)g_2(x,t) = 0, \tag{A7}$$

$$\frac{A_t(x,t)}{A(x,t)} + \frac{1}{2A^2(x,t)}[A^2(x,t)\phi_x(x,t)]_x = 0, \tag{A8}$$

$$\frac{A_{xx}(x,t)}{2A(x,t)} - \frac{\phi_x^2(x,t)}{2} - \phi_t(x,t) - \frac{1}{2}E\eta_x^2(x,t) - V(x) = 0. \tag{A9}$$

Further, with a little algebra, we obtain the amplitude, phase, and MF/BMF:

$$A(x,t) = \sqrt{\frac{b(t)}{\eta_x(x,t)}},\ \phi_z = -\frac{\eta_t(x,t)}{\eta_x(x,t)},\ g_1(x,t) = G_1\frac{\eta_x^2(x,t)}{2A(x,t)},\ g_2(x,t) = G_2\frac{\eta_x^2(x,t)}{2A^2(x,t)}, \tag{A10}$$

where $b(t)$ is an integration constant.

Next, we take $\eta(x,t) = \gamma \int_0^z exp[p_1 \cos^2(kx) + p_2 \cos^2(2kx)]\partial x$ and calculate the exact analytical form of the amplitude, phase, and nonlinearities:

$$A(x,t) = \sqrt{\frac{b(t)}{\gamma \times exp[p_1 \cos^2(kx) + p_2 \cos^2(2kx)]}},\ \theta(x,t) = [2p_1^2k^2 + 8p_2^2k^2 - 8k^2(p_1 + 4p_2)]t, \tag{A11}$$

$$g_1(x,t) = \frac{G_1\gamma^{3/2}}{2b(t)} exp[p_1 \cos^2(kx) + p_2 \cos^2(2kx)]^{\frac{3}{2}},\ g_2(x,t) = \frac{G_2\gamma^3}{2b(t)} exp[p_1 \cos^2(kx) + p_2 \cos^2(2kx)]^3. \tag{A12}$$

Thus, the complete solution of Equation (3) can be written as:

$$\psi(x,t) = \sqrt{\frac{b}{\gamma \times exp[p_1 \cos^2(kx) + p_2 \cos^2(2kx)]}} \times exp\left[i(2p_1^2k^2 + 8p_2^2k^2 - 8k^2(p_1 + 4p_2)t)\right] \frac{\frac{3E}{G_1}}{1 + \sqrt{1 - \frac{E}{\mu_0}\frac{G_2}{G_1^2}} \cosh(\sqrt{-E}(\gamma \int_0^x exp[p_1 \cos^2(kx) + p_2 \cos^2(2kx)]))}. \tag{A13}$$

References

1. Lewenstein, M.; Sanpera, A.; Ahufinger, V. *Ultracold Atoms in Optical Lattices: Simulating Quantum Many-Body Systems*; Oxford University Press: Oxford, UK, 2012.
2. Schäfer, F.; Fukuhara, T.; Sugawa, S.; Takasu, Y.; Takahashi, Y. Tools for quantum simulation with ultracold atoms in optical lattices. *Nat. Rev. Phys.* **2020**, *2*, 411–425. [CrossRef]
3. Gross, C.; Bloch, I. Quantum simulations with ultracold atoms in optical lattices. *Science* **2017**, *357*, 995–1001. [CrossRef] [PubMed]
4. Hauke, P.; Cucchietti, F.M.; Tagliacozzo, L.; Deutsch, I.; Lewenstein, M. Can one trust quantum simulators? *Rep. Prog. Phys.* **2012**, *75*, 082401. [CrossRef] [PubMed]
5. Qiu, X.; Zou, J.; Qi, X.; Li, X. Precise programmable quantum simulations with optical lattices. *NPJ Quantum Inf.* **2020**, *6*, 87. [CrossRef]
6. Haller, E.; Hart, R.; Mark, M.J.; Danzl, J.G.; Reichsöllner, L.; Gustavsson, M.; Dalmonte, M.; Pupillo, G.; Nägerl, H.-C. Pinning quantum phase transition for a Luttinger liquid of strongly interacting bosons. *Nature (Lond.)* **2010**, *466*, 597–600. [CrossRef] [PubMed]
7. Greiner, M.; Mandel, O.; Esslinger, T.; Hänsch, T.W.; Bloch, I. Quantum phase transition from a superfluid to a Mott insulator in a gas of ultracold atoms. *Nature (Lond.)* **2002**, *415*, 39–44. [CrossRef] [PubMed]
8. Billy, J.; Josse, V.; Zuo, Z.; Bernard, A.; Hambrecht, B.; Lugan, P.; Clement, D.; Sanchez-Palencia, L.; Bouyer, P.; Aspect, A. Direct observation of Anderson localization of matter waves in a controlled disorder. *Nature* **2008**, *453*, 891–894. [CrossRef] [PubMed]
9. Braun, S.; Ronzheimer, J.P.; Schreiber, M.; Hodgman, S.S.; Rom, T.; Bloch, I.; Schneider, U. Negative Absolute Temperature for Motional Degrees of Freedom. *Science* **2013**, *339*, 52–55. [CrossRef]
10. Nath, A.; Bera, J.; Ghosh, S.; Roy, U. Exact Analytical Model for Bose–Einstein Condensate at Negative Temperature. *Sci. Rep.* **2020**, *10*, 9016. [CrossRef]
11. Kundu, N.; Nath, A.; Bera, J.; Ghosh, S.; Roy, U. Synergy between the negative absolute temperature and the external trap for a Bose–Einstein condensate under optical lattices. *Phys. Lett. A* **2022**, *427*, 127922. [CrossRef]
12. Li, J.R.; Lee, J.; Huang, W.; Burchesky, S.; Shteynas, B.; Top, F.Ç.; Jamison, A.O.; Ketterle, W. A stripe phase with supersolid properties in spin-orbit coupled Bose–Einstein condensates. *Nature* **2017**, *543*, 91–94. [CrossRef] [PubMed]
13. Wang, Z.M.; Wu, L.A.; Modugno, M.; Byrd, M.S.; Yu, T.; You, J.Q. Fault-tolerant breathing pattern in optical lattices as a dynamical quantum memory. *Phys. Rev. A* **2014**, *89*, 042326. [CrossRef]
14. Brennen, G.K.; Pupillo, G.; Rey, A.M.; Clark, C.W.; Williams, C.J. Scalable register initialization for quantum computing in an optical lattice. *J. Phys. B At. Mol. Opt. Phys.* **2005**, *38*, 1687. [CrossRef]
15. Katori, H. Optical lattice clocks and quantum metrology. *Nat. Photonics* **2011**, *5*, 203–210. [CrossRef]
16. Yang, B.; Sun, H.; Huang, C.J.; Wang, H.Y.; Deng, Y.; Dai, H.N.; Yuan, Z.S.; Pan, J.W. Cooling and entangling ultracold atoms in optical lattices. *Science* **2020**, *369*, 550–553. [CrossRef]
17. Windpassinger, P.; Sengstock, K. Engineering novel optical lattices. *Rep. Prog. Phys.* **2013**, *76*, 086401. [CrossRef]
18. Yamamoto, D.; Fukuhara, T.; Danshita, I. Frustrated quantum magnetism with Bose gases in triangular optical lattices at negative absolute temperatures. *Nat. Phys.* **2020**, *3*, 56. [CrossRef]

19. Kohlert, T.; Scherg, S.; Li, X.; Lüschen, H.P.; Sarma, S.D.; Bloch, I.; Aidelsburger, M. Observation of many-body localization in a one dimensional system with a single-particle mobility edge. *Phys. Rev. Lett.* **2019**, *122*, 170403. [CrossRef]
20. Messer, M.; Desbuquois, R.; Uehlinger, T.; Jotzu, G.; Huber, S.; Greif, D.; Esslinger, T. Exploring Competing Density Order in the Ionic Hubbard Model with Ultracold Fermions. *Phys. Rev. Lett.* **2015**, *115*, 115303. [CrossRef]
21. Richaud, A.; Ferraretto, M.; Capone, M. Interaction-resistant metals in multicomponent Fermi systems. *Phys. Rev. B* **2021**, *103*, 205132. [CrossRef]
22. Kartashov, Y.V.; Astrakharchik, G.E.; Malomed, B.A.; Torner, L. Frontiers in multidimensional self-trapping of nonlinear fields and matter. *Nat. Rev. Phys.* **2019**, *1*, 185. [CrossRef]
23. Böttcher, F.; Schmidt, J.-N.; Hertkorn, J.; Ng, K.S.H.; Graham, S.D.; Guo, M.; Langen, T.; Pfau, T. New states of matter with fine-tuned interactions: Quantum droplets and dipolar supersolids. *Rep. Prog. Phys.* **2021**, *84*, 012403. [CrossRef] [PubMed]
24. Luo, Z.-H.; Pang, W.; Liu, B.; Li, Y.-L.; Malomed, B. A. A new form of liquid matter: Quantum droplets. *Front. Phys.* **2021**, *16*, 32201. [CrossRef]
25. Petrov, D.S. Quantum Mechanical Stabilization of a Collapsing Bose–Bose Mixture. *Phys. Rev. Lett.* **2015**, *115*, 155302. [CrossRef] [PubMed]
26. Cabrera, C.R.; Tanzi, L.; Sanz, J.; Naylor, B.; Thomas, P.; Cheiney, P.; Tarruell, L. Quantum liquid droplets in a mixture of Bose–Einstein condensates. *Science* **2018**, *359*, 301–304. [CrossRef] [PubMed]
27. Semeghini, G.; Ferioli, G.; Masi, L.; Mazzinghi, C.; Wolswijk, L.; Minardi, F.; Modugno, M.; Modugno, G.; Inguscio, M.; Fattori, M. Self-Bound Quantum Droplets of Atomic Mixtures in Free Space. *Phys. Rev. Lett.* **2018**, *120*, 235301. [CrossRef]
28. Ferrier-Barbut, I.; Kadau, H.; Schmitt, M.; Wenzel, M.; Pfau, T. Observation of quantum droplets in a strongly dipolar bose gas. *Phys. Rev. Lett.* **2016**, *116*, 215301. [CrossRef]
29. Edler, D.; Mishra, C.; Wächtler, F.; Nath, R.; Sinha, S.; Santos, L. Quantum fluctuations in quasi- one-dimensional dipolar bose-einstein condensates. *Phys. Rev. Lett.* **2017**, *119*, 050403. [CrossRef]
30. Ferrier-Barbut, I.; Pfau, T. Quantum liquids get thin. *Science* 2018, *359*, 274–275 . [CrossRef]
31. Astrakharchik, G.E.; Malomed, B.A. Dynamics of one-dimensional quantum droplets. *Phys. Rev. A* **2018**, *98*, 013631. [CrossRef]
32. Tylutki, M.; Astrakharchik, G.E.; Malomed, B.A.; Petrov, D.S. Collective excitations of a one-dimensional quantum droplet. *Phys. Rev. A* **2020**, *101*, 051601. [CrossRef]
33. Pathak, R.M.; Nath, A. Dynamics of Quantum Droplets in an External Harmonic Confinement. *Sci. Rep.* **2022**, *12*, 6904. [CrossRef] [PubMed]
34. Morera, I.; Astrakharchik, G.E.; Polls, A.; Juliá-Díaz, B. Universal Dimerized Quantum Droplets in a One-Dimensional Lattice. *Phys. Rev. Lett.* **2021**, *126*, 023001. [CrossRef] [PubMed]
35. Morera, I.; Astrakharchik, G.E.; Polls, A.; Juliá-Díaz, B. Quantum droplets of bosonic mixtures in a one-dimensional optical lattice. *Phys. Rev. Res.* **2020**, *2*, 022008. [CrossRef]
36. Chomaz, L.; Petter, D.; Ilzhöfer, P.; Natale, G.; Trautmann, A.; Politi, C.; Durastante, G.; van Bijnen, R.M.W.; Patscheider, A.; Sohmen, M.; et al. Long-Lived and Transient Supersolid Behaviors in Dipolar Quantum Gases. *Phys. Rev. X* **2019**, *9*, 021012. [CrossRef]
37. Norcia, M.A.; Politi, C.; Klaus, L.; Poli, E.; Sohmen, M.; Mark, M.J.; Bisset, R.; Santos, L.; Ferlaino, F. Two-dimensional supersolidity in a dipolar quantum gas. *Nature (Lond.)* **2021**, *596*, 357–361. [CrossRef] [PubMed]
38. Young, S.L.E.; Adhikari, S.K. Supersolid-like square- and honeycomb-lattice crystallization of droplets in a dipolar condensate. *Phys. Rev. A* **2022**, *105*, 033311. [CrossRef]
39. Otajonov, S.R.; Tsoy, E.N.; Abdullaev, F.K. Stationary and dynamical properties of one-dimensional quantum droplets. *Phys. Lett. A* **2019**, *383*, 34. [CrossRef]
40. Nath, A.; Roy, U. Bose–Einstein condensate in a bichromatic optical lattice: An exact analytical model. *Laser Phys. Lett.* **2014**, *11*, 115501. [CrossRef]
41. Halder, B.; Ghosh, S.; Basu, P.; Bera, J.; Malomed, B.; Roy, U. Exact Solutions for Solitary Waves in a Bose–Einstein Condensate under the Action of a Four-Color Optical Lattice. *Symmetry* **2022**, *14*, 49. [CrossRef]
42. Inouye, S.; Andrews, M.R.; Stenger, J.; Miesner, H.-J.; Stamper-Kurn, D.M.; Ketterle, W. Observation of Feshbach resonances in a Bose–Einstein condensate. *Nature* **1998**, *392*, 151–154. [CrossRef]
43. Viebahn, K.; Sbroscia, M.; Carter, E.; Yu, J.-C.; Schneider, U. Matter-Wave Diffraction from a Quasicrystalline Optical Lattice. *Phys. Rev. Lett.* **2019**, *122*, 110404. [CrossRef] [PubMed]
44. Khaykovich, L.; Schreck, F.; Ferrari, G.; Bourdel, T.; Cubizolles, J.; Carr, L.D.; Castin, Y.; Salomon, C. Formation of a Matter-Wave Bright Soliton. *Science* **2002**, *296*, 1290–1293. [CrossRef] [PubMed]
45. Zhang, T.; Jo, G.B. One-dimensional sawtooth and zigzag lattices for ultracold atoms. *Sci. Rep.* **2015**, *5*, 16044. [CrossRef]
46. Kengne, E.; Liu, W.-M.; Malomed, B.A. Spatiotemporal engineering of matter-wave solitons in Bose–Einstein condensates. *Phys. Rep.* **2021**, *899*, 1–62. [CrossRef]
47. Nath, A.; Bera, J.; Ghosh, S.; Panigrahi, P.K.; Roy, U. Soliton dynamics for an ingenious trap combination in a Bose–Einstein condensate. *Eur. Phys. J. D* **2020**, *74*, 27. [CrossRef]
48. Mithun, T.; Maluckov, A.; Kasamatsu, K.; Malomed, B.A.; Khare, A. Modulational Instability, Inter-Component Asymmetry, and Formation of Quantum Droplets in One-Dimensional Binary Bose Gases. *Symmetry* **2020**, **12**, 174. [CrossRef]

49. Kaur, P.; Gautam, S.; Adhikari, S.K. Supersolid-like solitons in a spin-orbit-coupled spin-2 condensate. *Phys. Rev. A* **2022**, *105*, 023303. [CrossRef]
50. Vakhitov, N.; Kolokolov, A.A. Stationary solutions of the wave equation in a medium with nonlinearity saturation. *Radiophys. Quantum Electron.* **1973**, *16*, 783. [CrossRef]
51. Sardanashvily, G. *Noether's Theorems. Applications in Mechanics and Field Theory*; Springer: Berlin/Heidelberg, Germany, 2016.
52. Owyed, S.; Abdou, M.A.; Abdel-Aty, A.-H.; Ibraheem, A.A.; Nekhili, R.; Baleanu, D. New optical soliton solutions of space-time fractional nonlinear dynamics of microtubules via three integration schemes. *J. Intell. Fuzzy Syst.* **2020**, *38*, 2859–2866. [CrossRef]

symmetry
MDPI

Article

Josephson-like Oscillations in Toroidal Spinor Bose–Einstein Condensates: A Prospective Symmetry Probe

Mário H. Figlioli Donato and **Sérgio R. Muniz** *

Instituto de Física de São Carlos, Universidade de São Paulo, IFSC-USP Caixa Postal 369, São Carlos 13560-970, SP, Brazil; mario.donato@usp.br
* Correspondence: srmuniz@ifsc.usp.br

Abstract: Josephson junctions are essential ingredients in the superconducting circuits used in many existing quantum technologies. Additionally, ultracold atomic quantum gases have also become essential platforms to study superfluidity. Here, we explore the analogy between superconductivity and superfluidity to present an intriguing effect caused by a thin finite barrier in a quasi-one-dimensional toroidal spinor Bose–Einstein condensate (BEC). In this system, the atomic current density flowing through the edges of the barrier oscillates, such as the electrical current through a Josephson junction in a superconductor, but in our case, there is no current circulation through the barrier. We also show how the nontrivial broken-symmetry states of spinor BECs change the structure of this Josephson-like current, creating the possibility to probe the spinor symmetry, solely using measurements of this superfluid current.

Keywords: Josephson effect; Bose–Einstein condensate; spinor BEC; toroidal superfluid

Citation: Donato, M.H.F.; Muniz, S.R. Josephson-like Oscillations in Toroidal Spinor Bose–Einstein Condensates: A Prospective Symmetry Probe. *Symmetry* **2022**, *14*, 867. https://doi.org/10.3390/sym14050867

Academic Editors: Rashid G. Nazmitdinov, Vyacheslav Yukalov and Vanderlei Salvador Bagnato

Received: 28 February 2022
Accepted: 19 April 2022
Published: 23 April 2022

1. Introduction

The Josephson effect [1,2] is one of the most relevant phenomena in *superconductivity*; for his theoretical predictions, Brian D. Josephson received the Nobel Prize in Physics in 1973. In a superconductor, the phenomenon is characterized by the tunneling of *Cooper pairs* through a junction, or a weak link, represented by a potential energy barrier between two superconductor regions [3–5]. When a voltage V is applied to these regions, the electrical current in the junction, I_j, oscillates according to the expression:

$$I_j = I_0 \sin(\phi); \quad \text{with } \frac{d\phi}{dt} = \frac{2eV}{\hbar}, \tag{1}$$

where I_0 is the current amplitude and ϕ is the angle representing the phase difference of the order parameter between each superconducting region [4]. There is a conceptual analogy between superconductivity and superfluidity [6], given by the lack of resistance in the flow of the electrical current, or, equivalently, the lack of viscosity in the flow of a superfluid. In fact, since superfluidity emerges naturally in Bose–Einstein condensates (BECs) produced in dilute ultracold Bose gases [7–15], the analogue of the Josephson effect has already been studied in experiments [16–19].

For *scalar* BECs [7,8] (i.e., condensates with a scalar order parameter: $\psi = \sqrt{n}e^{i\theta}$), the superfluid velocity is always proportional to the global phase gradient ($\nabla\theta$), and it represents the complete *structural symmetry* of the system. On the other hand, for *spinor* BECs [9–11] (i.e., condensates with multi-component order parameter: $[\boldsymbol{\psi}]_m = \psi_m$), the superfluid flow is related to the properties of their symmetry and topology, which can generate nontrivial spatial spin textures and current distributions not found in scalar BEC systems. In fact, there are many topological states in spinor BECs analogous to particles and structures studied across several areas of physics, such as *skyrmions*, *Dirac monopoles*, *knot solitons*, *vortices*, *half-vortices* [9,20–24], to name a few.

Here, based on the analogy between superfluidity and superconductivity, we study a Josephson-like oscillating current produced in toroidal spinor Bose–Einstein condensates due to the presence of a thin finite energy barrier (Figure 1), and we show that the current density at the edges of the barrier behaves similarly to I_j, and depends on the structural symmetry of the spinor condensate.

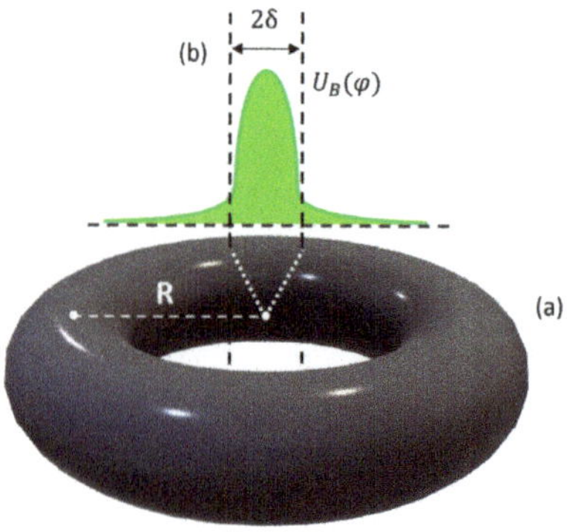

Figure 1. Schematic illustration of the external potentials applied to the condensate: (**a**) effective toroidal region $\mathcal{T}$ (with average radius R) where the potential U_{trap} confines the condensate; (**b**) repulsive barrier $U_B = U_B(\varphi)$, for φ the azimuthal angle in cylindrical coordinates (r, φ, z). Here, we consider thin angular widths $2\delta \ll 2\pi$.

2. General Modeling of Spinor BECs

For a system with atomic spin $f = 1$, the order parameter of a spinor BEC satisfies the following vector Gross–Pitaevskii equation (GPE) [9]:

$$i\hbar \frac{\partial}{\partial t}\boldsymbol{\psi} = \left[-\frac{\hbar^2\nabla^2}{2M} + U_{\text{total}} - p\mathbf{f}_z + q\mathbf{f}_z^2 + c_0 n + c_1 \vec{F} \cdot \vec{\mathbf{f}} \right] \boldsymbol{\psi} \,. \tag{2}$$

In the equation, the bold fonts indicate matrix quantities related to the vector order parameter $\boldsymbol{\psi} = (\psi_f(\vec{r}, t), \ldots, \psi_{-f}(\vec{r}, t))^T$, and where $n = \boldsymbol{\psi}^\dagger \boldsymbol{\psi} = \sum_m |\psi_m(\vec{r}, t)|^2 = \sum_m n_m(\vec{r}, t)$ is the number density, with spin components $\psi_m(\vec{r}, t) = \sqrt{n_m(\vec{r}, t)}\, e^{i\theta_m(\vec{r}, t)}$, $\vec{F} = \boldsymbol{\psi}^\dagger \vec{\mathbf{f}}\, \boldsymbol{\psi}$ is the spin density, and $\vec{\mathbf{f}} = (\mathbf{f}_x, \mathbf{f}_y, \mathbf{f}_z)$ is the spin operator (in the $f = 1$ representation). The constants $p \propto |\vec{B}|$ and $q \propto |\vec{B}|^2$ are related to the linear and quadratic Zeeman effects, respectively, when an external magnetic field, $\vec{B}$, is applied. The system's nonlinear interaction "strength" is modeled by the coefficients c_0 and c_1 (see Reference [9] for details).

In this work, we consider a toroidal trapping potential, U_{trap}, with a repulsive barrier, U_B, restricted to a region of small azimuthal angle. Therefore, the total external potential is given by $U_{\text{total}} = U_{\text{trap}} + U_B$, as sketched in Figure 1. We assume that U_{total} provides the necessary means to trap all the components of the spinor BEC in the torus. For instance, by using the appropriate optical trapping techniques [25]. Optical traps have the advantage of providing flexible control over the potential symmetry when combined, for example, with digital holography [26,27], direct imaging [28], or phase-mapping [29] of sharp optical patterns, to design almost arbitrary shapes. In fact, a toroidal optical potential with a controllable energy barrier similar to the one described here has been demonstrated in [30–32], and a toroidal potential with spinor BEC in [33].

The order parameter of a spinor BEC, represented as a column matrix, can always be factored into $\boldsymbol{\psi} = \psi(\vec{r}, t)\, \boldsymbol{\zeta}$, as the product of a *scalar part*, $\psi(\vec{r}, t) = \sqrt{n(\vec{r}, t)}\, e^{i\theta(\vec{r}, t)}$, and a normalized *spinor part*, $\boldsymbol{\zeta}$ (with $\boldsymbol{\zeta}^\dagger \boldsymbol{\zeta} = 1$) [9]. This vector description has a gauge symmetry with *artificial gauge fields* related to ζ. These fields emerge naturally from the description, and are analogous to the scalar and vector electromagnetic potentials [9,34]:

$$\Phi = -i\zeta^{\dagger}\frac{\partial}{\partial t}\zeta, \tag{3}$$

$$\vec{A} = i\zeta^{\dagger}\vec{\nabla}\zeta. \tag{4}$$

In particular, the superfluid number current density, $\vec{J} = n\,\vec{v}$, depends linearly on the vector potential $\vec{A}$, as the superfluid velocity $\vec{v}$ of a spinor BEC is given by:

$$\vec{v} = \frac{\hbar}{M}(\vec{\nabla}\theta - \vec{A}). \tag{5}$$

Note that the action of the artificial vector potential is analogous to the electromagnetic vector potential on the expression of the electrical current in a superconductor [4]. In addition, some of the topological properties of the condensate are determined by the *circulation* of $\vec{v}$ (i.e., $\oint \vec{v}\cdot\vec{dl}$), and, therefore, they may depend on *artificial magnetic fluxes*, given by a synthetic magnetic field $\vec{B} \equiv \hbar\vec{\nabla}\times\vec{A}$ [9].

In general, the current density of a spinor BEC is defined in the usual manner, replacing the scalar order parameter by its vector form

$$\vec{J} = \frac{\hbar}{M}\left(\boldsymbol{\psi}^{\dagger}\vec{\nabla}\boldsymbol{\psi} - (\vec{\nabla}\boldsymbol{\psi}^{\dagger})\boldsymbol{\psi}\right) = \frac{\hbar}{M}\mathrm{Im}\{\boldsymbol{\psi}^{\dagger}\vec{\nabla}\boldsymbol{\psi}\}, \tag{6}$$

therefore,

$$\vec{J} = \frac{\hbar}{M}\sum_{m=-f}^{f}\mathrm{Im}\{\psi_m^*(\vec{r},t)\vec{\nabla}\psi_m(\vec{r},t)\} \tag{7}$$

$$= \sum_{m=-f}^{f}\vec{J}_m(\vec{r},t) \tag{8}$$

Alternatively, using the spin components $\psi_m(\vec{r},t)$, one could write

$$\vec{J}_m = \frac{\hbar}{M}n_m(\vec{r},t)\vec{\nabla}\theta_m(\vec{r},t) = n_m(\vec{r},t)\,\vec{v}_m(\vec{r},t), \tag{9}$$

where the velocity spin components are $\vec{v}_m(\vec{r},t) = \frac{\hbar}{M}\vec{\nabla}\theta_m(\vec{r},t)$.

3. Modeling the Potential Barrier and Defining Some Approximations

This section explains in detail how we define the potential barrier, U_B, and its general properties. Because we will use the quasi-one-dimensional (*quasi-1D*) limit for the geometry of the condensate later, we consider the barrier $U_B(\varphi)$ is written in terms of a normalized angular distribution $f(\varphi)$, not depending on other spatial variables (r,z):

$$U_B(\varphi) = U_0 f(\varphi), \quad \text{with} \int_{-\pi}^{\pi} f(\varphi)d\varphi = 1. \tag{10}$$

We consider that the barrier is effectively restricted to an angular range $\varphi \in [-\delta,\delta]$, such that:

$$\int_{-\pi}^{\pi} f(\varphi)d\varphi \approx \int_{-\delta}^{\delta} f(\varphi)d\varphi \approx 1. \tag{11}$$

Integrating the GPE (2) along the arc $\varphi \in [-\delta,\delta]$, with fixed (r,z) in the thin barrier condition ($\delta \ll \pi$), leads to two dominant terms:

$$\frac{\hbar^2}{2Mr^2}\left[\partial_\varphi\boldsymbol{\psi}(\delta) - \partial_\varphi\boldsymbol{\psi}(-\delta)\right] \approx \int_{-\delta}^{\delta} U_B(\varphi)\boldsymbol{\psi}(\varphi)\,d\varphi, \tag{12}$$

where the left-hand side follows directly from the fundamental theorem of calculus, and it is related to the momentum along the $\hat{\varphi}$ direction ($p_\varphi \hat{\varphi} = -\hat{\varphi}(i\hbar\partial_\varphi)/r$). Note that we simplified the notation $\boldsymbol{\psi}(r,\varphi,z) \to \boldsymbol{\psi}(\varphi)$, given that we consider (r,z) fixed parameters.

The remaining terms not computed in the last equation were neglected because for small δ they are proportional to δ (i.e., these terms are $\mathcal{O}(\delta)$ while the dominant terms are proportional to the unit), and can be neglected as one takes the limit of $\delta \to 0$.

Additionally, in the limit of small δ, the integration of the barrier potential resembles the integration of a Dirac's delta (normalized) function, such that

$$\lim_{\delta\to 0}\left[\int_{-\delta}^{\delta} U_B(\varphi)\boldsymbol{\psi}\, d\varphi\right] = U_0 \lim_{\delta\to 0}\left[\int_{-\delta}^{\delta} f(\varphi)\boldsymbol{\psi}\, d\varphi\right] \approx U_0\, \boldsymbol{\psi}(0). \tag{13}$$

Here, it is important to emphasize that, in practice, the derivative of the potential barrier must not exceed a limit given by the Landau critical velocity, otherwise, the flow is dissipated and the system loses its superfluid properties [30,35]. To avoid such situations, one may always choose a larger radius R (see Figure 1) so that the barrier $U_B(\varphi)$ is spatially "smooth" enough. We discuss this condition in Section 6.

Finally, we consider one last approximation. If the number of atoms in the condensate is kept the same, it is expected that the average cross-section area of $\mathcal{T}$ (at fixed φ) becomes smaller when R is larger. So, for R big enough, the condensate would be trapped in small Δr and Δz ranges, and the number density would become approximately a function of only one variable (*quasi-1D approximation*):

$$n(r,\varphi,z) \to \begin{cases} n(\varphi), & \text{if } (r,\varphi,z) \in \mathcal{T}; \\ 0, & \text{otherwise.} \end{cases} \tag{14}$$

Moreover, these conditions also restrict the direction of $\vec{J}$, such that:

$$\vec{J} \approx J_\varphi \hat{\varphi} = \frac{\hbar}{Mr}\text{Im}\{\boldsymbol{\psi}^\dagger \partial_\varphi \boldsymbol{\psi}\}\hat{\varphi}. \tag{15}$$

In the next section, we discuss how the barrier potential, U_B, acts on the current density and the consequences of choosing an order parameter with *defined parity*.

4. Current Density and the Parity of the Order Parameter

Based on the discussion in the last section, we expect that, for relatively large R, a toroidal condensate under the presented conditions, will behave similar to a quasi-1D BEC with the current density effectively restricted to the $\hat{\varphi}$ direction. Now, we are interested in finding an expression for the current density $J_\varphi(\delta)$ at the edges of the barrier U_B (i.e., at $\varphi = \pm\delta$). To help us find such an expression, we define the following quantity:

$$\Delta J(\delta) := J_\varphi(\delta) - J_\varphi(-\delta). \tag{16}$$

Its interpretation is quite simple: if we multiply it by the area, A_c, of the cross-section of the condensate, it returns the rate of change in the number of atoms N_B located inside the barrier region:

$$\Delta J(\delta) \cdot A_c = -\frac{d}{dt}N_B. \tag{17}$$

Using (15) and (16), one finds the following expression for $\Delta J(\delta)$:

$$\Delta J(\delta) \approx \frac{\hbar}{MR}\text{Im}\Big\{\boldsymbol{\psi}^\dagger(\delta)\partial_\varphi\boldsymbol{\psi}(\delta) - \boldsymbol{\psi}^\dagger(-\delta)\partial_\varphi\boldsymbol{\psi}(-\delta)\Big\}. \tag{18}$$

To simplify it, we use symmetry arguments observing the *parity* of the order parameter $\boldsymbol{\psi}$. It is noticeable that, in general, the GPE (2) admits both *even* and *odd* solutions (with respect to the variable φ), and if we choose an even distribution $f(-\varphi) = f(\varphi)$, the order parameter follows

$$\begin{cases} \boldsymbol{\psi}(-\varphi) &= \boldsymbol{\psi}(\varphi) \quad \rightarrow \quad \text{even}, \\ \boldsymbol{\psi}(-\varphi) &= -\boldsymbol{\psi}(\varphi) \quad \rightarrow \quad \text{odd}. \end{cases} \tag{19}$$

Moreover, it is easy to show that J_φ is always an *odd* function if the parity of $\boldsymbol{\psi}$ is well defined. This way, it is clear that $\Delta J(\delta)$ allows us to calculate $J_\varphi(\delta)$ directly:

$$\Delta J(\delta) = 2J_\varphi(\delta). \tag{20}$$

In particular, if we choose $\boldsymbol{\psi}(\varphi)$ to be even, one finds that:

$$\boldsymbol{\psi}^\dagger(\delta)\partial_\varphi\boldsymbol{\psi}(\delta) - \boldsymbol{\psi}^\dagger(-\delta)\partial_\varphi\boldsymbol{\psi}(-\delta) = \boldsymbol{\psi}^\dagger(\delta)\left[\partial_\varphi\boldsymbol{\psi}(\delta) - \partial_\varphi\boldsymbol{\psi}(-\delta)\right], \tag{21}$$

and, using the Equations (12) and (13), we finally obtain:

$$J_\varphi(\delta) \approx \frac{U_0 R}{\hbar}\mathrm{Im}\{\boldsymbol{\psi}^\dagger(\delta)\boldsymbol{\psi}(0)\}. \tag{22}$$

Note that making $\boldsymbol{\psi}$ an even function is not the only way of finding such an approximation for the current difference $\Delta J(\delta)$. For instance, choosing RU_0 big enough in the complete expression for $\Delta J(\delta)$,

$$\Delta J(\delta) \approx \frac{\hbar}{MR}\mathrm{Im}\left\{\left[\boldsymbol{\psi}^\dagger(\delta) - \boldsymbol{\psi}^\dagger(-\delta)\right]\partial_\varphi\boldsymbol{\psi}(\delta) + \frac{2MR^2}{\hbar^2}U_0\boldsymbol{\psi}^\dagger(-\delta)\boldsymbol{\psi}(0)\right\}, \tag{23}$$

should effectively return the same $\Delta J(\delta)$ value from the even case, regardless of whether the parity of $\boldsymbol{\psi}$ is defined or not. However, choosing the even parity case for $\boldsymbol{\psi}$ leads to a symmetrical constraint of the BEC global phase, as we show in the next section (see Equations (31) and (38)).

Recalling what we mentioned before, it is always possible to factorize the order parameter $\boldsymbol{\psi} = \psi\boldsymbol{\zeta}$, and the scalar term can be written as $\psi = \sqrt{n}\,e^{i\theta}$. Applying such factorization in (22), we find:

$$J_\varphi(\delta) \approx \frac{U_0 R}{\hbar}\sqrt{n(\delta)n(0)}\,\mathrm{Im}\left\{\boldsymbol{\zeta}^\dagger(\delta)\boldsymbol{\zeta}(0)e^{i[\theta(0)-\theta(\delta)]}\right\}. \tag{24}$$

According to reference [9], $\boldsymbol{\zeta}$ is highly dependent on the symmetrical and topological properties of the condensate. Therefore, we expect that J_φ will also depend on such properties. Nevertheless, for now, to keep the analogy between the Josephson current (1) and $J_\varphi(\delta)$ as simple as possible, we consider the case of a scalar BEC (i.e., $\boldsymbol{\zeta} \equiv 1$):

$$J_\varphi(\delta) \approx \left\{\frac{U_0 R}{\hbar}\sqrt{n(\delta)n(0)}\right\} \times \sin[\theta(0) - \theta(\delta)] \tag{25}$$

Note that the current density J_φ at the edges of the barrier ($\varphi = \pm\delta$) is proportional to the *sine* of the phase difference between the *barrier region* ($\varphi \sim 0$) and the *remaining region* ($\varphi \gtrsim \delta$ or $\varphi \lesssim -\delta$), and the analogy to the Josephson current is (mathematically) clear.

It is also possible to derive the time-evolution equation for the global phase difference $\Delta\theta = \theta(0) - \theta(\delta)$, similarly to the second equation in (1). Here, we use the following relation between $\partial_t\theta$ and the local average energy per particle $\delta E/\delta n(\vec{r})$ [7]:

$$-\hbar\frac{\partial}{\partial t}\theta = \frac{\delta E}{\delta n(\vec{r})}, \tag{26}$$

where $\delta E/\delta n(\vec{r})$ is the functional derivative of the average energy of the condensate, E, with respect to the number density [7], $n(\vec{r})$,

$$E = \int \psi^* \left[-\frac{\hbar^2 \nabla^2}{2M} + U_{\text{total}} \right] \psi + \frac{c_0 n^2}{2} \, d^3 r. \tag{27}$$

If we consider the phase difference $\Delta\theta = \theta(0) - \theta(\delta)$,

$$\frac{\partial}{\partial t}[\Delta\theta] = \frac{1}{\hbar} \frac{\delta E}{\delta n} \bigg|_{\varphi=0}^{\varphi=\delta}, \tag{28}$$

we find that the difference of the local energy per particle (right-hand side of (28)) has the same role to $J_\varphi(\delta)$ as $2eV$ has to I_j in Equation (1).

5. The Broken-Symmetry Spinor BEC Case

This section discusses the current $J_\varphi(\delta)$ beyond the scalar BEC case. For that, it is convenient to define the function $C(\delta)$ as the quantity that characterizes such current according to the spinor nature of the condensate:

$$C(\delta) := \text{Im}\left\{ \zeta^\dagger(\delta) \zeta(0) e^{i[\theta(0) - \theta(\delta)]} \right\}. \tag{29}$$

Note that, in general,

$$C(\delta) \neq \sin(\Delta\theta), \tag{30}$$

because the *spinor* part $\boldsymbol{\zeta}$ may not be trivial (i.e., $\zeta \not\equiv 1$). Before discussing specific cases, we recall some useful properties of spinor BECs applied to the system analyzed here.

Firstly, due to the defined parity of $\boldsymbol{\psi}$, it is simple to show that the superfluid velocity in the $\hat{\varphi}$ direction (v_φ) is an odd function. Therefore, its circulation must always be zero and the *global phase circulation* is proportional to the *artificial magnetic flux*:

$$\oint d\theta = \oint \vec{A} \cdot d\vec{l} = \frac{1}{\hbar} \iint \vec{B} \cdot d\vec{S}, \tag{31}$$

which, in several situations [9], might be proportional to integer multiples of 2π.

In addition, the time-evolution equation for $\Delta\theta$ in the spinor BEC case is given by:

$$\frac{\partial}{\partial t}[\Delta\theta] = \left(\frac{1}{\hbar} \frac{\delta E}{\delta n} + \Phi \right) \bigg|_{\varphi=0}^{\varphi=\delta}, \tag{32}$$

where the average energy E is (for atomic spin $f = 1$):

$$E = \int \boldsymbol{\psi}^\dagger \left[-\frac{\hbar^2 \nabla^2}{2M} + U_{\text{total}} - p\mathbf{f}_z + q\mathbf{f}_z^2 \right] \boldsymbol{\psi} + \frac{c_0 n^2}{2} + \frac{c_1 \vec{F}^2}{2} \, d^3 r, \tag{33}$$

and it implicitly depends on the gauge field $\vec{A}$.

Moreover, according to reference [9], spinor BECs are naturally described by *broken-symmetry* states in the long-wavelength limit (i.e., when the characteristic dimensions of the condensate are much larger than its *healing length*). Furthermore, the spinor part of these states is characterized by the following type of expansion:

$$\boldsymbol{\zeta} = e^{-i\alpha\mathbf{f}_z} e^{-i\beta\mathbf{f}_y} e^{-i\gamma\mathbf{f}_z} \boldsymbol{\zeta}_0, \tag{34}$$

where the parameters (α, β, γ) are functions of time and space, and represent an arbitrary unitary transformation $\cong SO(3)$ (i.e., they are Euler angles). For $f = 1$ condensates, such transformation is equivalent to the following matrix:

$$e^{-i\alpha \mathbf{f}_z} e^{-i\beta \mathbf{f}_y} e^{-i\gamma \mathbf{f}_z} = \begin{pmatrix} e^{-i(\alpha+\gamma)}\cos^2\frac{\beta}{2} & -\frac{e^{-i\alpha}}{\sqrt{2}}\sin\beta & e^{-i(\alpha-\gamma)}\sin^2\frac{\beta}{2} \\ \frac{e^{-i\gamma}}{\sqrt{2}}\sin\beta & \cos\beta & -\frac{e^{i\gamma}}{\sqrt{2}}\sin\beta \\ e^{i(\alpha-\gamma)}\sin^2\frac{\beta}{2} & \frac{e^{i\alpha}}{\sqrt{2}}\sin\beta & e^{i(\alpha+\gamma)}\cos^2\frac{\beta}{2} \end{pmatrix}. \tag{35}$$

The *constant generator* of the spinor part ζ_0 is directly related to the order parameter of a uniform condensate, whose possible states (ferromagnetic, polar, anti-ferromagnetic, and others) have been studied and classified [9,10,36]. For generators with average spin in the $\hat{z}$ direction ($\zeta_0^\dagger \vec{\mathbf{f}} \zeta_0 = \zeta_0^\dagger \mathbf{f}_z \zeta_0 \hat{z} = f_0 \hat{z}$), the artificial electromagnetic potentials are [9]:

$$\vec{A} = f_0\Big(\cos(\beta)\vec{\nabla}\alpha + \vec{\nabla}\gamma\Big), \tag{36}$$

$$\Phi = -f_0\left(\cos(\beta)\frac{\partial}{\partial t}\alpha + \frac{\partial}{\partial t}\gamma\right). \tag{37}$$

So, we can rewrite (31) as:

$$\oint d\theta = f_0 \oint (\cos(\beta) d\alpha + d\gamma), \tag{38}$$

which is a constrain between θ and the Euler angles (given that the parity of ψ is defined).

Now, we will show two examples of how such broken-symmetry states affect the current $J_\varphi(\delta)$and how it differs from the scalar BEC case.

5.1. Ferromagnetic States

There are two families of *ferromagnetic* states in spinor ($f = 1$) BECs, called *positive* and *negative*, generated by $\zeta_0^{\text{ferro}+} = (1,0,0)^T$ and $\zeta_0^{\text{ferro}-} = (0,0,1)^T$, respectively. Using (35), one derives [9] the following spinor parts of the ferromagnetic states $\zeta_{\text{ferro}\pm}$:

$$\zeta_{\text{ferro}+} = e^{-i\gamma}\begin{pmatrix} e^{-i\alpha}\cos^2\frac{\beta}{2} \\ \frac{1}{\sqrt{2}}\sin\beta \\ e^{i\alpha}\sin^2\frac{\beta}{2} \end{pmatrix} \text{ and } \zeta_{\text{ferro}-} = e^{i\gamma}\begin{pmatrix} e^{-i\alpha}\sin^2\frac{\beta}{2} \\ \frac{-1}{\sqrt{2}}\sin\beta \\ e^{i\alpha}\cos^2\frac{\beta}{2} \end{pmatrix}, \tag{39}$$

from which one can compute all the *symmetrical* and *topological* properties. With these expressions, we directly find how the parameters (α, β, γ) act on $C(\delta)$:

$$\begin{aligned} C_{\text{ferro}\pm}(\delta) &= \cos^2\frac{\beta(0)}{2}\cos^2\frac{\beta(\delta)}{2}\sin[\Delta\theta \mp (\Delta\gamma + \Delta\alpha)] + \frac{1}{2}\sin\beta(0)\sin\beta(\delta)\sin[\Delta\theta \mp \Delta\gamma] \\ &+ \sin^2\frac{\beta(0)}{2}\sin^2\frac{\beta(\delta)}{2}\sin[\Delta\theta \mp (\Delta\gamma - \Delta\alpha)], \end{aligned} \tag{40}$$

for $\Delta G := G(0) - G(\delta)$, when $G = \theta, \alpha$ or γ.

Note that $C_{\text{ferro}\pm}(\delta)$ is not equal to $\sin[\Delta\theta]$ in general. This shows that the current $J_\varphi(\delta)$ in BECs is sensitive to their nontrivial structural spinor properties. However, we are still able to access the scalar limit, because $C_{\text{ferro}\pm}(\delta) = \sin[\Delta\theta]$, if (α, β, γ) are constant parameters. This situation corresponds to spinor BECs with trivial spinor parts (for instance, in the ferromagnetic case, $\zeta = (1,0,0)^T$ or $(0,0,1)^T$), representing an effective scalar BEC in a rotated reference frame.

5.2. Polar State

The *polar* states are generated by $\zeta_0^{\text{polar}} = (0,1,0)^T$. Such a family of states is unique because they are closely related to scalar BECs (given that $f_0 = 0$ and $\vec{F} = 0$), while their symmetrical and topological properties might be nontrivial [9]. Using (35), one finds the spinor part of the polar states:

$$\zeta_{\text{polar}} = \begin{pmatrix} -\frac{e^{-i\alpha}}{\sqrt{2}}\sin\beta \\ \cos\beta \\ \frac{e^{i\alpha}}{\sqrt{2}}\sin\beta \end{pmatrix}. \tag{41}$$

Applying this expression in (29), we obtain $C_{\text{polar}}(\delta)$:

$$\begin{aligned} C_{\text{polar}}(\delta) &= \frac{1}{2}\sin\beta(0)\sin\beta(\delta)\sin[\Delta\theta - \Delta\alpha] + \cos\beta(0)\cos\beta(\delta)\sin[\Delta\theta] \\ &+ \frac{1}{2}\sin\beta(0)\sin\beta(\delta)\sin[\Delta\theta + \Delta\alpha]. \end{aligned} \tag{42}$$

With a little algebraic work, we find that $C_{\text{polar}}(\delta) \propto \sin[\Delta\theta]$ in general. In this sense, we might interpret such a relation as a signature of the *partial analogy* between *polar states* and *scalar* BECs, because the current density $J_\varphi(\delta)$ can be written in the following way:

$$J_\varphi(\delta) = J_{\max}(\alpha,\beta)\sin[\Delta\theta], \tag{43}$$

where the parameters (α,β) modulate the maximum value of $J_\varphi(\delta)$.

6. Landau Critical Velocity, Lower Bound for R, and Experimental Protocol

This section discusses the effects of the Landau critical velocity on the parameters of our model and shows that it defines a lower bound for R. We also introduce a simple experimental scheme to test the theory, proposing an upper limit to how fast the barrier height, $U_0(t)$, can be turned on. In addition, we discuss some relevant considerations and practical suggestions for the experiments, particularly related to time-lapsed measurements of the BEC density distribution, using specialized imaging techniques, from which the analysis of the time evolution would lead to $C(\delta)$.

The Landau critical velocity is an upper limit for the velocity of particles in a superfluid, before the appearance of dissipation (viscosity) [7,8,13,35]. It is set by the energy gap between the ground state and the lowest elementary excitation leading to dissipation in the fluid. In BECs, this critical velocity is typically in the same order of magnitude as the *sound velocity*, being identical to it for a weakly-interacting homogeneous (uniform) scalar BEC. According to the Bogoliubov theory for spin-1 spinor BECs [9], similarly to the scalar case, this velocity v_c is

$$v_c \sim \sqrt{\frac{gn}{M}}, \tag{44}$$

where here g is usually a linear combination of the constants c_0 and c_1 from Equation (2), and it depends on the broken-symmetry state of the condensate. For example, in a ferromagnetic state, according to Sec. 5.2.1 in Ref. [9], $g = c_0 + c_1$.

Because $J_\varphi(\varphi)$ is an odd (anti-symmetrical) function in our formulation, from symmetry arguments alone, we expect $J_\varphi(0) = J_\varphi(\pi) = J_\varphi(-\pi) = 0$, with the maximum amplitudes symmetrically occurring in the interval $|\varphi| \in [0,\pi]$. Therefore, assuming that the dominant contribution to the current occurs at $\varphi = \pm\delta$, because we are interested in the superfluid regime, we impose $|\vec{v}(\delta)| < v_c(\delta) = \sqrt{gn(\delta)/M}$, where $n(\delta)$ is the total (local) density at $\varphi = \pm\delta$. Moreover, we use Equation (5) to estimate $|\vec{v}(\delta)|$:

$$|\vec{v}(\delta)| \sim \frac{\hbar}{MR}|\partial_\varphi\theta(\delta) - i\zeta^\dagger(\delta)\partial_\varphi\zeta(\delta)| \sim \frac{\hbar}{MR}\frac{1}{\delta}. \tag{45}$$

Considering δ is a given angle (i.e., we chose δ in an experiment), we can estimate a lower bound for R, such that the condensate is in the superfluid regime:

$$R > \sqrt{2}\frac{\hbar}{\sqrt{2Mgn(\delta)}}\frac{1}{\delta} = \sqrt{2}\frac{\xi}{\delta}; \quad \text{with } \xi \equiv \sqrt{\hbar^2/(2Mgn(\delta))}. \tag{46}$$

Or, more intuitively:

$$R \gg \frac{\text{(Healing Length)}}{\delta} \tag{47}$$

if one wants the system to be far from the critical velocity.

As indicated previously in Equation (17), the currents $J_\varphi(\pm\delta)$ are helpful to model the rate of change in the number of atoms N_B leaving the region inside the barrier. Conversely, measuring this rate of change, i.e., monitoring $N_B(t)$, is a way to measure $C(\delta)$ as a function of time. In principle, one could use non-destructive imaging techniques [37,38] to take multiple snapshots of the condensate density distribution at different times. Depending on the experimental parameters, such as the number of atoms and the radius R, the cross-section A_c could be so small that the optical density of the atoms would make dispersive (phase contrast) imaging [37] difficult, but this is exactly the conditions for which partial-transfer absorption imaging (PTAI) [38] was developed. Therefore, in principle, one can make these measurements even in the deep quasi-1D limit (i.e., extending the radius R as necessary to fulfill the condition in Section 4).

According to Section 5, different types of spinor BECs (ferromagnetic, polar, and scalar, with equivalent initial conditions for the density and velocity) should respond differently to the application of the same potential barrier $U_B(\varphi)$. The details will depend on the specific case, as well as the external magnetic field. Therefore, further theoretical, numerical, and experimental studies are necessary, but they may unveil new ways of sensing the symmetry of spinor BEC states, solely using measurements of $C(\delta)$.

As a simple example, we can imagine an experimental protocol starting initially with the barrier turned off and a homogeneous condensate in equilibrium ($n(0) = n(\delta)$ and $\partial_t n(\varphi) = 0$). At $t = 0$, the barrier is turned on and ramped up following a sufficiently fast time protocol $U_0 = U_0(t)$. The barrier causes a perturbation on the densities (and also on N_B and $C(\delta)$), and, after a time $t = \tau$, the system reaches a new equilibrium ($\partial_t N_B = 0$ and $C(\delta) = 0, \forall t > \tau$). For scalar BECs, the equilibrium is described by $C(\delta) = \sin(\Delta\theta) = 0$, with $\Delta\theta =$ constant, which (as expected) simply implies in:

$$\mu = \left(\frac{\delta E}{\delta n}\right)_{\varphi=\delta} = \left(\frac{\delta E}{\delta n}\right)_{\varphi=0}. \tag{48}$$

Finally, we briefly show here that the time protocol $U_0(t)$ must always have an upper limit that depends on the density's perturbation and $C(\delta)$ at any given time so that the condensate does not lose its superfluid properties. This upper limit is found using Equation (24) to estimate $|\vec{v}(\delta)|$ and it corresponds to

$$U_0(t) < v_c(\delta,t)\frac{\hbar}{R}\sqrt{\frac{n(\delta,t)}{n(0,t)}\frac{1}{|C(\delta,t)|}}. \tag{49}$$

7. Conclusions

In this manuscript, we used the parallel between superconductors and superfluids to present a curious new effect in a superfluid, similar to the current oscillations in the Josephson effect, but it happens without current flowing through the barrier. The result was derived for BECs in toroidal traps with a thin finite repulsive barrier, and it works both for scalar and spinor BECs.

Using a thin barrier approximation, in Equations (12) and (13), and assuming a defined parity for the order parameter, we derived analytical expressions for the current density.

We showed that the current at the edges of the barrier oscillates in a similar fashion to the current flowing through a Josephson junction (1), but in our case, it happens without a net circulation of the superfluid current. We also showed how the nontrivial symmetry properties of spinor BECs could generate other current structures, beyond the typical $C(\delta) = \sin[\Delta\theta]$ case, indicating that such current is sensitive to these properties, suggesting that it could be used to probe spinor symmetry or, perhaps, provide precision measurements related to the superfluid flow.

Throughout this work, we assumed the parity of $\boldsymbol{\psi}$ was defined, which is a strong mathematical imposition to guarantee that Equation (24) is a reasonable approximation for the current density. Moreover, this imposition implies that the circulation of the condensates considered here is always zero, excluding several $(\theta, \alpha, \beta, \gamma)$ configurations with nontrivial symmetry and topology. However, as we discussed in Section 4, ensuring that the last term in (23) is dominant should keep the structure of $\Delta J(\delta)$ unchanged, regardless of the parity of $\boldsymbol{\psi}$. Therefore, it might be possible to derive a similar expression (24) for BECs with nonzero circulation, such as the ones shown in experiments [30–33,39], and for spinor BECs with nontrivial structure or topology [9,40–48].

Here, we also neglected the effects of fluctuations, either classical or quantum, which can be relevant in the quasi-1D limit [49–51]. However, as we have shown previously in Ref. [52], one can always play with the number of atoms and the aspect ratio of the trapping potential to place oneself within the best range of parameters for the experiments. Exercising this ability deliberately allows one to control the influence of thermal phase fluctuations at finite temperatures. Therefore, interesting future directions would be to explore the effects of finite temperature and thermal fluctuations, and the non-equilibrium effects caused by quickly turning on the repulsive barrier in our proposed experimental scheme. In Section 6, we discussed some considerations for an idealized protocol, but the general spinor case is more complex and deserves a detailed analysis. For instance, our simple estimate for the Landau critical velocity does not take into account other relevant mechanisms of decay [33,39,53], especially at finite temperatures [54].

Therefore, further numerical, experimental, and theoretical studies are necessary and may improve, expand, and, perhaps, help to classify (maybe in symmetry terms) the formulation of this intriguing Josephson-like effect in toroidal spinor Bose–Einstein condensates.

Author Contributions: Conceptualization, M.H.F.D. and S.R.M.; methodology, M.H.F.D. and S.R.M.; writing—original draft preparation, M.H.F.D.; writing—review and editing, S.R.M.; supervision, S.R.M.; funding acquisition, S.R.M. All authors have read and agreed to the published version of the manuscript.

Funding: We acknowledge the financial support provided by FAPESP (Fundação de Amparo à Pesquisa do Estado de São Paulo), the São Paulo Research Foundation, under Grant no. 2013/07276-1 and 2019/27471-0, and also by CAPES Grant nos. 88887.338144/2019-00 and 88887.616990/2021-00.

Institutional Review Board Statement: Not applicable.

Informed Consent Statement: Not applicable.

Conflicts of Interest: The authors declare no conflict of interest.

References

1. Josephson, B.D. Possible new effects in superconductive tunnelling. *Phys. Lett.* **1962**, *1*, 251–253. [CrossRef]
2. Josephson, B.D. The discovery of tunnelling supercurrents. *Rev. Mod. Phys.* **1974**, *46*, 251–254. [CrossRef]
3. Barone, A.; Paterno, G. *Physics and Applications of the Josephson Effect*; John Wiley & Sons, Ltd.: Hoboken, NJ, USA, 1982. [CrossRef]
4. Tinkham, M. *Introduction to Superconductivity*; Dover Books on Physics Series; Dover Publications: Mineola, NY, USA, 2004.
5. Bardeen, J.; Cooper, L.N.; Schrieffer, J.R. Theory of Superconductivity. *Phys. Rev.* **1957**, *108*, 1175–1204. [CrossRef]
6. Cummings, F.W.; Johnston, J.R. Theory of Superfluidity. *Phys. Rev.* **1966**, *151*, 105–112. [CrossRef]
7. Pethick, C.; Smith, H. *Bose–Einstein condensation in Dilute Gases*; Cambridge University Press: Cambridge, UK, 2008.
8. Pitaevskii, L.; Stringari, S. *Bose-Einstein Condensation and Superfluidity*; Oxford University Press: Oxford, UK, 2016; p. 576. [CrossRef]
9. Kawaguchi, Y.; Ueda, M. Spinor Bose–Einstein condensates. *Phys. Rep.* **2012**, *520*, 253–381. [CrossRef]

10. Stamper-Kurn, D.M.; Ueda, M. Spinor Bose gases: Symmetries, magnetism, and quantum dynamics. *Rev. Mod. Phys.* **2013**, *85*, 1191–1244. [CrossRef]
11. Yukalov, V.I. Dipolar and spinor bosonic systems. *Laser Physics* **2018**, *28*, 053001. [CrossRef]
12. Penrose, O.; Onsager, L. Bose-Einstein Condensation and Liquid Helium. *Phys. Rev.* **1956**, *104*, 576–584. [CrossRef]
13. Bogoliubov, N. On the Theory of Superfluidity. *J. Phys.* **1947**, *11*, 3–20. [CrossRef]
14. Zapata, I.; Sols, F.; Leggett, A.J. Josephson effect between trapped Bose-Einstein condensates. *Phys. Rev. A* **1998**, *57*, R28–R31. [CrossRef]
15. Williams, J.; Walser, R.; Cooper, J.; Cornell, E.; Holland, M. Nonlinear Josephson-type oscillations of a driven, two-component Bose-Einstein condensate. *Phys. Rev. A* **1999**, *59*, R31–R34. [CrossRef]
16. Hall, D.S.; Matthews, M.R.; Wieman, C.E.; Cornell, E.A. Measurements of Relative Phase in Two-Component Bose-Einstein Condensates. *Phys. Rev. Lett.* **1998**, *81*, 1543–1546. [CrossRef]
17. Albiez, M.; Gati, R.; Fölling, J.; Hunsmann, S.; Cristiani, M.; Oberthaler, M.K. Direct Observation of Tunneling and Nonlinear Self-Trapping in a Single Bosonic Josephson Junction. *Phys. Rev. Lett.* **2005**, *95*, 010402. [CrossRef] [PubMed]
18. Levy, S.; Lahoud, E.; Shomroni, I.; Steinhauer, J. The a.c. and d.c. Josephson effects in a Bose–Einstein condensate. *Nature* **2007**, *449*, 579–583. [CrossRef] [PubMed]
19. Ryu, C.; Blackburn, P.W.; Blinova, A.A.; Boshier, M.G. Experimental Realization of Josephson Junctions for an Atom SQUID. *Phys. Rev. Lett.* **2013**, *111*, 205301. [CrossRef] [PubMed]
20. Sadler, L.E.; Higbie, J.M.; Leslie, R.; Vengalattore, M.; Stamper-Kurn, D.M. Spontaneous symmetry breaking in a quenched ferromagnetic spinor Bose–Einstein condensate. *Nature* **2006**, *443*, 312–315. [CrossRef]
21. Faddeev, L.; Niemi, A.J. Stable knot-like structures in classical field theory. *Nature* **1997**, *387*, 58–61. [CrossRef]
22. Kawaguchi, Y.; Nitta, M.; Ueda, M. Knots in a Spinor Bose-Einstein Condensate. *Phys. Rev. Lett.* **2008**, *100*, 180403. [CrossRef] [PubMed]
23. Savage, C.M.; Ruostekoski, J. Dirac monopoles and dipoles in ferromagnetic spinor Bose-Einstein condensates. *Phys. Rev. A* **2003**, *68*, 043604. [CrossRef]
24. Khawaja, U.A.; Stoof, H. Skyrmions in a Ferromagnetic Bose-Einstein Condensate. *Nature* **2001**, *411*, 918–920. [CrossRef] [PubMed]
25. Grimm, R.; Weidemüller, M.; Ovchinnikov, Y.B. Optical Dipole Traps for Neutral Atoms. In *Advances In Atomic, Molecular, and Optical Physics*; Academic Press: Cambridge, MA, USA, 2000; pp. 95–170 [CrossRef]
26. Pasienski, M.; DeMarco, B. A high-accuracy algorithm for designing arbitrary holographic atom traps. *Opt. Express* **2008**, *16*, 2176–2190. [CrossRef] [PubMed]
27. Gaunt, A.L.; Hadzibabic, Z. Robust Digital Holography for Ultracold Atom Trapping. *Sci. Rep.* **2012**, *2*, 721. [CrossRef] [PubMed]
28. Gauthier, G.; Lenton, I.; Parry, N.M.; Baker, M.; Davis, M.J.; Rubinsztein-Dunlop, H.; Neely, T.W. Direct imaging of a digital-micromirror device for configurable microscopic optical potentials. *Optica* **2016**, *3*, 1136–1143. [CrossRef]
29. Silva, P.F.; Muniz, S.R. Generating arbitrary laser beam shapes through phase-mapped designed beam splitting. In Proceedings of the 2021 SBFoton International Optics and Photonics Conference (SBFoton IOPC), Sao Carlos, Brazil, 31 May–2 June 2021; pp. 1–4. [CrossRef]
30. Ramanathan, A.; Wright, K.C.; Muniz, S.R.; Zelan, M.; Hill III, W.T.; Lobb, C.J.; Helmerson, K.; Phillips, W.D.; Campbell, G.K. Superflow in a Toroidal Bose-Einstein Condensate: An Atom Circuit with a Tunable Weak Link. *Phys. Rev. Lett.* **2011**, *106*, 130401. [CrossRef]
31. Wright, K.C.; Blakestad, R.B.; Lobb, C.J.; Phillips, W.D.; Campbell, G.K. Driving Phase Slips in a Superfluid Atom Circuit with a Rotating Weak Link. *Phys. Rev. Lett.* **2013**, *110*, 025302. [CrossRef]
32. Eckel, S.; Lee, J.G.; Jendrzejewski, F.; Murray, N.; Clark, C.W.; Lobb, C.J.; Phillips, W.D.; Edwards, M.; Campbell, G.K. Hysteresis in a quantized superfluid 'atomtronic' circuit. *Nature* **2014**, *506*, 200–203. [CrossRef] [PubMed]
33. Beattie, S.; Moulder, S.; Fletcher, R.J.; Hadzibabic, Z. Persistent Currents in Spinor Condensates. *Phys. Rev. Lett.* **2013**, *110*, 025301. [CrossRef] [PubMed]
34. Goldman, N.; Juzeliūnas, G.; Öhberg, P.; Spielman, I.B. Light-induced gauge fields for ultracold atoms. *Rep. Prog. Phys.* **2014**, *77*, 126401. [CrossRef] [PubMed]
35. Baym, G.; Pethick, C.J. Landau critical velocity in weakly interacting Bose gases. *Phys. Rev. A* **2012**, *86*, 023602. [CrossRef]
36. Zhang, W.; Yi, S.; You, L. Mean field ground state of a spin-1 condensate in a magnetic field. *New J. Phys.* **2003**, *5*, 77. [CrossRef]
37. Andrews, M.R.; Mewes, M.O.; van Druten, N.J.; Durfee, D.S.; Kurn, D.M.; Ketterle, W. Direct, Nondestructive Observation of a Bose Condensate. *Science* **1996**, *273*, 84–87. [CrossRef] [PubMed]
38. Ramanathan, A.; Muniz, S.R.; Wright, K.C.; Anderson, R.P.; Phillips, W.D.; Helmerson, K.; Campbell, G.K. Partial-transfer absorption imaging: A versatile technique for optimal imaging of ultracold gases. *Rev. Sci. Instr.* **2012**, *83*, 083119. [CrossRef] [PubMed]
39. Moulder, S.; Beattie, S.; Smith, R.P.; Tammuz, N.; Hadzibabic, Z. Quantized supercurrent decay in an annular Bose-Einstein condensate. *Phys. Rev. A* **2012**, *86*, 013629. [CrossRef]
40. Leanhardt, A.E.; Görlitz, A.; Chikkatur, A.P.; Kielpinski, D.; Shin, Y.; Pritchard, D.E.; Ketterle, W. Imprinting Vortices in a Bose-Einstein Condensate using Topological Phases. *Phys. Rev. Lett.* **2002**, *89*, 190403. [CrossRef] [PubMed]

41. Leanhardt, A.E.; Shin, Y.; Kielpinski, D.; Pritchard, D.E.; Ketterle, W. Coreless Vortex Formation in a Spinor Bose-Einstein Condensate. *Phys. Rev. Lett.* **2003**, *90*, 140403. [CrossRef]
42. Kumakura, M.; Hirotani, T.; Okano, M.; Takahashi, Y.; Yabuzaki, T. Topological formation of a multiply charged vortex in the Rb Bose-Einstein condensate: Effectiveness of the gravity compensation. *Phys. Rev. A* **2006**, *73*, 063605. [CrossRef]
43. Wright, K.C.; Leslie, L.S.; Bigelow, N.P. Optical control of the internal and external angular momentum of a Bose-Einstein condensate. *Phys. Rev. A* **2008**, *77*, 041601. [CrossRef]
44. Wright, K.C.; Leslie, L.S.; Hansen, A.; Bigelow, N.P. Sculpting the Vortex State of a Spinor BEC. *Phys. Rev. Lett.* **2009**, *102*, 030405. [CrossRef] [PubMed]
45. Leslie, L.S.; Hansen, A.; Wright, K.C.; Deutsch, B.M.; Bigelow, N.P. Creation and Detection of Skyrmions in a Bose-Einstein Condensate. *Phys. Rev. Lett.* **2009**, *103*, 250401. [CrossRef] [PubMed]
46. Juliá-Díaz, B.; Guilleumas, M.; Lewenstein, M.; Polls, A.; Sanpera, A. Josephson oscillations in binary mixtures of $F = 1$ spinor Bose-Einstein condensates. *Phys. Rev. A* **2009**, *80*, 023616. [CrossRef]
47. Juliá-Díaz, B.; Melé-Messeguer, M.; Guilleumas, M.; Polls, A. Spinor Bose-Einstein condensates in a double well: Population transfer and Josephson oscillations. *Phys. Rev. A* **2009**, *80*, 043622. [CrossRef]
48. Muñoz Mateo, A.; Gallemí, A.; Guilleumas, M.; Mayol, R. Persistent currents supported by solitary waves in toroidal Bose-Einstein condensates. *Phys. Rev. A* **2015**, *91*, 063625. [CrossRef]
49. Petrov, D.S.; Shlyapnikov, G.V.; Walraven, J.T.M. Phase-Fluctuating 3D Bose-Einstein Condensates in Elongated Traps. *Phys. Rev. Lett.* **2001**, *87*, 050404. [CrossRef] [PubMed]
50. Al Khawaja, U.; Proukakis, N.P.; Andersen, J.O.; Romans, M.W.J.; Stoof, H.T.C. Dimensional and temperature crossover in trapped Bose gases. *Phys. Rev. A* **2003**, *68*, 043603. [CrossRef]
51. Gerbier, F. Quasi-1D Bose-Einstein condensates in the dimensional crossover regime. *Europhys. Lett. (EPL)* **2004**, *66*, 771–777. [CrossRef]
52. Mathey, L.; Ramanathan, A.; Wright, K.C.; Muniz, S.R.; Phillips, W.D.; Clark, C.W. Phase fluctuations in anisotropic Bose-Einstein condensates: From cigars to rings. *Phys. Rev. A* **2010**, *82*, 033607. [CrossRef]
53. Piazza, F.; Collins, L.A.; Smerzi, A. Critical velocity for a toroidal Bose–Einstein condensate flowing through a barrier. *J. Phys. B At. Mol. Opt. Phys.* **2013**, *46*, 095302. [CrossRef]
54. Mathey, A.C.; Clark, C.W.; Mathey, L. Decay of a superfluid current of ultracold atoms in a toroidal trap. *Phys. Rev. A* **2014**, *90*, 023604. [CrossRef]

symmetry MDPI

Review

Non-Thermal Fixed Points in Bose Gas Experiments

Lucas Madeira [1,*] and **Vanderlei S. Bagnato** [1,2]

1 Instituto de Física de São Carlos, Universidade de São Paulo, CP 369, São Carlos 13560-970, Brazil; vander@ifsc.usp.br
2 Hagler Fellow, Hagler Institute for Advanced Study, Texas A&M University, College Station, TX 77843, USA
* Correspondence: madeira@ifsc.usp.br

Abstract: One of the most challenging tasks in physics has been understanding the route an out-of-equilibrium system takes to its thermalized state. This problem can be particularly overwhelming when one considers a many-body quantum system. However, several recent theoretical and experimental studies have indicated that some far-from-equilibrium systems display universal dynamics when close to a so-called non-thermal fixed point (NTFP), following a rescaling of both space and time. This opens up the possibility of a general framework for studying and categorizing out-of-equilibrium phenomena into well-defined universality classes. This paper reviews the recent advances in observing NTFPs in experiments involving Bose gases. We provide a brief introduction to the theory behind this universal scaling, focusing on experimental observations of NTFPs. We present the benefits of NTFP universality classes by analogy with renormalization group theory in equilibrium critical phenomena.

Keywords: non-thermal fixed points; out-of-equilibrium system; Bose–Einstein condensate; turbulence

Citation: Madeira, L.; Bagnato, V.S. Non-Thermal Fixed Points in Bose Gas Experiments. *Symmetry* **2022**, *14*, 678. https://doi.org/10.3390/sym14040678

Academic Editor: Sergei D. Odintsov

Received: 27 February 2022
Accepted: 24 March 2022
Published: 25 March 2022

1. Introduction

Although studying closed interacting quantum many-body systems is challenging, some theories have successfully described aspects of these systems [1]. Quantum statistical mechanics provides a good description of many physical systems in thermodynamic equilibrium. However, most natural phenomena occur under conditions outside equilibrium. This brings us to a new challenge in searching for descriptions of out-of-equilibrium conditions that allow a reasonable understanding of the observations.

Some observables are insensitive to initial conditions and system parameters in certain situations, leading to universal phenomena. In the case of far-from-equilibrium initial conditions, this can be observed well before an equilibrium or a quasi-stationary state is reached. This has been investigated in a wide variety of systems such as the inflation in the early Universe [2–6], heavy-ion collisions producing quark–gluon matter [7–10], and cold-gas systems, in both theory [5,11–17] and experiments [18–21]. A universal spatio-temporal scaling emerges, independent of the initial state or microscopic parameters.

It has been suggested that the universal scaling observed in these far-from-equilibrium isolated systems is due to the presence of non-thermal fixed points (NTFPs) [4,5,9,13,22]. In the literature, it is possible to find evidence of non-thermal universality classes encompassing a variety of systems. These are far away from any phase transitions; hence, their mechanism cannot be the same as that of the well-known critical phenomena and their characteristic exponents. Another difference is that no fine-tuning of the parameters is required to observe the characteristic scaling provided by nearby NTFPs, unlike in equilibrium critical phenomena, where system variables need to be tuned to specific critical values. This is evidenced by the universal scaling for several different initial states in NTFP investigations.

Recently, with the advent of Bose–Einstein condensates (BECs) [23], the possibility of carrying out controlled experiments has become a reality, opening a window of opportunity to investigate intrinsic properties of these systems or even classify them according to their pattern of time evolution.

The production of an out-of-equilibrium BEC can be achieved experimentally in several ways, one of them being via so-called "quantum quenching". The system, initially in equilibrium, is represented by a Hamiltonian $H(\alpha_0)$, where α_0 is a set of system parameters. Then, an abrupt change is made to a new situation described by a Hamiltonian $H(\alpha_1)$, where α_1 is a new collection of parameters. In this case, the system leaves the equilibrium condition, having its previous state determined by $H(\alpha_0)$ as its initial condition. Its temporal evolution is unitary and governed by the new Hamiltonian $H(\alpha_1)$. The operator $U = \exp[-iH(\alpha_1)t/\hbar]$ determines what happens until the system can decay to a new state of $H(\alpha_1)$. One example corresponds to abrupt potential changes, when only a few degrees of freedom are present, allowing a simplified system evolution. This simplicity can change when dealing with a system with many degrees of freedom, which is the case in a many-body quantum system. Each subsystem can see the rest as a reservoir, and the final time evolution can take quite complicated routes. There are many open questions, including the fundamental question of how the system evolves temporally toward equilibrium, or even how to quantify the out-of-equilibrium state and identify which system components are determinant in the typical relaxation time for a new state of equilibrium.

Since there are many quench protocols that can be employed to produce out-of-equilibrium BECs, some even preserving symmetries present in the equilibrium state, the question of whether a system can approach a NTFP arises. In general, this depends strongly on the chosen initial conditions, which should correspond to extreme out-of-equilibrium configurations [24]. For example, in the case of a dilute Bose gas, this could be achieved by populating only modes below a certain momentum scale Q. The initial state, a constant momentum distribution for $k < Q$ that drops abruptly at $k = Q$, is strongly overpopulated at low momenta (compared to the final equilibrium distribution). Hence, the subsequent dynamics is of particle transport toward lower momenta and energy migration to high momenta. If the spatio-temporal scaling presented in Section 2 is observed, then this is a "smoking gun" indicating a nearby NTFP [24].

In experiments, preparing and maintaining a closed system can be very challenging. Cold-atom systems are, in this sense, very close to the ideal. Isolation allows experimental access by external agents in a controlled manner. The isolation of trapped condensates and the excellent control of quantum states are fundamental elements required to achieve long coherence times, an essential feature in out-of-equilibrium studies since the coherence time must be of at least the same order as the temporal scale of the dynamics involved.

Some clarification about the scope of this review is in order. We chose to focus our attention on experiments involving Bose gases [18–21]. Although this corresponds only to a small part of all research being conducted on NTFPs, we consider these examples to be illustrative of the field. Notably, it is possible to present straightforward applications of the theory without exploring its inner workings. For readers that wish to do so, we provide references that may be helpful [17,24,25]. One of the motivations is to use cold-atom setups to simulate the dynamics of currently inaccessible systems; for example, the inflation of the early Universe. This is counter-intuitive because cold-atom systems are in the low-energy regime, whereas the early Universe is at the other end of the energy spectrum. It is only possible through the universal scaling due to the presence of NTFPs.

This paper is organized as follows. In Section 2, we introduce the universal scaling provided by NTFPs and related quantities of interest. Section 3 presents four experiments dealing with NTFPs in Bose gas systems: a quasi-one-dimensional Bose gas in Section 3.1; a spin-1 condensate in Section 3.2; a homogeneous condensate in three-dimensions in Section 3.3; and a turbulent harmonically trapped BEC in Section 3.4. Finally, a summary is given in Section 4.

2. Universal Scaling and NTFPs

Our goal in this section is to present the universal scaling function and its associated exponents. The motivations, details, and derivations regarding the NTFP theory can be found in [17,24,25] and the references therein. The theory developed for NTFPs is inspired by equilibrium renormalization group theory where, in the vicinity of a phase transition, the correlations are self-similar (independent of the resolution). Thus, scaling the spatial resolution by a parameter s creates a correlation function that depends on the distance x between two points behave according to $C(x;s) = s^{\zeta} f(x/s)$. In this way, the correlation is characterized exclusively by a universal exponent ζ and a function f. A fixed point in equilibrium critical phenomena corresponds to a situation where varying s does not change $C(x;s)$, meaning that the universal function is a power law, $f(x) \propto x^{-\zeta}$. In physical systems, this behavior is approximate, and information may be retained about characteristic scales.

Non-thermal fixed points in the dynamics of far-from-equilibrium systems are analogous to fixed points in critical equilibrium phenomena, with the distinction that the former uses the time t as the scale parameter. Near NTFPs, the correlations take the form $C(x,t) = t^{\alpha} f(t^{-\beta} x)$, now with two universal exponents. Universality classes, in this context, correspond to the same exponents α and β and a scaling function f. Figure 1 shows a schematic time evolution for a system that encounters an NTFP.

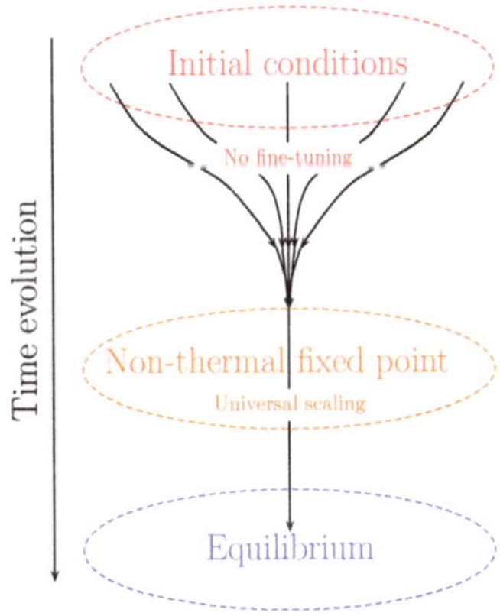

Figure 1. Illustration of the dynamics of a system passing near an NTFP. For several initial conditions (the key idea being that no fine-tuning is needed), the system can pass near a non-thermal fixed point. The correlation functions show a spatio-temporal scaling with a universal function when that occurs. After some time, the system leaves the vicinity of the NTFP and reaches equilibrium.

In the case of cold bosonic gases, the momentum distribution can be used to characterize the time evolution of the system. Consider an isotropic momentum distribution, i.e., a distribution that is only a function of $k = |\mathbf{k}|$ and time, $n(k,t)$. The total number of particles $N(t)$ is obtained by integrating the momentum distribution,

$$N(t) = \int d^d k \, n(k,t). \tag{1}$$

We provide the d-dimensional expression, since we cover the cases $d = 1, 2$, and 3 in this review. The fluctuations in the total number of particles in cold-gas experiments are relatively small due to the low temperatures involved and the high levels of control that one can exert over these systems. However, unwanted losses occur mainly due to the heating of the sample. For this reason, several experimental studies report the momentum distribution normalized by the total number of particles at a given time, $\bar{n}(k,t) \equiv n(k,t)/N(t)$. In theoretical and numerical investigations, N is a fixed number and this distinction is irrelevant.

The universal scaling displayed by a system in the vicinity of an NTFP is given by

$$\bar{n}(k,t) = \left(\frac{t}{t_0}\right)^{\alpha} f\left[\left(\frac{t}{t_0}\right)^{\beta} k\right]. \tag{2}$$

where α and β are scaling exponents, f is a universal scaling function, and t_0 is an arbitrary reference time within the time interval where the scaling takes place. From this definition of the scaling exponents, it is possible to see that the sign of α is related to the direction of the particle transport. Positive values indicate particles migrating toward lower momenta, whereas negative values increase the population of the high-momentum components.

Global Observables

Besides the important result of Equation (2), there are also global observables of interest when dealing with systems that display universal scaling due to the presence of nearby NTFPs. The main difference is that these quantities are computed in the infrared (IR) scaling region, $k_D \leqslant (t/t_0)^{\beta} k \leqslant k_c$, where k_D is the smallest wave vector that can be measured (inversely proportional to the largest length scale of the system D, typically its size) and k_c defines the cut-off where the universal scaling takes place.

One of the observables is the number of particles in the scaling region,

$$\bar{N}(t) = \int\limits_{k \leqslant \left(\frac{t}{t_0}\right)^{-\beta} k_c} d^d k\, \bar{n}(k,t) \propto \left(\frac{t}{t_0}\right)^{\alpha - d\beta}. \tag{3}$$

A necessary condition for Equation (2) to hold is that $\bar{N}(t)$ is constant during the time interval where the universal scaling is observed. The time dependence $\bar{N}(t) \propto t^{\alpha - d\beta}$ can be derived straightforwardly by inserting Equation (2) into Equation (3) and changing the variables of integration.

The moments of the momentum distribution can also be computed in the region of the universal scaling. The second moment has the physical interpretation of the mean kinetic energy per particle in the scaling region,

$$\bar{M}_2(t) = \int\limits_{k \leqslant \left(\frac{t}{t_0}\right)^{-\beta} k_c} d^d k\, \frac{k^2 \bar{n}(k,t)}{\bar{N}(t)} \propto \left(\frac{t}{t_0}\right)^{-2\beta}. \tag{4}$$

Its time dependence, $\bar{M}_2 \propto t^{-2\beta}$, can also be derived using Equations (2) and (3). Hence, the sign of β is related to the direction of the energy transport. For $\beta > 0$, the energy leaves the scaling region and migrates to higher momenta, whereas a negative sign indicates an energy increase in the IR scaling interval.

These are not the only global observables that can be defined. For example, in Section 3.2 we encounter another quantity, relevant in the context of a spinor condensate.

3. Experiments

3.1. One-Dimensional Bose gas

The authors in [18] produced a one-dimensional Bose gas by strongly quenching a three-dimensional one. They reported a universal scaling in the time-dependent momentum distribution due to the presence of an NTFP.

These authors employed ^{87}Rb atoms, which correspond to a repulsively interacting Bose gas in a very elongated (quasi-1D) harmonic trap. At the beginning of the experiment, the thermal gas was just above the critical temperature. During the quench, the trap depth was ramped linearly, such that its final value was below the first radially excited state. This causes the evaporation of atoms occupying higher energy states. Finally, the depth of the trap was raised to close the trap. The resulting Bose gas was in a far-from-equilibrium condition, and its evolution was recorded after a time t. They were able to probe the system using both the in situ density and the momentum distribution, which is the quantity of interest for the universal scaling in the form of Equation (2).

Some aspects of the momentum distribution at early and late times were known. The authors provided a quantitative description of the initial state in terms of solitonic defects [14,26]. Since the quenching procedure is almost instantaneous, the initial state had a large population of high-energy modes, which makes the observation of the universal dynamics associated with NTFPs possible when the system relaxes. During the time evolution of the system, a peak appears at low momenta, indicative of the formation of the quasi-condensate. The system reaches a thermal quasi-condensate state at late times, which is described by a Lorentzian function (its width is inversely proportional to the temperature).

The normalized momentum distributions are shown in Figure 2a. As time progresses, the distribution increases (decreases) at low (high) momenta, signaling the formation of the quasi-condensate. In Figure 2b, the authors provide the scaled curves using Equation (2) with $\alpha = 0.09(5)$ and $\beta = 0.10(4)$. All curves, below a cut-off value k_c indicated by a dashed line in the figure, collapse into a single universal function. Both exponents are positive, indicating that particles migrate toward the low-momenta region, and energy flows in the high-momenta direction. This is consistent with the formation of a quasi-condensate after the quenching procedure is performed. The theoretical study in [5] predicts a value of $\beta = 1/2$, independent of the dimension d, for far-from-equilibrium dynamics in an isolated Bose gas following a strong quench. The authors of [18] provide arguments for why this theory does not apply fully to their system.

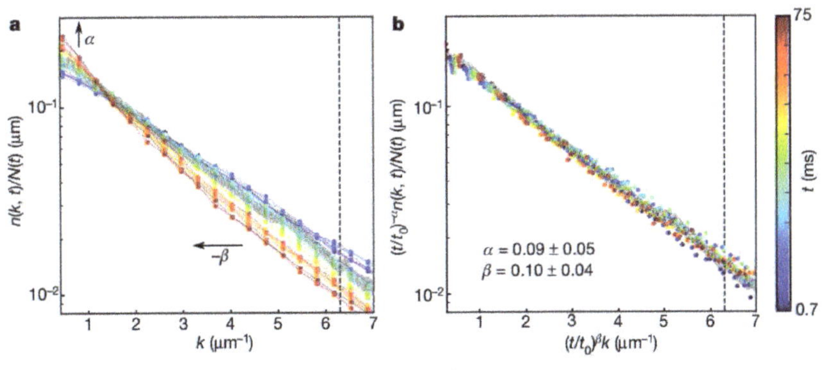

Figure 2. Universal scaling dynamics observed by the authors of [18]. (**a**) Time evolution of the normalized momentum distributions. (**b**) Momentum distributions scaled according to Equation (2). All the curves collapse into a single function, signaling the universal scaling. Reprinted by permission from Springer Nature Customer Service Centre GmbH: Springer Nature, Nature, Erne et al., Universal dynamics in an isolated one-dimensional Bose gas far from equilibrium, © 2018.

As well as analyzing the momentum distributions, the authors also computed the number of particles and the mean kinetic energy per particle in the scaling region (Equations (3) and (4)). The quantity $\bar{N}$ was approximately constant in that region, which is consistent with the time-dependence prediction of Equation (3), i.e., $\bar{N}(t) \propto t^{\alpha - d\beta}$, since $d = 1$ and the authors found $\alpha \approx \beta$. Moreover, the $\bar{M}_2 \propto t^{-2\beta}$ behavior was also verified in the region of interest. As expected, both quantities showed a different time dependence outside the scaling region.

3.2. Spinor Bose Gas

The authors in [19] observed universal dynamics in a quasi-one-dimensional spinor Bose gas [27] by analyzing spin correlations. They employed a ^{87}Rb gas in the $F = 1$

hyperfine state, which has three possible magnetic sublevels: $m_F = -1, 0, 1$. Hence, it behaves as a spin-1 system with ferromagnetic interactions [28]. Initially, all atoms are in the $m_F = 0$ state. The system is driven far from equilibrium by a sudden change in the energy splitting of the sublevels, producing excitations in the $F_x - F_y$ plane.

Although both this system and the one presented in Section 3.1 are quasi-one-dimensional systems, the degrees of freedom and underlying physics are quite different. Hence, the function entering into the scaling described by Equation (2) is not as straightforward as being simply the momentum distribution. For details regarding the function and the measurements, the reader is referred to [19]. Here, we present only a brief overview.

First, the authors computed the mean spin length, $\langle |F_\perp(t)| \rangle$, where $F_\perp = F_x + iF_y$. A local angle was defined and extracted using $\theta(y,t) = \arcsin(F_x(y,t)/\langle |F_\perp(t)| \rangle)$. Fluctuations were probed by means of a two-point correlation function, $C(y,y';t) = \langle \theta(y,t)\theta(y',t) \rangle$. Finally, the desired function is the structure factor, which is the averaged Fourier transform:

$$f_\theta(k,t) = \int dy \int d\bar{y}\, C(y+\bar{y}, y; t) e^{-2\pi i k \bar{y}}. \tag{5}$$

Although the derivation of the function defined in Equation (5) may seem more intricate than the more familiar momentum distribution of Section 3.1, both are functions of momentum and time and are determined by experimental parameters and the initial state.

In Figure 3a the authors present the time evolution of the structure factor. It is possible to see a shift toward lower momenta as time passes. When the scaling of Equation (2) is applied to the data, all the points collapse into a single universal curve, as shown in Figure 3b. The exponents employed were $\alpha = 0.33(8)$ and $\beta = 0.54(6)$.

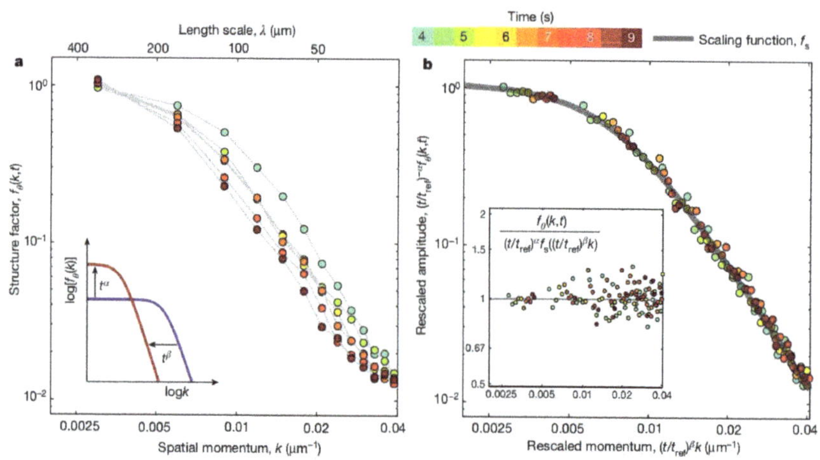

Figure 3. Universal scaling in a spinor Bose gas [19]. (**a**) Time evolution of the structure factor (Equation (5)). (**b**) After the scaling of Equation (2) has been applied, the curves collapse into a universal function. Reprinted by permission from Springer Nature Customer Service Centre GmbH: Springer Nature, Nature, Prüfer et al., Observation of universal dynamics in a spinor Bose gas far from equilibrium, © 2018.

The authors also find a relevant global observable $\int dk f_\theta(k,t)$, which is an approximately conserved quantity during the evolution. The arguments used in Section 2 can be employed to derive its time dependence $\propto t^{\alpha-\beta}$. The observed migration of the conserved quantity toward lower momenta is consistent with $\beta > 0$.

Although the values of the exponents do not strictly correspond to $\alpha \approx \beta$, as one would expect from $\alpha - d\beta = 0$, we should keep in mind that this system is composed of $N_s = 3$ unidimensional Bose gases, for which there are no theoretical predictions. For other systems described by $\mathcal{O}(N_s)$ symmetric models and $d \geqslant 2$, the universal value of $\beta \approx 0.5$ has been predicted [5].

3.3. Homogeneous Three-Dimensional Bose Gas

The authors of [20] observed a bidirectional dynamical scaling in a homogeneous three-dimensional Bose gas [29]. While the systems described in Sections 3.1 and 3.2 only report universal scaling in the IR region, the authors of [20] also observed scaling in the ultraviolet (UV) region of the momentum distribution, hence the term bidirectional. Although the authors found a different set of exponents for the universal scaling in each region, these do not correspond to two different NTFPs. Instead, they are a consequence of particle transport toward small momenta and energy transport toward large momenta, in agreement with NTFP theory [24].

In their experiment, the authors employed ^{39}K atoms in a cylindrical box, producing a homogeneous 3D Bose gas. Their experimental protocol depended on the depth of the optical box trap and the *s*-wave two-body scattering length a, which characterizes the interparticle interactions. First, a cloud containing the atoms just above the condensation temperature was prepared. The idea was to quickly remove atoms and energy from the system to produce a far-from-equilibrium state. This was achieved by turning off the interactions ($a \to 0$) and lowering the depth of the trap so that high-energy atoms evaporated. Since there are no interactions, the system does not thermalize. This step removed ≈77% of the atoms and ≈98% of the energy such that, if it were in equilibrium, a significant fraction of the system would condense. Next, the system was closed again by increasing the depth of the trap potential. At $t = 0$, interactions were turned on again ($a \neq 0$) and thermalization began to take place. After a variable time t, the momentum distribution $n(k,t)$ was obtained through absorption images. For long enough times, the system reaches equilibrium with both condensed and thermal components.

During thermalization, the total number of particles N and total energy E are conserved, but there are two distinctive flows in opposite directions as time progresses. The majority of particles migrate toward the IR region, consistent with the condensate formation, but a small fraction of the atoms transfer the energy in the UV direction.

The presence of a nearby NTFP allows for the universal scaling of Equation (2) in a time interval between the initial and equilibrium states. In Figure 4a, the authors show the unscaled $n(k,t)$ profiles. The UV region is described well by the exponents $\alpha = -0.70(7)$ and $\beta = -0.14(2)$ (Figure 4b), while the IR region collapses into a universal function with $\alpha = 1.15(8)$ and $\beta = 0.34(5)$ (Figure 4c).

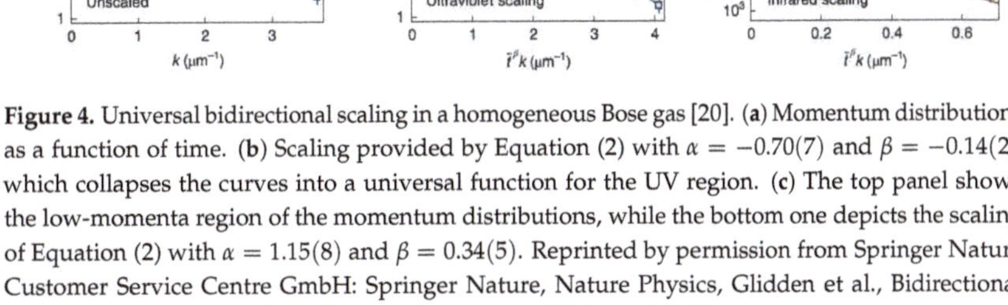

Figure 4. Universal bidirectional scaling in a homogeneous Bose gas [20]. (**a**) Momentum distributions as a function of time. (**b**) Scaling provided by Equation (2) with $\alpha = -0.70(7)$ and $\beta = -0.14(2)$, which collapses the curves into a universal function for the UV region. (**c**) The top panel shows the low-momenta region of the momentum distributions, while the bottom one depicts the scaling of Equation (2) with $\alpha = 1.15(8)$ and $\beta = 0.34(5)$. Reprinted by permission from Springer Nature Customer Service Centre GmbH: Springer Nature, Nature Physics, Glidden et al., Bidirectional dynamic scaling in an isolated Bose gas far from equilibrium, © 2021.

For a d-dimensional system with a dispersion relation of the form $\omega(k) \propto k^z$, energy density conservation requires $\alpha/\beta = d + z$ [17,24]. Hence, the UV scaling, where $\alpha/\beta \approx 5$, is consistent with energy-conserving transport with a quadratic dispersion relation in 3D. Moreover, weak-wave turbulence predicts a value of $\beta = -1/6$ for the UV dynamics [30,31], close to that observed in [20].

For the IR region, $\alpha/\beta \approx 3$, consistent with particle conservation in $d = 3$ dimensions. While several theoretical investigations predict $\beta = 1/2$ for the IR region [4,5,17], certain conditions may yield $\beta = 1/3$ [24,32].

The authors of [20] also investigated a dynamical scaling depending on the interactions $t \rightarrow ta/a_0$, where a_0 is a reference scattering length. The exponents obtained by this other scaling were similar to those previously obtained.

3.4. *Harmonically Trapped Three-Dimensional Bose gas*

The authors in [21] investigated the emergence of universal scaling due to the presence of NTFPs in a harmonically trapped three-dimensional Bose gas, driven to a turbulent state.

The experiment began with a cigar-shaped ^{87}Rb BEC with a condensed fraction of ≈70% in equilibrium. The production of a far-from-equilibrium state was achieved by a sinusoidal time-varying magnetic field gradient, such that it was not aligned with the principal axes of the trap. This corresponds to rotations and distortions of the original trap shape. The amplitude, frequency, and duration of the excitation could be varied and controlled. At $t = 0$ the excitation was turned off, and the system was left to evolve in the trap. After a time t, absorption images were taken to obtain the momentum distribution of the system.

The emergence of a turbulent state depends on the parameters of the excitation protocol, as identified by a characteristic power-law behavior. Figure 5a shows the time evolution of the momentum distribution of a turbulent state. As time passes, the distribution shifts toward high momenta, indicating the depletion of the condensate. The scaling employing Equation (2) with $\alpha = -0.50(8)$ and $\beta = -0.2(4)$ is shown in Figure 5b, which collapses all curves into a single function.

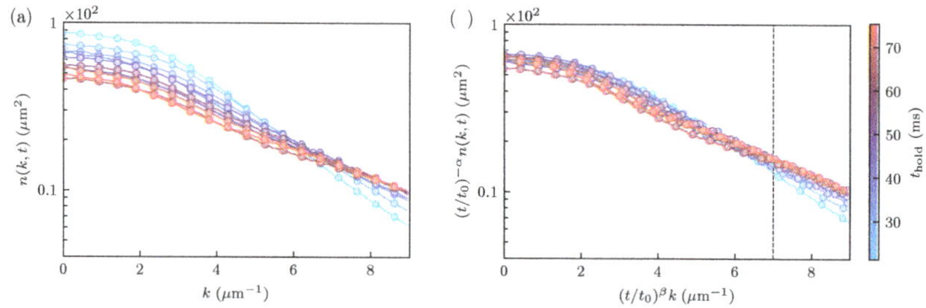

Figure 5. Universal scaling in a turbulent harmonically trapped Bose gas [21]. (**a**) Momentum distributions of the turbulent state. (**b**) Scaled momentum distributions with $\alpha = -0.50(8)$ and $\beta = -0.2(4)$. Figure taken from [21].

The momentum distributions of Figure 5a were obtained via absorption images of the cloud, corresponding to two-dimensional projections of a three-dimensional system. Using the inverse Abel transform [33] and some assumptions, the authors in [21] were able to reconstruct the three-dimensional momentum distributions (see Figure 6a). The exponents determined using the two-dimensional projections did not provide the universal scaling, as evidenced by Figure 6b. The authors showed that the exponents obtained through the scaling of a projection were related to those of the isotropic three-dimensional distribution through $\alpha_{3D} = 3\alpha/2$ (β remains the same). Using $\alpha = -0.75$ and $\beta = -0.2$, the collapse into a universal function was much better adjusted, as shown in Figure 6c.

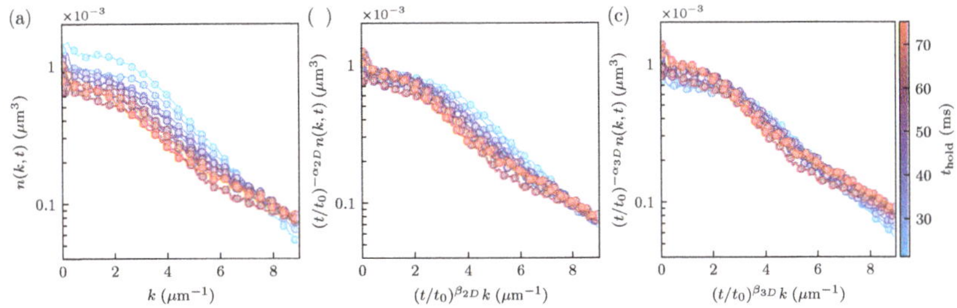

Figure 6. (**a**) Three-dimensional momentum distribution reconstructed with the inverse Abel transform. (**b**) Scaling provided by Equation (2) with $\alpha = -0.50$ and $\beta = -0.2$, which are the same exponents as those employed in the two-dimensional projection. (**c**) Scaling using $\alpha = -0.75$ and $\beta = -0.2$, corresponding to the prediction of the exponents for the three-dimensional case. The collapse is much better than that shown in panel (**b**). The figure is taken from [21].

The global observables of Equations (3) and (4) were also computed in [21]. The authors observed a slight decrease in the particle number in the scaling region, $\bar{N} \propto t^{-0.1}$, and an increase in the mean kinetic energy consistent with $\bar{M}_2 \propto t^{-2\beta}$. These observations are consistent with the energy leaving the IR region and the depletion of the condensate. The authors also presented the benefits of merging quantum turbulence phenomena into a universality class of dynamically scaling systems characteristic of NTFPs [5,13,17,22,24–26,31,34,35].

4. Final Remarks

In this review, we focused on experiments with Bose gases that, due to the presence of NTFPs, display the spatio-temporal scaling of Equation (2). Table 1 contains a summary of the systems and related exponents. Although we discussed these experiments only in terms of dimensionality, the number of components, and the signs of the exponents, these systems cover a wide range of scenarios.

Table 1. Summary of the exponents α and β found in the experiments covered in this review [18–21]. We indicate the dimensionality d, the number of components N_s, and the two different sets of exponents found in [20,21].

Bose Gas		d	N_s	α	β
1D [18]		1	1	0.09(5)	0.10(4)
Spinor [19]		1	3	0.33(8)	0.54(6)
3D Homogeneous [20]	IR region	3	1	1.15(8)	0.34(5)
	UV region	3	1	−0.70(7)	−0.14(2)
Turbulent, harmonically trapped [21]	2D projection	2	1	−0.50(8)	−0.2(4)
	3D reconstruction	3	1	−0.75	−0.2

A common aspect of the experiments presented in this review is the relation followed by the scaling exponents, $\alpha \approx d\beta$, in the IR region. This is not surprising, since according to Equation (3), $\bar{N} \propto t^{\alpha - d\beta}$ and $\bar{N}(t)$ must be conserved during the interval where the scaling occurs. Nevertheless, verifying these theoretical predictions experimentally in systems that are so different from each other strengthens the claim regarding NTFPs.

In equilibrium critical phenomena, renormalization group theory and fixed points lead to critical exponents [36], which provide a unified description in terms of universality classes sharing the same exponents. The experimental evidence provided by the studies presented in this review is crucial to providing something similar in isolated far-from-equilibrium systems. Hopefully, the interplay between theory and experiments may lead to classification schemes based on universal properties, which would be important for various systems. This would allow, for example, cold-atom experiments to be employed to simulate different systems of the same universality class.

Author Contributions: Conceptualization, V.S.B.; methodology, L.M.; software, L.M.; validation, L.M.; formal analysis, L.M.; investigation, L.M.; resources, V.S.B.; data curation, L.M.; writing—original draft preparation, L.M.; writing—review and editing, V.S.B.; visualization, L.M.; supervision, V.S.B.; project administration, L.M.; funding acquisition, V.S.B. All authors have read and agreed to the published version of the manuscript.

Funding: This work was supported by the São Paulo Research Foundation (FAPESP) under the grants 2013/07276-1, 2014/50857-8, and 2018/09191-7, and by the National Council for Scientific and Technological Development (CNPq) under the grant 465360/2014-9.

Institutional Review Board Statement: Not applicable.

Informed Consent Statement: Not applicable.

Data Availability Statement: No new data were created or analyzed in this study. Data sharing is not applicable to this article.

Conflicts of Interest: The authors declare no conflict of interest.

Abbreviations

The following abbreviations are used in this manuscript:

NTFP	non-thermal fixed point
BEC	Bose–Einstein condensate
IR	infrared
UV	ultraviolet

References

1. Polkovnikov, A.; Sengupta, K.; Silva, A.; Vengalattore, M. Colloquium : Nonequilibrium dynamics of closed interacting quantum systems. *Rev. Mod. Phys.* **2011**, *83*, 863–883. [CrossRef]
2. Kofman, L.; Linde, A.; Starobinsky, A.A. Reheating after Inflation. *Phys. Rev. Lett.* **1994**, *73*, 3195–3198. [CrossRef] [PubMed]
3. Micha, R.; Tkachev, I.I. Relativistic Turbulence: A Long Way from Preheating to Equilibrium. *Phys. Rev. Lett.* **2003**, *90*, 121301. [CrossRef]
4. Berges, J.; Rothkopf, A.; Schmidt, J. Nonthermal Fixed Points: Effective Weak Coupling for Strongly Correlated Systems Far from Equilibrium. *Phys. Rev. Lett.* **2008**, *101*, 041603. [CrossRef]
5. Piñeiro Orioli, A.; Boguslavski, K.; Berges, J. Universal self-similar dynamics of relativistic and nonrelativistic field theories near nonthermal fixed points. *Phys. Rev. D* **2015**, *92*, 025041. [CrossRef]
6. Moore, G.D. Condensates in relativistic scalar theories. *Phys. Rev. D* **2016**, *93*, 065043. [CrossRef]
7. Baier, R.; Mueller, A.; Schiff, D.; Son, D. "Bottom-up" thermalization in heavy ion collisions. *Phys. Lett. B* **2001**, *502*, 51–58. [CrossRef]
8. Berges, J.; Boguslavski, K.; Schlichting, S.; Venugopalan, R. Turbulent thermalization process in heavy-ion collisions at ultrarelativistic energies. *Phys. Rev. D* **2014**, *89*, 074011. [CrossRef]
9. Berges, J.; Boguslavski, K.; Schlichting, S.; Venugopalan, R. Universality Far from Equilibrium: From Superfluid Bose Gases to Heavy-Ion Collisions. *Phys. Rev. Lett.* **2015**, *114*, 061601. [CrossRef]
10. Berges, J.; Boguslavski, K.; Mace, M.; Pawlowski, J.M. Gauge-invariant condensation in the nonequilibrium quark-gluon plasma. *Phys. Rev. D* **2020**, *102*, 034014. [CrossRef]
11. Lamacraft, A. Quantum Quenches in a Spinor Condensate. *Phys. Rev. Lett.* **2007**, *98*, 160404. [CrossRef] [PubMed]
12. Barnett, R.; Polkovnikov, A.; Vengalattore, M. Prethermalization in quenched spinor condensates. *Phys. Rev. A* **2011**, *84*, 023606. [CrossRef]
13. Schole, J.; Nowak, B.; Gasenzer, T. Critical dynamics of a two-dimensional superfluid near a nonthermal fixed point. *Phys. Rev. A* **2012**, *86*, 013624. [CrossRef]
14. Schmidt, M.; Erne, S.; Nowak, B.; Sexty, D.; Gasenzer, T. Non-thermal fixed points and solitons in a one-dimensional Bose gas. *New J. Phys.* **2012**, *14*, 075005. [CrossRef]
15. Hofmann, J.; Natu, S.S.; Das Sarma, S. Coarsening Dynamics of Binary Bose Condensates. *Phys. Rev. Lett.* **2014**, *113*, 095702. [CrossRef]
16. Williamson, L.A.; Blakie, P.B. Universal Coarsening Dynamics of a Quenched Ferromagnetic Spin-1 Condensate. *Phys. Rev. Lett.* **2016**, *116*, 025301. [CrossRef]

17. Chantesana, I.; Orioli, A.P.; Gasenzer, T. Kinetic theory of nonthermal fixed points in a Bose gas. *Phys. Rev. A* **2019**, *99*, 043620. [CrossRef]
18. Erne, S.; Bücker, R.; Gasenzer, T.; Berges, J.; Schmiedmayer, J. Universal dynamics in an isolated one-dimensional Bose gas far from equilibrium. *Nature* **2018**, *563*, 225–229. [CrossRef]
19. Prüfer, M.; Kunkel, P.; Strobel, H.; Lannig, S.; Linnemann, D.; Schmied, C.M.; Berges, J.; Gasenzer, T.; Oberthaler, M.K. Observation of universal dynamics in a spinor Bose gas far from equilibrium. *Nature* **2018**, *563*, 217–220. [CrossRef]
20. Glidden, J.A.P.; Eigen, C.; Dogra, L.H.; Hilker, T.A.; Smith, R.P.; Hadzibabic, Z. Bidirectional dynamic scaling in an isolated Bose gas far from equilibrium. *Nat. Phys.* **2021**, *17*, 457–461. [CrossRef]
21. García-Orozco, A.D.; Madeira, L.; Moreno-Armijos, M.A.; Fritsch, A.R.; Tavares, P.E.S.; Castilho, P.C.M.; Cidrim, A.; Roati, G.; Bagnato, V.S. Universal dynamics of a turbulent superfluid Bose gas. *arXiv* **2021**, arXiv:2107.07421.
22. Nowak, B.; Sexty, D.; Gasenzer, T. Superfluid turbulence: Nonthermal fixed point in an ultracold Bose gas. *Phys. Rev. B* **2011**, *84*, 020506. [CrossRef]
23. Pethick, C.J.; Smith, H. *Bose-Einstein Condensation in Dilute Gases*; Cambridge University Press: Cambridge, UK, 2008; Volume 9780521846, pp. 1–569. [CrossRef]
24. Schmied, C.M.; Mikheev, A.N.; Gasenzer, T. Non-thermal fixed points: Universal dynamics far from equilibrium. *Int. J. Mod. Phys. A* **2019**, *34*, 1941006. [CrossRef]
25. Nowak, B.; Erne, S.; Karl, M.; Schole, J.; Sexty, D.; Gasenzer, T. Non-thermal fixed points: Universality, topology, & turbulence in Bose gases. *arXiv* **2013**, arXiv:1302.1448.
26. Karl, M.; Gasenzer, T. Strongly anomalous non-thermal fixed point in a quenched two-dimensional Bose gas. *New J. Phys.* **2017**, *19*, 093014. [CrossRef]
27. Sadler, L.E.; Higbie, J.M.; Leslie, S.R.; Vengalattore, M.; Stamper-Kurn, D.M. Spontaneous symmetry breaking in a quenched ferromagnetic spinor Bose–Einstein condensate. *Nature* **2006**, *443*, 312–315. [CrossRef]
28. Stamper-Kurn, D.M.; Ueda, M. Spinor Bose gases: Symmetries, magnetism, and quantum dynamics. *Rev. Mod. Phys.* **2013**, *85*, 1191–1244. [CrossRef]
29. Gaunt, A.L.; Schmidutz, T.F.; Gotlibovych, I.; Smith, R.P.; Hadzibabic, Z. Bose-Einstein Condensation of Atoms in a Uniform Potential. *Phys. Rev. Lett.* **2013**, *110*, 200406. [CrossRef]
30. Zakharov, V.E.; L'vov, V.S.; Falkovich, G. *Kolmogorov Spectra of Turbulence I*; Springer Series in Nonlinear Dynamics; Springer: Berlin/Heidelberg, Germany, 1992. [CrossRef]
31. Dyachenko, S.; Newell, A.; Pushkarev, A.; Zakharov, V. Optical turbulence: Weak turbulence, condensates and collapsing filaments in the nonlinear Schrödinger equation. *Phys. D Nonlinear Phenom.* **1992**, *57*, 96–160. [CrossRef]
32. Mikheev, A.N.; Schmied, C.M.; Gasenzer, T. Low-energy effective theory of nonthermal fixed points in a multicomponent Bose gas. *Phys. Rev. A* **2019**, *99*, 063622. [CrossRef]
33. Bracewell, R.N.; Bracewell, R.N. *The Fourier Transform and Its Applications*; McGraw-Hill Series in Electrical Engineering; McGraw-Hill: New York, NY, USA, 1986.
34. Nowak, B.; Schole, J.; Sexty, D.; Gasenzer, T. Nonthermal fixed points, vortex statistics, and superfluid turbulence in an ultracold Bose gas. *Phys. Rev. A* **2012**, *85*, 043627. [CrossRef]
35. Scheppach, C.; Berges, J.; Gasenzer, T. Matter-wave turbulence: Beyond kinetic scaling. *Phys. Rev. A* **2010**, *81*, 033611. [CrossRef]
36. Wilson, K.G. The renormalization group: Critical phenomena and the Kondo problem. *Rev. Mod. Phys.* **1975**, *47*, 773–840. [CrossRef]

symmetry

MDPI

Article

Spontaneous Symmetry Breaking: The Case of Crazy Clock and Beyond

Maja C. Pagnacco [1,*], Jelena P. Maksimović [2], Marko Daković [2], Bojana Bokic [3], Sébastien R. Mouchet [4,5], Thierry Verbiest [6], Yves Caudano [7] and Branko Kolaric [3,8,*]

1 Center for Catalysis and Chemical Engineering, Institute of Chemistry, Technology and Metallurgy, University of Belgrade, Njegoševa 12, 11000 Belgrade, Serbia
2 Faculty of Physical Chemistry, University of Belgrade, Studentski Trg 12, 11158 Belgrade, Serbia; jelena.maksimovic@ffh.bg.ac.rs (J.P.M.); marko@ffh.bg.ac.rs (M.D.)
3 Center for Photonics, Institute of Physics, University of Belgrade, Pregrevica 118, 11080 Belgrade, Serbia; bojana@ipb.ac.rs
4 Department of Physics & Namur Institute of Structured Matter (NISM), University of Namur, Rue de Bruxelles 61, 5000 Namur, Belgium; sebastien.mouchet@unamur.be
5 School of Physics, University of Exeter, Stocker Road, Exeter EX4 4QL, UK
6 Department of Chemistry, Molecular Imaging and Photonics, KU Leuven, 3001 Heverlee, Belgium; thierry.verbiest@fys.kuleuven.be
7 Department of Physics & Namur Institute of Structured Matter (NISM) & Namur Institute for Complex Systems (naXys), University of Namur, Rue de Bruxelles 61, 5000 Namur, Belgium; yves.caudano@unamur.be
8 Micro- and Nanophotonic Materials Group, University of Mons, Place du Parc 20, 7000 Mons, Belgium
* Correspondence: maja.pagnacco@nanosys.ihtm.bg.ac.rs (M.C.P.); branko.kolaric@umons.ac.be (B.K.)

Abstract: In this work, we describe the crazy-clock phenomenon involving the state I (low iodide and iodine concentration) to state II (high iodide and iodine concentration with new iodine phase) transition after a Briggs–Rauscher (BR) oscillatory process. While the BR crazy-clock phenomenon is known, this is the first time that crazy-clock behavior is linked and explained with the symmetry-breaking phenomenon, highlighting the entire process in a novel way. The presented phenomenon has been thoroughly investigated by running more than 60 experiments, and evaluated by using statistical cluster K-means analysis. The mixing rate, as well as the magnetic bar shape and dimensions, have a strong influence on the transition appearance. Although the transition for both mixing and no-mixing conditions are taking place completely randomly, by using statistical cluster analysis we obtain different numbers of clusters (showing the time-domains where the transition is more likely to occur). In the case of stirring, clusters are more compact and separated, revealed new hidden details regarding the chemical dynamics of nonlinear processes. The significance of the presented results is beyond oscillatory reaction kinetics since the described example belongs to the small class of chemical systems that shows intrinsic randomness in their response and it might be considered as a real example of a classical liquid random number generator.

Keywords: crazy clock; Briggs–Rauscher reaction; state I to state II transition; symmetry breaking; iodine; K-means analysis; random number generator

Citation: Pagnacco, M.C.; Maksimović, J.P.; Daković, M.; Bokic, B.; Mouchet, S.R.; Verbiest, T.; Caudano, Y.; Kolaric, B. Spontaneous Symmetry Breaking: The Case of Crazy Clock and Beyond. *Symmetry* **2022**, *14*, 413. https://doi.org/10.3390/sym14020413

Academic Editor: Stefano Profumo

Received: 10 January 2022
Accepted: 14 February 2022
Published: 19 February 2022

1. Introduction

The presence of symmetry around us inspired many scientists to search for beauty, harmony, order, and regularity in nature and her fundamental laws [1,2]. Additionally, phase transitions with and without spontaneously broken symmetries are widespread concepts through different areas of physics and physical chemistry. The applications of spontaneously broken symmetries cover a wide range of condensed matter science topics, such as superconductivity, super-fluidity, Bose–Einstein condensation, nucleation physics, self-assembly processes, morphogenesis, and chemical kinetics. In this account, we describe spontaneous symmetry breaking in the case of the nonlinear Briggs–Rauscher reaction. We highlight the importance of symmetry breaking, in non-equilibrium and pattern formation processes, which is of vital meaning to the understanding of the morphogenesis process and for applications in several areas of biomimetics and nanoscience.

The Briggs–Rauscher (BR) [3] reaction is a hybrid oscillating reaction formed by coupling two chemical oscillators, Bray–Liebhafsky [4,5] and Belousov–Zhabotinskii [6]. Since its discovery in 1973, the Briggs–Rauscher oscillating reaction has been one of the most investigated oscillatory systems. It is probably due to its simplicity and exciting colour alternation caused by changes in reaction kinetics (when starch is used as an indicator) [7].

BR reaction typically occurs within mixtures of H_2O_2, H_2SO_4, and KIO_3. Additionally, Mn(II) ions are added as a metal catalyst and malonic acid (H_2MA) as an organic substrate. Substitutions of chemicals are possible; different acids, organic substrates, and ions, such as Ce(III) instead of Mn(II) catalyst, can be used to generate BR oscillations [7–10]. However, the oscillatory behavior is not the only one that attracted the attention of non-linear scientists in the Briggs–Rauscher reaction [11–15].

Indeed, as described elsewhere [16], after the well-controlled initial oscillatory behavior, the reaction becomes chaotic. Depending on the initial conditions, particularly on the ratio $[H_2MA]_0/[IO_3^-]_0$ [16,17], the reaction exhibits a sudden and unpredictable phase transition. This transition, from state I (low concentration of iodide and iodine) to state II (high concentration of iodide and iodine), happens randomly in practice, as the time spent by the system in the state I is irreproducible (see Figure 1). The transition is characterized by a "sharp and sudden" increase of iodine and iodide concentration, followed by the formation of solid iodine. The observed stochastic feature, called a crazy clock (due to the unpredictable time needed to provoke the transition), is linked to imperfect mixing that affects convection and diffusion dynamics. The imperfect mixing results in extremely complicated phenomena, which occur on multiple length and time scales [18]. Possible kinetical consequences are the appearance of bifurcation, chaos, intermittent behavior, and symmetry-breaking [18,19]. In the experiments of our previous paper [16], the mixing was stopped after an intensive homogenization (stirring at 900 rpm) of the Briggs–Rauscher solution in the oscillatory period. Herein, the experiments carried out with or without specific mixing were maintained all the time. Additionally, we apply the statistical cluster K-means analysis for the first time, by processing more than 60 experiments. Therefore, this paper further studies the mixing effects in connection to the crazy-clock phenomenon in the Briggs–Rauscher oscillatory reaction. It compares and processes statistically more than #60 experiments obtained under identical initial concentrations of all reactants. Although the BR crazy-clock phenomenon was previously detected [16], this behavior is linked for the first time to symmetry-breaking, highlighting the entire process in a novel way. Furthermore, the investigated crazy clock exhibits a truly random behavior that might be considered as an example of a classical, liquid random number generator. Additionally, the investigated system also belongs to the particular class of classical systems that shows intrinsic randomness in their response (as also observed in colloid particles placed on an oscillating surface) [20,21].

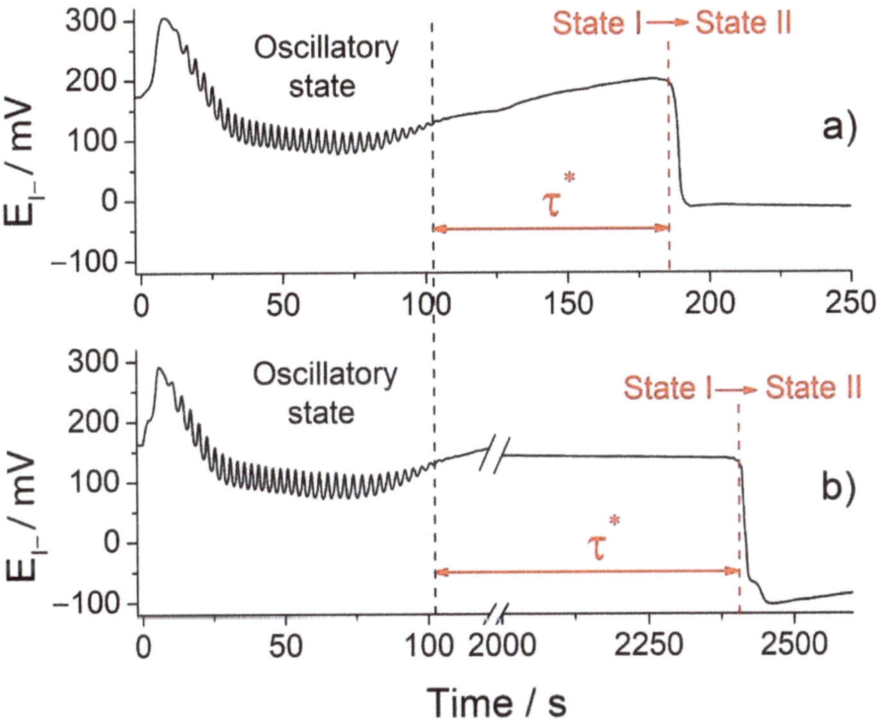

Figure 1. Two independent measurements (**a**,**b**) of iodide potential vs. time obtained for BR reaction under experimental conditions: $[H_2MA]_0$ = 0.0789 mol/dm^3, $[MnSO_4]_0$ = 0.00752 mol/dm^3, $[HClO_4]_0$ = 0.03 mol/dm^3, $[KIO_3]_0$ = 0.0752 mol/dm^3, $[H_2O_2]_0$ = 1.176 mol/dm^3, T = 37.0 °C. The experiments were performed without stirring and without protection from light. τ^* denotes the time from the end of the oscillatory mode to the occurrence of state I→state II transition.

2. Materials and Methods

2.1. Briggs–Rauscher Experimental Setup

Since the time of the transition between state I and II is unpredictable, great attention must be paid to the experimental procedure. Only analytically graded reagents without further purification were used for preparing the solutions. Malonic acid was obtained from Acrōs Organics (Geel, Belgium), manganese sulphate from Fluka (Buchs, Switzerland), perchloric acid, potassium iodate, and hydrogen peroxide from Merck (Darmstadt, Germany). The solutions were prepared in deionized water with specific resistance 18 MΩ/cm (Milli-Q, Millipore, Bedford, MA, USA).

All experiments were done in a container not protected from light. Reactions were monitored electrochemically (unless specified in the text). An I-ion-sensitive electrode (Metrohm 6.0502.160) was used as the working electrode and an Ag/AgCl electrode (Metrohm 6.0726.100), as the reference. During the experiments, the temperature of the reaction container was regulated by a circulating thermostat (JULABO GmbH, Seelbach, Germany) and maintained constant at 37 °C. The reaction mixture was stirred by magnetic stirrer (Ingenieurbüro, M. Zipperrer GmbH, Cat-ECM5, Staufen, Denmark).

Five independent series of measurements were carried (they differed in stirring bar size and shape, as well as mixing rate) with the identical solution composition $[H_2MA]_0$ = 0.0789 mol/dm^3, $[MnSO_4]_0$ = 0.00752 mol/dm^3, $[HClO_4]_0$ = 0.03 mol/dm^3, $[KIO_3]_0$ = 0.0752 mol/dm^3 and $[H_2O_2]_0$ = 1.176 mol/dm^3 in 25 mL volume:

(1) without mixing (number of conducted experiments #30);
(2) with mixing 100 rpm using cylindrical stirring bar 10 mm length, 4 mm diameter (BRAND magnetic stirring bar, PTFE-coated cylindrical), (number of conducted experiments #30);
(3) with mixing 300 rpm using cylindrical stirring bar 10 mm length, 4 mm diameter (BRAND magnetic stirring bar, PTFE-coated cylindrical), (in triplicate);
(4) with mixing 100 rpm using cylindrical stirring bar 20 mm length, 6 mm diameter (BRAND magnetic stirring bar, PTFE-coated cylindrical) (in triplicate);
(5) with mixing 100 rpm using triangular stirring bar 12 mm length, 6 mm diameter (BRAND magnetic stirring bar, PTFE-coated triangular) (in triplicate).

2.2. Statistical Processing and Cluster Analysis

The obtained experimental results were analyzed in the open-source statistic software "R" using "hclust" algorithm for the hierarchical cluster analysis (HCA) [22,23].

3. Results and Discussion

3.1. Effects of the Stirring Bar Shape and Dimensions on the State I→State II Transition

The BR oscillatory period is strongly reproducible, while the transition from state I to state II occurred practically randomly (Figure 1), as previously reported by our research group [16].

It is imperative to emphasize that in our measurements (Figure 1), unlike in other crazy-clock reactions found in the literature [18,19,24], large time fluctuations (the order of magnitude could be more than two hours) occur after a highly reproducible oscillatory period. Two independent measurements and consequently obtained BR oscillograms exhibit identical trends in oscillation amplitude and time between two neighboring oscillation maxima $\tau_{n-(n-1)} = t_n - t_{n-1}$, as it can be observed in Figures 1 and 2a,b. Conversion from higher to lower potential of iodine electrode (or from low to high iodide concentration) marks the transition from state I to state II (state I→state II). Furthermore, the choice of the working electrode affects only the transition shape. However, the transition itself is very noticeable due to the intense color change of the system from colorless to yellow accompanied by solid iodine formation. This allows monitoring the state I→state II transition with the naked eye.

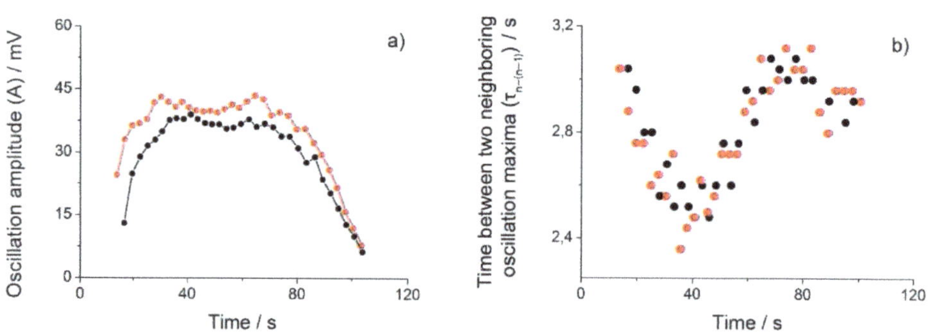

Figure 2. Briggs–Rauscher oscillation amplitude (**a**) and time between two neighboring oscillation maxima $\tau_{n-(n-1)} = t_n - t_{n-1}$ (**b**) in the two independent measurements presented at Figure 1. The resulting BR oscillograms have the same number of oscillations (N_{osc} = 33) and identical oscillation period, however the time of state I→state II transition differs more than 10 times (as shown at Figure 1).

The cause of this unexpected transition is still unknown. Previous work highlighted the significance of mixing conditions for the appearance of state I→state II transition and crazy-clock behavior [13,16]. Therefore, we want to reveal in detail the effect of mixing on the transition, by using stirring bars of different sizes and shapes and applying various mixing rates (Figure 3).

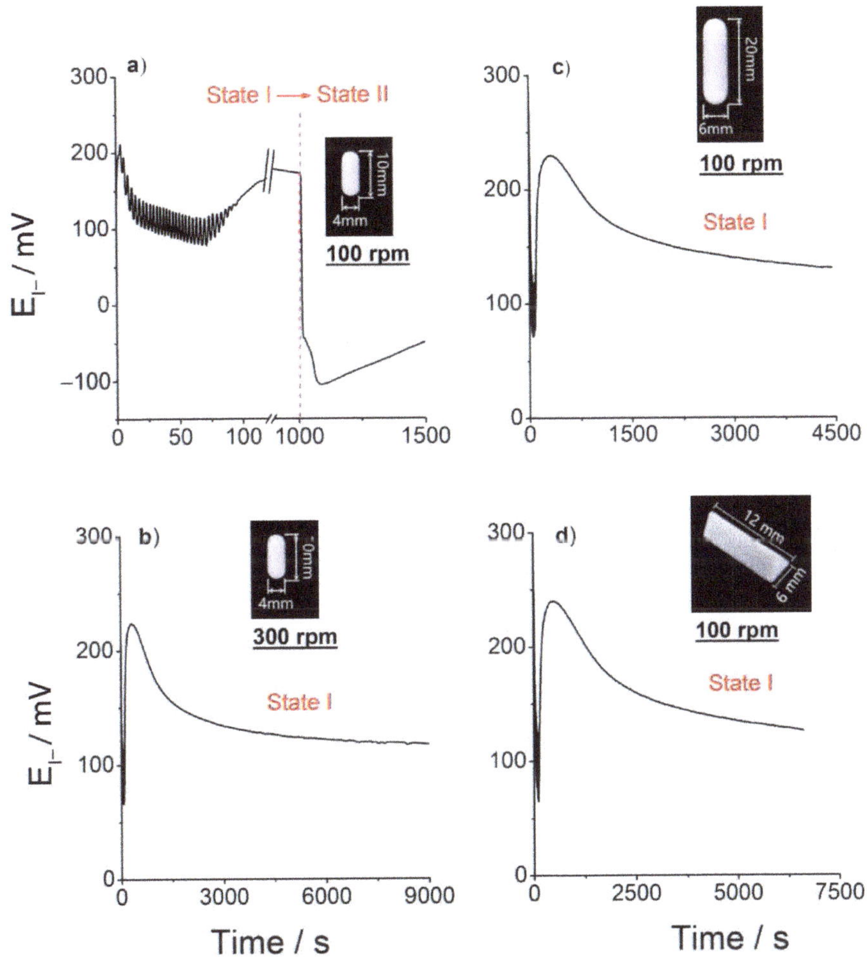

Figure 3. Typical measurements with iodide-sensing and reference electrodes with different mixing rates and different shapes of magnetic bar: (**a**) 100 rpm with magnetic stirring bar, PTFE-coated cylindrical, 10 mm length, 4 mm diameter (in inset), (**b**) 300 rpm with magnetic stirring bar, PTFE-coated cylindrical, 10 mm length, 4 mm diameter (in inset), (**c**) 100 rpm with magnetic stirring bar, PTFE-coated cylindrical, 20 mm length, 6 mm diameter (in inset), and (**d**) 100 rpm with magnetic stirring bar, PTFE-coated triangular, 12 mm length, 6 mm diameter (in inset). The reactant concentrations are identical as in Figure 1.

The transition from the state I to state II occurs only with a low stirring rate and a stirring bar of small dimensions (namely, 100 rpm and a magnetic stirring bar made of PTFE-coated cylindrical with a length of 10 mm and a diameter of 4 mm, Figure 3a). The results also underline the importance of the particular magnetic bar shape and mixing rate that was used (Figure 3a–d). Even with a low stirring rate (100 rpm), the transition does not occur with a bar exhibiting a triangular section (bar 12 mm length, 6 mm diameter) (Figure 3d). This result implies that the transition is strongly connected a particular diffusion conditions and vortex type behavior created by using specific stirring rods. Furthermore, we perform a detailed statistical analysis to reveal the connection between the state I→state II transition and the mixing rate (using the 10-mm long and 4-mm large cylindrical stirring bar). A set of 60 experiments were performed without stirring and with stirring at a 100-rpm mixing rate (30 experiments, each). The results (τ_{osc} and τ^*) are tabulated (Tables S1 and S2) and presented in Supplementary Materials. The τ^* mean value with 95% confidence limit is for no-mixing $\tau^*_{no\ mix}$ = (12 ± 4) min and for mixing conditions, τ^*_{mix} = (17 ± 5) min.

3.2. Statistical Analysis of Experimental Results and Evidence of Clustering

Is there a connection between the time that the system spends in the oscillatory regime (τ_{osc}) and the time when the state I to state II transition occurs (τ^*)? Or, in other words, are the minor differences in oscillatory period duration responsible for a significant deviation in transition appearance? The detailed exploration of the relation between BR oscillatory time, τ_{osc}, and the time τ^* of the occurrence of the state I→state II transition (Figure 1), with and without stirring of the solutions, was performed by statistical cluster analysis (CA). Cluster analysis performs subdivision of datasets based on the relationships among their members (in our case datasets of τ_{osc} and τ^*). The application of CA allows the separation of data in clusters (namely, in groups) based on mutual distances, which reflect a degree of similarity among data [25]. The greater the similarity in the cluster, the higher the distance between the clusters, and hence the better the clustering. Our results combine a total of 60 experiments, obtained with no-mixing conditions (30 experiments) and with a mixing rate of 100 rpm (30 experiments). They were analyzed in the open-source statistic software "R" using "hclust" algorithm for the hierarchical cluster analysis (HCA) [22,23]. The HCA divided the ratios τ_{osc}/τ^* into three clusters for both mixing and no-mixing measurements (Figure 4a,b). It can be noticed that the number of members of a particular cluster slightly changes upon alteration of experimental conditions (Table 1).

It appears that the stirring effect causes an increase in the members of Cluster 1 and Cluster 3, as well as decreasing in members in Cluster 2. The increase of members in Cluster 3 suggested that stirring has prolonged time for the state I→state II transition taking place (Figure 4b.). It can be also seen from the τ^* mean value for no-mixing $\tau^*_{no\ mix}$ = (12 ± 4) min and mixing conditions, τ^*_{mix} = (17 ± 5) min. Furthermore, all clusters are more compact and separated, in the case of stirring (if compared with those obtained without stirring).

The results presented in Table 1 clearly indicate that the investigated crazy-clock exhibits a truly random behavior. The shift (Figure 4a,b, Table 1) in the position of the cluster centroids towards higher τ^* values (i.e., a delaying time to transition to happen) can be observed for the case of applied stirring conditions. Due to a small number of members, the centroid of Cluster 3, in both cases, was not calculated. Furthermore, there is no significant change in the cluster's centroid position regarding τ_{osc} oscillatory coordinate for clusters. That leads to the conclusion that the transition from state I to state II is independent of oscillatory time duration or, in other words, that the minor differences in oscillatory period duration are not responsible for a significant deviation in transition appearance. Therefore, we calculated the optimal number of clusters by a one parameter (by parameter τ^*) K-means analysis, for no-mixing and mixing conditions (Figure 5a,b, respectively). The determination of optimal numbers of clusters was performed using gap method in fviz_nbclust alghoritm [26]. In brief, the gap statistic compares the total within intra-cluster variation for different values of k with their expected values under null reference distribution of the data. The estimate of the optimal clusters will be value that maximizes the gap statistic (i.e., that yields the largest gap statistic) [27].

As it can be seen from the Figure 5 the optimal number of clusters obtained by using one parameter K-means analysis is changed. The BR system which is not stirred has one cluster (Figure 5a), while the stirring induces differentiation of two clusters (see Figures 5b and 6). The dimensions refer to the first two components. The fitviz_cluster function has been used to analyze the main components, after which the cluster is represented in the dimensions of the first two Principal Component Analysis (PCA) components.

The clusters are well separated. This indicates that mixing introduced additional effects responsible for a significant cluster separation revealing the existence of time domains where the state I to state II transition is more likely to occur.

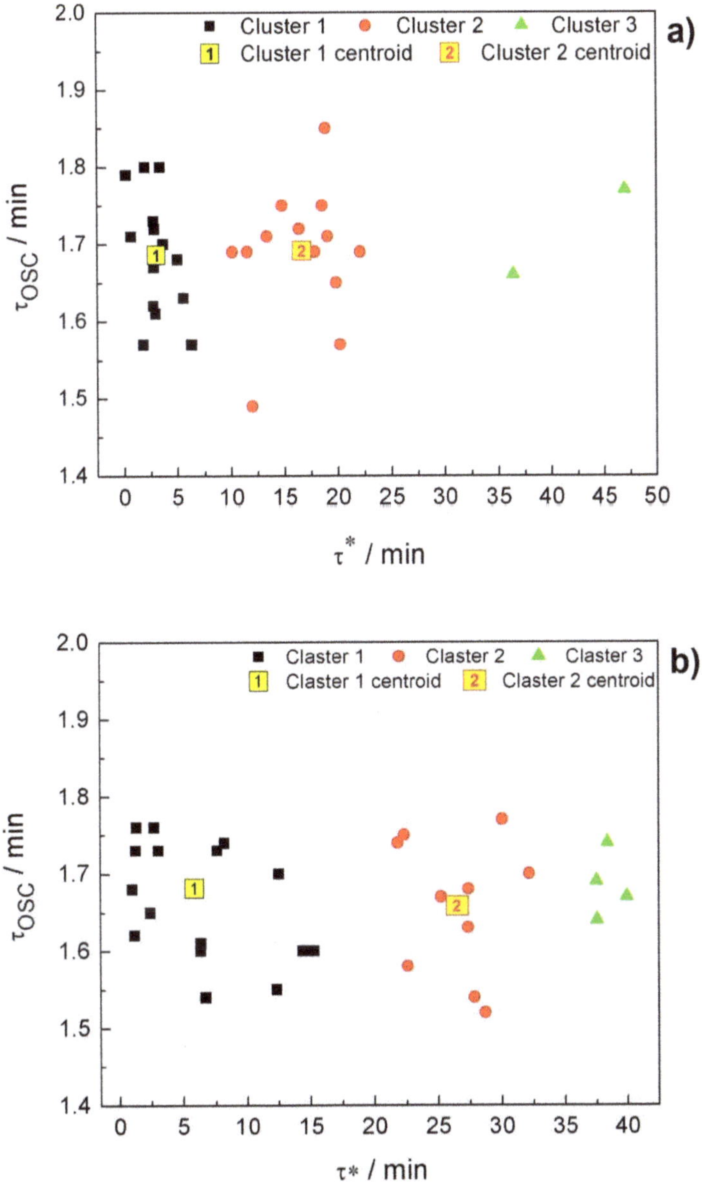

Figure 4. Clusters obtained for thirty experiments run without stirring of the reaction mixture (**a**) and clusters obtained for thirty experiments run with stirring of reaction mixture in the vessel, stirring rate was 100 rpm (**b**). τ_{osc} is the BR oscillatory time, while τ^* denotes the time from the end of the oscillatory mode to the occurrence of state I→state II transition (see Figure 1).

Table 1. Clusters and Cluster centroids.

Exp. Condition	Cluster	Number of Cluster Members	Cluster Centroids in Minute	
			τ_{osc}	τ^*
without stirring	1	15	1.686	3.036
	2	13	1.691	16.672
	3	2	/	/
with stirring	1	16	1.681	5.893
	2	10	1.658	26.536
	3	4	/	/

3.3. The State I→State II Phenomenon and Its Relation to (Spontaneous) Symmetry Breaking

The appearance of clusters indicates the existence of time domains where the state I to state II transition is more likely to occur. The stirring of the reaction mixture has a strong indirect influence on the state I→state II transition, delaying the crazy-clock behavior, shifting cluster centroids toward higher τ^* values, and increasing the cluster separation (i.e., time domain separation), as well. The existence of clusters could be connected to different nucleation and growth mechanisms of iodine crystals in the case of mixing [28], and further examination of solid iodine products would be the subject of future work.

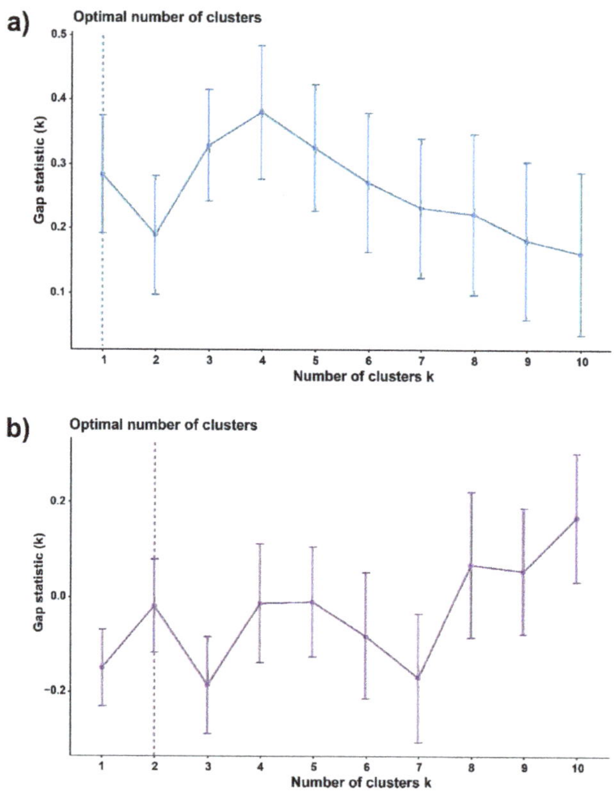

Figure 5. The optimal number of clusters for one parameter analysis by parameter τ^* (time when state I to state II transition occurs) for non-mixing conditions (**a**) and mixing conditions 100 rpm (**b**).

However, if we assumed that state I (low iodide and iodine concentration) is the symmetric state, which under some conditions becomes absolutely unstable, then, reaching the state II (high iodide and iodine concentration, with a new I_2 solid phase) could be considered as spontaneous symmetry breaking [29], see Figure 7. Such an observation of state I to state II transition could also explain the persistence of the BR system "indefinitely" in the state I, as obtained for strong mixing condition (see Figure 3).

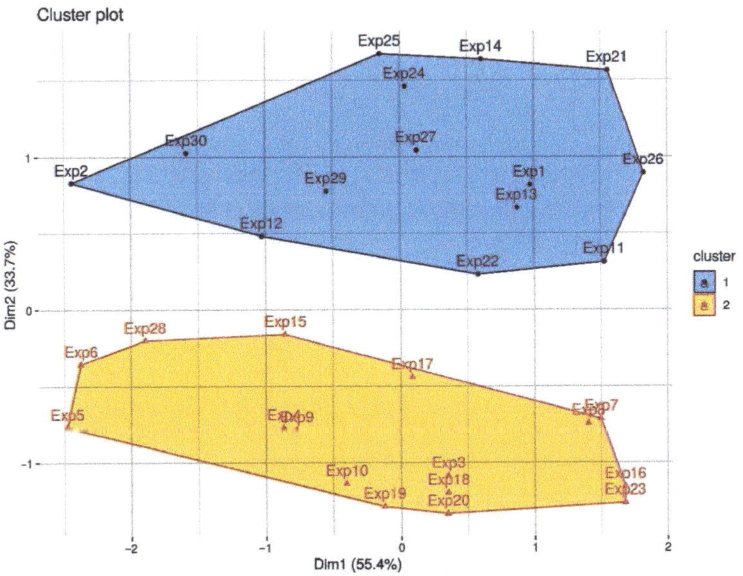

Figure 6. The evidence of two clusters for state I→state II transition obtained for mixing condition in BR system. Clusters are plotted in two dimensions (Dim1 and Dim2).

Namely, in the case of the symmetry-breaking process, the system must overcome a sufficiently large energy barrier (as shown in Figure 7). Therefore, the system's state (state I, low iodide and iodine concentration) will remain unchanged until a sufficiently large perturbation throws it over the energy barrier, ΔV, which separates the states. We assume that ΔV corresponds to the energy threshold of the formation of solid iodine from chemical reactions that principally occur in the BR solution after the oscillatory period. It is well-known that the post-oscillatory period could be excitable [30,31], meaning that a nonlinear system can be shifted (perturbed) from one state to another. It is usually achieved by the addition of some stable intermediate or reactants, playing the role of an external perturbation, to the reaction mixture [32]. Since, in our case, there is no external perturbants/stimulus (all experiments are conducted under identical conditions, and the system remained under constant temperature), the BR system should find an internal stimulus to overcome the energy barrier. The mixing itself should not influence activation energies of chemical reactions responsible for state I to state II transition. In other words, the energy threshold (ΔV) should be identical for mixing and no-mixing condition. However, the system has behaved differently, and the state I to state II transition strongly depends on the mixing conditions. Therefore, some phenomena related to mixing are accountable for the obtained behavior. Figure 7 is a cartoon view that links thermodynamics (far from equilibrium) with kinetic processes driven by the gradient of diffusion/concentration. The changes in time of these local gradients are probably a source of fluctuations. When fluctuations reach a certain threshold (such as a critical number of interacted dissipative structures), a new order/phase spontaneously appears. Cluster analysis (Figures 4–6) allows us to group different dissipation architectures as a function of time.

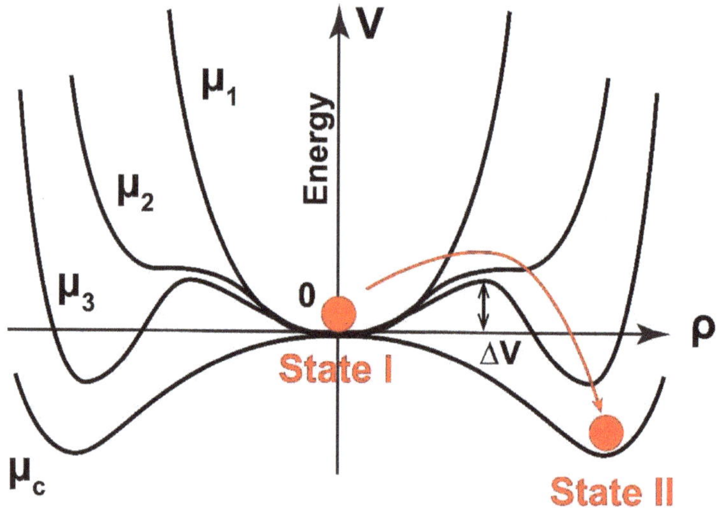

Figure 7. Possible symmetry breaking during the transition from the state I (low iodide and iodine concentration) to state II (high iodide and iodine, with segregation of solid iodine) in Briggs–Rauscher oscillatory reaction after an oscillatory period. Energetic consideration and symmetry breaking depending on control parameter, μ and ordering parameter, ρ. ΔV corresponds to the energy threshold of the formation of solid iodine from chemical reactions that occur in the BR solution after oscillatory period.

The diffusion-driven instability combined with nonlinear chemical reactions (with autocatalytic steps and radical reactions) is a broad concept and it could be responsible for the "internal stimulus" necessary for passing the energy barrier. The diffusion-driven instability is intensified by gaseous oxygen and carbon dioxide/carbon monoxide, which are released in the BR solution during the oscillatory period [33]. Additionally, the possible energetic coupling between physical and chemical processes, such as the nucleation of gaseous phase (O_2 and/or CO_2), nucleation of solid iodine and particular chemical reactions, could also be considered as an "internal stimulus" necessary for overcoming the energy barrier and breaking symmetry [34–36].

This proposal is actually a reformulated original idea of Turing [37], where the interplay of chemical reactions and diffusion are responsible for pattern formation in living systems. As suggested by Prigogine, the spontaneous appearance of a spatial organization via diffusion-driven instability can be considered as a spontaneous symmetry-breaking transition [38]. In the presented work, we deal with a bulk solution and there is no visible spatial organization, but the spatial organization (spatiotemporal patterns) of the identical process (state I→state II transition in Briggs–Rauscher reaction) in a thin layer, is very recently found by Li and coworkers [15]. Therefore, this work indirectly links spontaneous symmetry breaking and crazy-clock behavior (stochastic nature) in the bulk. The stochastic nature of state I to state II transition and its relation to symmetry breaking (and pattern formation), introduced a new approach in the investigation of crazy-clock behavior. On the other hand, the investigation of chemical systems with stochastic nature and symmetry breaking could improve our understanding of more complex phenomena in living organisms, such as morphogenesis.

Additionally, this paper nominates state I→state II transition as an easily available chemical system for intrinsic random number generator and thus, expands the potential application of this crazy-clock reaction.

4. Conclusions

In this work, we further investigated the crazy-clock phenomenon (state I to state II transition) which occurs after a strongly reproducible Briggs–Rauscher oscillatory reaction. The mixing rate, as well as the magnetic bar shape and dimensions, have a strong influence on the transition appearance. In order to better understand the stochasticity of the mentioned process, we ran more than 60 experiments (30 experiments with mixing and 30 experiments with no-mixing conditions), and we applied the statistical cluster K-means analysis. Although the transition for both mixing and no-mixing conditions are taking place completely randomly, by using statistical cluster analysis, we obtained different number of clusters pointing to different time-domains where the transition is more likely to occur. Two-parameter analysis (by oscillatory time duration, τ_{osc} and by the moment when the transition occurs, τ^*) suggests that the state I$\rightarrow$state II is independent of the oscillatory time duration. Therefore, we performed a one parameter analysis (by τ^*). In the case of no-mixing, we found one cluster, while the statistical analysis of the results for mixing conditions revealed two compact and well-separated clusters. The clustering method reveals new hidden details regarding the chemical dynamics of nonlinear processes. The state I to state II transition could be explained through a symmetry breaking approach, and the necessity of the BR system to overcome a sufficiently large energy barrier. This is the first link of the crazy-clock behavior to the symmetry breaking phenomenon. The investigation of chemical systems with stochastic nature and symmetry breaking could improve our understanding of more complex phenomena in living organisms and therefore the scope of the presented results goes beyond oscillatory reaction kinetics. Furthermore, the described example belongs to the small class of chemical systems that shows intrinsic randomness in their response and it might be considered as a real example of a classical liquid random number generator.

Supplementary Materials: The following supporting information can be downloaded at: https://www.mdpi.com/article/10.3390/sym14020413/s1, Table S1: The Briggs-Rauscher oscillogram duration (τ_{osc}), the numbers of oscillations, and state I to state II transition time (τ^*) when no-mixing conditions is applied: [Malonic acid]0 = 0.0789 M, [MnSO4]0 = 0.00752 M, [HClO4]0 = 0.03 M, [KIO3]0 = 0.0752 M, [H2O2]0 = 1.176 M, T= 37.0 °C, reaction volume 25 mL, Table S2: The Briggs-Rauscher oscillogram duration (τ_{osc}), the numbers of oscillations, and state I to state II transition time (τ^*) when 100 rpm mixing conditions is applied: [Malonic acid]$_0$ = 0.0789 M, [$MnSO_4$]$_0$ = 0.00752 M, [$HClO_4$]$_0$ = 0.03 M, [KIO_3]$_0$ = 0.0752 M, [H_2O_2]$_0$ = 1.176 M, T= 37.0 °C, reaction volume 25 mL.

Author Contributions: Conceptualization, M.C.P. and B.K.; methodology, M.C.P., J.P.M., Y.C., and B.K.; statistical K-means analysis M.D.; investigation, M.C.P. and J.P.M.; writing—original draft preparation, M.C.P. and B.K.; writing—review and editing, M.C.P., B.K., Y.C., S.R.M. and T.V.; visualization, J.P.M. and B.B. All authors have read and agreed to the published version of the manuscript.

Funding: This research was funded by the Ministry of Education, Science and Technological Development of the Republic of Serbia, Contract numbers: 451-03-68/2022-14/200026 and 451-03-68/2022-14/200146; BEWARE Fellowship of the Walloon Region (Convention n°2110034); the Fund for Scientific Research F.R.S.-FNRS; the Ministry of Education, Science and Technological Development of the Republic of Serbia (grant III 45016); the Office of Naval Research Global (Research Grant N62902-22-1-2024).

Institutional Review Board Statement: Not applicable.

Informed Consent Statement: Not applicable.

Data Availability Statement: Data underlying the results presented in this paper are not publicly available at this time but may be obtained from the authors upon reasonable request. Please contact for this corresponding authors.

Acknowledgments: This work was supported by the Ministry of Education, Science and Technological Development of the Republic of Serbia Contract numbers: 451-03-68/2022-14/200026 and 451-03-68/2022-14/200146. S.R.M. was supported by a BEWARE Fellowship of the Walloon Region (Convention n°2110034), as a postdoctoral researcher. Y.C. is a research associate of the Fund for

Scientific Research F.R.S.-FNRS. B.K. and B.B. acknowledge financial support of the Ministry of Education, Science and Technological Development of the Republic of Serbia (grant III 45016). Additionally, B.K. acknowledges support from F.R.S.-FNRS. M.P., B.B., and B.K. acknowledge the support of the Office of Naval Research Global through the Research Grant N62902-22-1-2024.

Conflicts of Interest: The authors declare no conflict of interest.

References

1. Gross, D.J. The role of symmetry in fundamental physics. *Proc. Natl. Acad. Sci. USA* **1996**, *93*, 14256–14259. [CrossRef] [PubMed]
2. Smits, J.; Stoof, H.T.C.; Straten, P.V.D. Spontaneous symmetry breaking in a driven-dissipative system. *Phys. Rev. A* **2021**, *104*, 023318. [CrossRef]
3. Briggs, T.S.; Rauscher, W.C. An oscillating iodine clock. *J. Chem. Educ.* **1973**, *50*, 496. [CrossRef]
4. Bray, W.C. A periodic reaction in homogeneous solution and its relation to catalysis. *J. Am. Chem. Soc.* **1921**, *43*, 1262–1267. [CrossRef]
5. Bray, W.C.; Liebhafsky, H.A. Reaction involving hydrogen peroxide, iodine and iodate ion. *I. Introd. J. Am. Chem. Soc.* **1931**, *53*, 38–43. [CrossRef]
6. Belousov, B.P. A periodic reaction and its mechanism, in collection of short papers on radiation medicine for 1958. *Medgiz* **1959**, 145–147.
7. Furrow, S.D. A modified recipe and variations for the briggs–rauscher oscillating reaction. *J. Chem. Educ.* **2012**, *89*, 1421–1424. [CrossRef]
8. Furrow, S.D.; Noyes, R.M. The oscillatory Briggs-Rauscher reaction. 1. Examination of subsystems. *J. Am. Chem. Soc.* **1982**, *104*, 38–42. [CrossRef]
9. Noyes, R.M.; Furrow, S.D. The oscillatory Briggs-Rauscher reaction. 3. A skeleton mechanism for oscillations. *J. Am. Chem. Soc.* **1982**, *104*, 45–48. [CrossRef]
10. Schmitz, G.; Furrow, S. Kinetics of the iodate reduction by hydrogen peroxide and relation with the Briggs–Rauscher and Bray–Liebhafsky oscillating reactions. *Phys. Chem. Chem. Phys.* **2012**, *14*, 5711–5717. [CrossRef]
11. Furrow, S.D.; Cervellati, R.; Greco, E. A study of the cerium-catalyzed briggs-rauscher oscillating reaction. *Z. Nat. B* **2012**, *67*, 89–97.
12. Vanag, V.K.; Alfimov, M.V. Light-induced nonequilibrium phase transition between quasistationary states of the Briggs-Rauscher reaction under batch conditions. *J. Phys. Chem.* **1993**, *97*, 1878–1883. [CrossRef]
13. Vanag, V.K.; Alfimov, M.V. Effects of stirring on photoinduced phase transition in a batch-mode Briggs-Rauscher reaction. *J. Phys. Chem.* **1993**, *97*, 1884–1890. [CrossRef]
14. Pagnacco, M.C.; Mojović, M.D.; Popović-Bijelić, A.D.; Horváth, A.K. Investigation of the halogenate–hydrogen peroxide reactions using the electron paramagnetic resonance spin trapping technique. *J. Phys. Chem. A* **2017**, *121*, 3207–3212. [CrossRef] [PubMed]
15. Li, Z.; Yuan, L.; Liu, M.; Cheng, Z.; Zheng, J.; Epstein, I.R.; Gao, Q. The briggs–rauscher reaction: A demonstration of sequential spatiotemporal patterns. *J. Chem. Educ.* **2021**, *98*, 665–668. [CrossRef]
16. Pagnacco, M.C.; Maksimović, J.P.; Potkonjak, N.I.; Božić, B.; Horváth, A.K. The transition from low to high iodide and iodine concentration state in the briggs rauscher reaction-evidence on crazy clock behavior. *J. Phys. Chem. A* **2018**, *122*, 482–491. [CrossRef]
17. Furrow, S.D.; Cervellati, R.; Greco, E. Study of the transition to higher iodide in the malonic acid Briggs–Rauscher oscillator. *React. Kinet. Mech. Catal.* **2015**, *118*, 59–71. [CrossRef]
18. Epstein, I.R. The consequences of imperfect mixing in autocatalytic chemical and biological systems. *Nature* **1995**, *374*, 321–327. [CrossRef]
19. Valkai, L.; Cseko, G.; Horvath, A.K. Initial Inhomogeneity-induced crazy-clock behavior in the iodate-arsenous acid reaction in a buffered medium under stirred batch conditions. *Phys. Chem. Chem. Phys.* **2015**, *17*, 22187–22194. [CrossRef]
20. Andersen, A.; Madsen, J.; Reichelt, C.; Ahl, S.R.; Lautrup, B.; Ellegaard, C.; Levinsen, M.T.; Bohr, T. Double-slit experiment with single wave-driven particles and its relation to quantum mechanics. *Phys. Rev. E* **2015**, *92*, 013006. [CrossRef]
21. Ellegaard, C.; Levinsen, M.T. Interaction of wave-driven particles with slit structures. *Phys. Rev. E* **2020**, *102*, 023115. [CrossRef] [PubMed]
22. Murtagh, F. *Multidimensional Clustering Algorithms in COMPSTAT Lectures 4*; Physica-Verlag: Wuerzburg, Germany, 1985.
23. R Core Team. *R: A Language and Environment for Statistical Computing*; R Foundation for Statistical Computing: Vienna, Austria, 2018. Available online: https://www.R-project.org/ (accessed on 20 November 2021).
24. Horváth, A.K.; Nagypál, I. Classification of Clock Reactions. *ChemPhysChem* **2014**, *16*, 588–594. [CrossRef] [PubMed]
25. Kaufman, L.; Rousseeuw, P.J. *Finding Groups in Data: An Introduction to Cluster Analysis*; John Wiley & Sons: New York, NY, USA, 2009.
26. Available online: https://cran.r-project.org/web/packages/factoextra/factoextra.pdf (accessed on 20 November 2021).
27. Tibshirani, R.; Walther, G.; Hastiee, T. Estimating the number of clusters in a data set via the gap statistic. *J. R. Stat. Soc. Ser. B (Stat. Methodol.)* **2001**, *63*, 411–423. [CrossRef]

28. Pagnacco, M.C.; Maksimovic, J.P.; Jankovic, B.Z. Analysis of transition from low to high iodide and iodine state in the briggs–rauscher oscillatory reaction containing malonic acid using kolmogorov–johnson–mehl–avrami (kjma) theory. *React. Kinet. Mech. Catal.* **2018**, *123*, 61–80. [CrossRef]
29. Genz, H. Symmetry and symmetry breaking in nature. *Interdiscip. Sci. Rev.* **1999**, *24*, 129–138. [CrossRef]
30. Ruoff, P. Excitability in a closed stirred Belousov—Zhabotinskii system. *Chem. Phys. Lett.* **1982**, *90*, 76–80. [CrossRef]
31. Rouff, P. Excitability created by oxygen inhibition in the stirred malonic acid belousov-zabotinskii system. *Chem. Phys. Lett.* **1982**, *92*, 239–244. [CrossRef]
32. Vukojevic, V.; Anic, S.; Kolar-Anic, L. Investigation of dynamic behavior of the bray–liebhafsky reaction in the CSTR. Properties of the system examined by pulsed perturbations with I^-. *Phys. Chem. Chem. Phys.* **2002**, *4*, 1276–1283. [CrossRef]
33. Muntean, N.; Szabo, G.; Wittmann, M.; Lawson, T.; Fulop, J.; Noszticzius, Z.; Onel, L. Reaction routes leading to CO_2 and CO in the briggs−rauscher oscillator: Analogies between the oscillatory br and bz reactions. *J. Phys. Chem. A* **2009**, *113*, 9102–9108. [CrossRef]
34. Bowers, P.G.; Noyes, R.M. Chemical oscillations and instabilities. 51. Gas evolution oscillators. 1. Some new experimental examples. *J. Am. Chem. Soc.* **1983**, *105*, 2572–2574. [CrossRef]
35. Sevcik, P.; Kissimonova, K.; Adamcikova, L. Oxygen Production in the oscillatory bray−liebhafsky reaction. *J. Phys. Chem. A* **2000**, *104*, 3958–3963. [CrossRef]
36. Stevanovic, K.Z.; Bubanja, I.N.M.; Stanisavljev, D.R. Is iodine oxidation with hydrogen peroxide coupled with nucleation processes? *J. Phys. Chem. C* **2019**, *123*, 16671–16680. [CrossRef]
37. Turing, A.M. The chemical basis of morphogenesis. *Philos. Trans. R. Soc. B Biol. Sci.* **1952**, *237*, 37–72.
38. Prigogine, I.; Nicolis, G. On symmetry-breaking instabilities in dissipative systems. *J. Chem. Phys.* **1967**, *46*, 3542–3550. [CrossRef]

symmetry

Article

Exact Solutions for Solitary Waves in a Bose-Einstein Condensate under the Action of a Four-Color Optical Lattice

Barun Halder [1], Suranjana Ghosh [2], Pradosh Basu [1], Jayanta Bera [1], Boris Malomed [3,4,*] and Utpal Roy [1,*]

1 Department of Physics, Indian Institute of Technology Patna, Patna 801103, India; barun.pph15@iitp.ac.in (B.H.); pradosh_2021ph23@iitp.ac.in (P.B.); jayanta.ppy14@iitp.ac.in (J.B.)
2 Department of Physics, Indian Institute of Science Education and Research Kolkata, Kolkata 741246, India; sghosh@iitp.ac.in or suranjana.ghosh15@gmail.com
3 Department of Physical Electronics, School of Electrical Engineering, Faculty of Engineering, Center for Light-Matter Interaction, Tel Aviv University, Ramat Aviv, Tel Aviv P.O. Box 39040, Israel
4 Instituto de Alta Investigación, Universidad de Tarapacá, Casilla 7D, Arica 1000000, Chile
* Correspondence: malomed@tauex.tau.ac.il (B.M.); uroy@iitp.ac.in (U.R.)

Abstract: We address dynamics of Bose-Einstein condensates (BECs) loaded into a one-dimensional four-color optical lattice (FOL) potential with commensurate wavelengths and tunable intensities. This configuration lends system-specific symmetry properties. The analysis identifies specific multi-parameter forms of the FOL potential which admits exact solitary-wave solutions. This newly found class of potentials includes more particular species, such as frustrated double-well superlattices, and bichromatic and three-color lattices, which are subject to respective symmetry constraints. Our exact solutions provide options for controllable positioning of density maxima of the localized patterns, and tunable Anderson-like localization in the frustrated potential. A numerical analysis is performed to establish dynamical stability and structural stability of the obtained solutions, which makes them relevant for experimental realization. The newly found solutions offer applications to the design of schemes for quantum simulations and processing quantum information.

Keywords: four-color optical lattice; Bose-Einstein condensate; soliton

Citation: Halder, B.; Ghosh, S.; Basu, P.; Bera, J.; Malomed, B.; Roy, U. Exact Solutions for Solitary Waves in a Bose-Einstein Condensate under the Action of a Four-Color Optical Lattice. *Symmetry* **2022**, *14*, 49. https://doi.org/10.3390/sym14010049

Academic Editors: V.I. Yukalov, V. S. Bagnato and Rashid G. Nazmitdinov

Received: 27 November 2021
Accepted: 28 December 2021
Published: 31 December 2021

1. Introduction

A suitably prepared standing wave of laser radiation can form an optical lattice (OL), which are broadly used for trapping and steering ultracold atoms [1–8]. Offering a versatile platform for research in the area of matter waves, OLs have become the most appropriate candidate for the realization of quantum simulations [9–11]. Further, ultracold atoms and Bose-Einstein condensates (BECs) trapped in an OL are used as a basis for the development of atomic clocks, quantum sensors, quantum computers, and a variety of other applications in quantum technologies [12–14].

In particular, the study of BEC under the action of geometrically frustrated OLs has drawn much interest [15–18]. Many complex phenomena have been found in this connection, including Anderson-like localization and negative absolute temperature [17,19–21]. Optical superlattices subjected to frustration offer potential for the development of tools which can hold and mould robust matter-wave states, such as solitons [22–26]. Theoretical studies in this direction are chiefly limited to a variety of bi-color optical lattices (BOL). A more general form of multi-color OLs may offer additional advantages, including the following points: (i) the color (wavelength) and intensity of the constituent beams, building the effective optical potential, greatly influence the manner in which the atoms are trapped; (ii) the formation of solitons requires a specific correlation between the nonlinearity and the trap parameters, which the multi-color OL may help to maintain; (iii) relations between intensities of the constituent beams may be used to optimize the creation of the self-trapped patterns. Thus, multi-color beams can be used to design potentials necessary for holding complex soliton patterns.

The aim of this work is to introduce a four-color OL (FOL) with commensurate wavelengths, which acts on a cigar-shaped (quasi-one-dimensional) BEC with the cubic nonlinearity. The corresponding Gross-Pitaevskii equation (GPE) is used to find appropriate relations between the nonlinearity and the potential parameters which help to support solitons. We produce analytical solutions which identify the specific form of the FOL and its parameter domain which provide tunability of the soliton-building scheme. Many exact condensate wave functions are obtained, and the results are illustrated by several characteristic examples. These solutions may be used for applications similar to those proposed in previous works [27–32]. Stability of the exact wave functions is addressed by means of direct simulations, adding random perturbations either to the underlying stationary solution, or to the external trap (the latter implies the consideration of the structural stability of the exact solutions). We thus find that our solutions are fully stable, both dynamically and structurally.

2. Exact Analytical Model for Obtaining the Solitary Excitations under the Novel FOL Trap

The FOL potential is produced by the combinations of four OLs with commensurate wave numbers, l, $2l$, $3l$, and $4l$, while the corresponding intensities of the laser beams, $V_{1,2,3,4}$, are treated as free parameters, with the intention to find appropriate relations between them. The corresponding effective potential acting on atoms is

$$V(z) = \sum_{j=1}^{4} V_j \cos(jlz). \tag{1}$$

The lattice depth may be compared to the recoil energy, $E_R = 2\pi^2\hbar^2/(M\lambda^2)$, and the scaled lattice wave-vector is given by $l = 2\pi a_\perp/\lambda$, where λ is the wavelength, M is the mass of the BEC atom, $a_\perp = (\hbar/(M\omega_\perp))^{1/2}$ and $\omega_\perp$ is the transverse frequency. The dimensionless 1D-GPE has the form

$$\left[i\frac{\partial}{\partial t} + \frac{1}{2}\frac{\partial^2}{\partial z^2} - g(z,t)|\psi(z,t)|^2 - V(z) - i\tau(z,t)\right]\psi(z,t) = 0. \tag{2}$$

Here, $g(z,t)$ is the nonlinearity coefficient, which may be made space- and time-modulated, while $\tau(z,t)$ represents the space- and time-modulated loss or gain of the condensate atoms. For illustration, we have exploited experimentally feasible parameters of Li^7 BEC in the quasi-1D trapping configuration with transverse frequency $\omega_\perp = 2\pi \times 710$ Hz, OL wavelength $\lambda = 10.62$ μm, and scattering length $a_s = -0.21$ nm corresponding to attractive interactions between atoms [33]. By varying the applied magnetic field and angle between the overlapping laser beams, it is possible to engineer the shape of the external potential [34,35].

To produce a spatially localized solution of Equation (2), following the general scheme used for engineering matter-wave configurations [36,37], we choose an ansatz,

$$\psi(z,t) = A(z,t)F(B(z,t))e^{i\theta(z,t)}, \tag{3}$$

such that the external potential is supposed to be found by precisely solving $B(z,t)$, amplitude $A(z,t)$, phase $\theta(z,t)$, and the condensate form factor $F[B(z,t)]$. We substitute this ansatz into the GPE (2) and separate out the real and imaginary parts. To establish relations between the physically relevant quantities like nonlinearity, amplitude, phase and external trap for a solitary wave solution, the real part can be mapped to the following nonlinear differential equation,

$$\frac{\partial^2 F[B(z,t)]}{\partial B(z,t)^2} - GF^3[B(z,t)] = 0, \tag{4}$$

which introduces a constant $G = 2g(z,t)A^2(z,t)/B_z^2(z,t)$ in the case of solitary wave excitations. G is -1 for attractive and 1 for repulsive inter-atomic interactions. The last consistency condition in Equation (4) is nothing but the elliptic equation, whose exact solutions are well-known in the form of 12 Jacobian elliptic functions ($cn[z,m]$, $sn[z,m]$, etc.), where m is the modulus parameter with $0 \leq m \leq 1$ [38]. One can choose various shapes of the elliptic functions from periodic ($m = 0$) to localized ($m = 1$), by varying the value of its modulus parameter. Here, we will focus only on the localized forms of the elliptic functions, that is, $cn[z,1] = sech[z]$, for the bright soliton case with attractive nonlinearity and $sn[z,1] = tanh[z]$ for the dark soliton case with repulsive nonlinearity. In addition to solving the above-mentioned equation, we also obtain the following consistency relations:

$$\begin{aligned} GB_z^2(z,t) - 2A^2(z,t)g(z,t) &= 0, \\ B_t(z,t) + B_z(z,t)\theta_z(z,t) = 0, [A^2(z,t)B_z(z,t)]_z &= 0 \\ 2A(z,t)A_t(z,t) + [A^2(z,t)\theta_z(z,t)]_z & \\ -2\tau(z,t)A^2(z,t) &= 0 \\ \frac{A_{zz}(z,t)}{2A(z,t)} - \frac{\theta_z^2(z,t)}{2} - \theta_t(z,t) - V(z) &= 0, \end{aligned} \tag{5}$$

where the subscripts stand, as usual, for partial derivatives. The above set of equations is solved simultaneously to produce

$$\begin{aligned} B(z,t) = \frac{c(t)}{A^2(z,t)}, \qquad \theta_z(z,t) = -\frac{A_t(z,t)}{A_z(z,t)}, \\ g(z,t) = GB_z^2(z,t)/2A^2(z,t), \end{aligned} \tag{6}$$

where $c(t)$ is an arbitrary positive definite function of time. These equations indicate a direct dependence of phase and nonlinearity on the amplitude of the system which will be determined by the trapping potential through the last equation of the system (5).

We substitute the expression of the external potential from Equation (4) into the set of Equations (5) and (6) and obtain the amplitude, phase and nonlinearity in the following exact forms:

$$\begin{aligned} A(z,t) &= \sqrt{\frac{c(t)}{\gamma \exp(b_1\cos(lz) + b_2\cos(2lz))}}, \\ \theta(z,t) &= \frac{1}{16}(l^2b_1^2 + l^2b_2^2)t, \ \tau(z,t) = \frac{1}{2}\frac{c'(t)}{c(t)}, \\ g(z,t) &= \frac{G\gamma^4}{2c^2(t)}\exp(4b_1\cos(lz) + 4b_2\cos(2lz)). \end{aligned} \tag{7}$$

We here introduce two real constants, b_1 and b_2, which help us to define the final form of the FOL amplitudes:

$$\begin{aligned} V_1 = (1+b_2)\frac{l^2b_1}{4}, \qquad V_2 = \left(\frac{-b_1^2}{16} + b_2\right)l^2, \\ V_3 = -\frac{l^2b_1b_2}{4}, \qquad V_4 = -\frac{l^2b_2^2}{4}. \end{aligned} \tag{8}$$

This is one of the essential results of the present work. Constants b_1 and b_2 are thus identified as the prime tuning parameters for controlling the trapping potential and condensate density. For the attractive and repulsive interactions, assuming the commonly known bright- or dark-soliton solutions of the elliptic equation (Equation (4)), the condensate wave functions take, severally, the following form:

$$\psi(z,t) = \sqrt{\frac{c(t)}{\gamma \exp[b_1 \cos(lz) + b_2 \cos(2lz)]}}$$
$$\times \text{sech}\left[\gamma \int_0^z \exp(b_1 \cos(lz) + b_2 \cos(2lz))dz'\right] \exp(i\theta(z,t)),$$
$$\psi(z,t) = \sqrt{\frac{c(t)}{\gamma \exp(b_1 \cos(lz) + b_2 \cos(2lz))}} \tag{9}$$
$$\times \tanh\left[\gamma \int_0^z \exp(b_1 \cos(lz) + b_2 \cos(2lz))dz'\right] \exp(i\theta(z,t)).$$

We are now in a position to analyze the relevant potential profiles and the corresponding condensate densities. Potential profiles are explained in Figures 1 and 2. Condensate densities will be delineated in Figures 3 and 4 for some parameter domains of b_1 and b_2 with $c = 0.1$, $g = 0.1$, and $l = 0.84$.

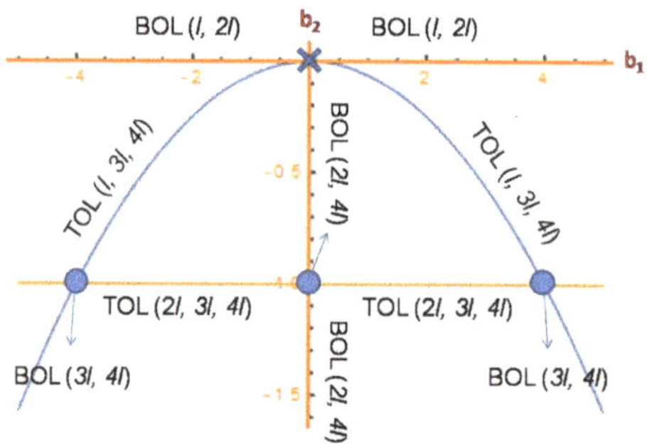

Figure 1. Curves and points for $l = 0.84$ where the potential is not a FOL, but a TOL or a BOL. '$\times$' signifies no potential for $b_1 = b_2 = 0$. All other points in the (b_1, b_2) plane correspond to FOLs.

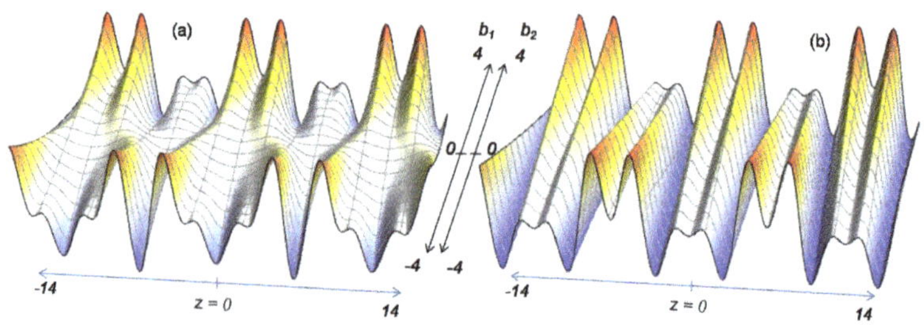

Figure 2. The variation of the FOL potential following the change of the tuning parameters for $l = 0.84$: (**a**) for fixed $b_2 = 2$, b_1 varies from -4 to $+4$; (**b**) for fixed $b_1 = 2$, b_2 varies from-4 to $+4$. Here and in the figures following below, the results are displayed in interval $-14 < z < +14$.

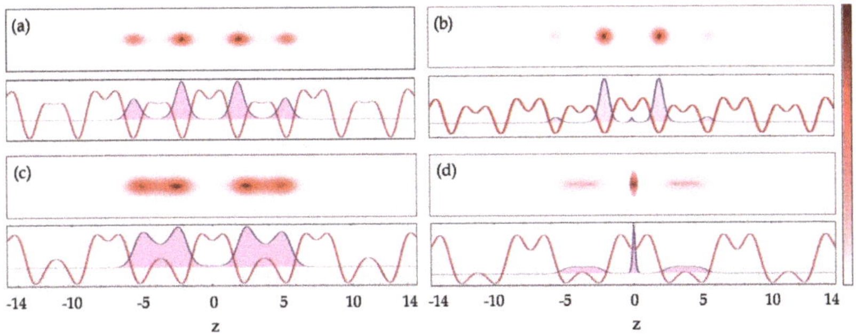

Figure 3. Condensate density patterns for $b_1 > 0$ and $b_2 > 0$: (**a**) $b_1 = 1; b_2 = 2$, (**b**) $b_1 = 1; b_2 = 3.5$, (**c**) $b_1 = 2; b_2 = 1$, and (**d**) $b_1 = 3.5; b_2 = 1$. Each plot of (**a–d**) has two panels: the upper panel shows the contour plot of the density and the lower panel consists of a 2D plot of the density combined with the corresponding potential profile.

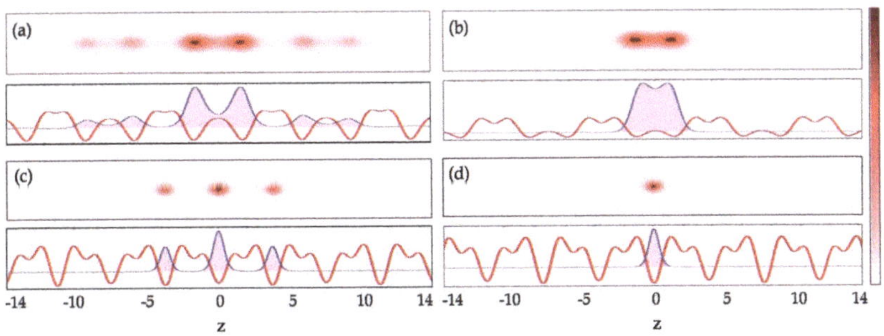

Figure 4. Condensate density patterns for $b_1 < 0$ or $b_2 < 0$ or both: (**a**) $b_1 = -1; b_2 = 1$, (**b**) $b_1 = -3; b_2 = 1$, (**c**) $b_1 = 1; b_2 = -3$, and (**d**) $b_1 = -1; b_2 = -3$. Each plot of (**a–d**) has two panels: the upper panel shows the contour plot of the density and the lower panel consists of a 2D plot of the density combined with the corresponding potential profile.

3. The Parameter Domain and Shape of the Tunable FOL

Figure 1 depicts the structure in the (b_1, b_2) space, produced by Equation (9), where one obtains, as particular cases, a tri-color optical lattice (TOL), or a BOL. On the contrary, FOL is obtained in the entire space, excluding the curves and points indicated in the figure.

The respective FOL potential, drawn in Figure 2, seems interesting enough. For $b_1 > 0$ and $b_2 > 0$, the FOL is a disordered double-well superlattice, featuring frustrations in terms of both inter- and intra-well separations. Figure 2a,b reduces to a BOL at $b_1 = b_2 = 0$. However, the transition to the domain of $b_1 < 0$ or $b_2 < 0$ makes the potential shapes quite different. In the former case, a triple-well superlattice gradually appears at $b_1 < 0$, whereas in the latter case, a translational shift of the double-wells in the superlattice by half a period is observed. The presently engineered FOL may be the most advanced trapping potential for BEC, derived as an ingredient of exact solutions. It may find applications to the design of quantum simulation, information and computation schemes [28–31]. We will further illustrate the results by displaying density patterns.

4. Density Patterns Supported by the Engineered FOL

The density patterns in the domain of $b_{1,2} > 0$ are displayed in Figure 3, along with the respective trapping profile, which help to understand the formation mechanism for the patterns. The presence of the inter- and intra-well potential frustration helps one to realize well-distinguished quantum clouds that may be employed for the design of enhanced atom-interferometry (Figure 3a–c). When the intra-well frustration disappears, the previously separated clouds inside the double well become indistinguishable and the condensate starts accumulating at the central frustrated site, causing Anderson-like localization (Figure 3d). The wide tunability of the FOL and the corresponding mesoscopic clouds make it possible to predict a variety of quantum states that may be useful for quantum technology [39–41].

In Figure 4, we illustrate the situation in the negative domain: $b_1 < 0$ in Figure 4a,b, $b_2 < 0$ in Figure 4c, and $b_{1,2} < 0$ in Figure 4d. It produces several aligned, well-separated spatial Schrödinger-cat states for $b_1 < 0$ [42]. More negative b_1 offers localization of the cat-state at the central double well. For $b_2 < 0$, the resulting triple-well super-lattice generates an odd number of well-separated clouds. Interestingly, changing the sign of b_1 at $b_2 < 0$ spatially translates the triple-well lattice by one period to create a single BEC cloud at the center (Figure 4d). Thus, a transition from Figure 4d to Figure 4c splits the single cloud into a set of three ones. In addition, a transition from Figure 4b to Figure 4a splits the Schrödinger-cat state from one to three. Along with the above-mentioned possibilities, this scheme of potential engineering offers an efficient scheme for designing quantum logic gates [31,32,43–46]. To illustrate the temporal dynamics of one of the obtained solutions, we choose the trap corresponding to $b_1 = 2$ and $b_2 = 1$ which shows a frustrated double-well super-lattice. Condensate, trapped in this potential, is allowed to evolve in time with a random noise of amplitude 10% of the maximum density. Condensate densities are depicted in Figure 5a–c for $t = 0$, $t = 10$ ms, and $t = 20$ ms, respectively. One can observe that the condensate is maintaining its shape after $t = 10$ ms, but getting distorted at $t = 20$ ms.

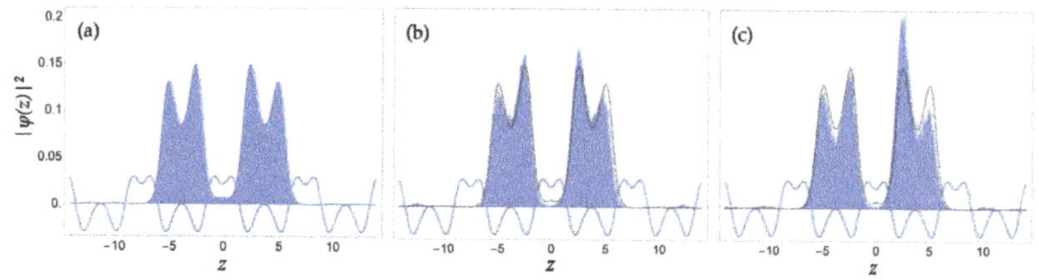

Figure 5. Condensate densities are depicted by filled plots at times (**a**) $t = 0$, (**b**) $t = 10$ ms, and (**c**) $t = 20$ ms, along with the potential energy profile (solid-line curve) for $b_1 = 2$ and $b_2 = 1$. Initial density (dotted curve) is merged with the densities in (**b**,**c**) for reference.

5. Dynamical Stability and Structural Stability of the Condensate

It is obviously necessary to check the dynamical and structural stability of the special analytical solution produced above. The dynamical stability pertains to disturbance added to the wave function, while the structural stability implies deformation of the external trap. We addressed these problems separately by numerically solving the GPE with the help of the split-step Fourier method [21,24,32]. The results are presented in Figure 6. In the former case, we have added random white noise $\Re_w$ to the analytically obtained wave function, while in the latter case, the noise is added to the external trap. The noisy form of the initial wave function and potential are represented as

$$\begin{aligned} \psi_{\text{noisy}}(z,t &= 0) = \psi(z,t=0) + \Re_w \\ V_{\text{noisy}}(z) &= V(z) + \Re_w. \end{aligned} \tag{10}$$

While the stability analysis was performed for a broad range of the parameters, here, we choose $b_1 = 2$ and $b_2 = 1$ for the purpose of illustration. In Figure 6, the condensate density profile, along with the external trap (not in scale), are depicted without the noise. The wave function is numerically evolved for both the noisy configurations defined as per Equation (10). Amplitude of noise $\Re_w$ varies from 0 to 5% of the maximum amplitude of the initial wave function. In the first scenario, we monitored the evolution of the wave functions, induced by the inputs $\psi_{\text{noisy}}(z, t = 0)$ and $\psi(z, t = 0)$, with our model potential,

$V(z)$. To observe the stability of the stationary state, we computed deviation of the evolving condensate density (D_W). In the latter case, we monitored the evolution of the input wave function ($\psi(z, t = 0)$) under the action of the potentials $V_{\text{noisy}}(z)$ and $V(z)$, to observe the structural deformation in the condensate density (D_P). We simulated the evolution for 10,000 time iterations with properly chosen space and time steps, $dz = 0.277$ µm and $dt = 0.22$ µs, respectively. The deviation (maximum relative error) of the evolved noisy data from their noise-free counterparts is shown for both kinds of the stability analyses in Figure 6 by the upper curve ($*$) and lower one, $\oplus$. Observing the noisy density profile after 10,000 iterations, we conclude that the density patterns retain their shapes with minimal deformation, which implies that the analytical solutions are indeed stable against both kinds of the random perturbations (Figure 6). The observed relative perturbation in the final configurations is near to 1% when the noise is initially added to the wave function, and near to 2% when it is added to the trapping potential. Thus, the presented model and its analytical solutions are physically relevant ones.

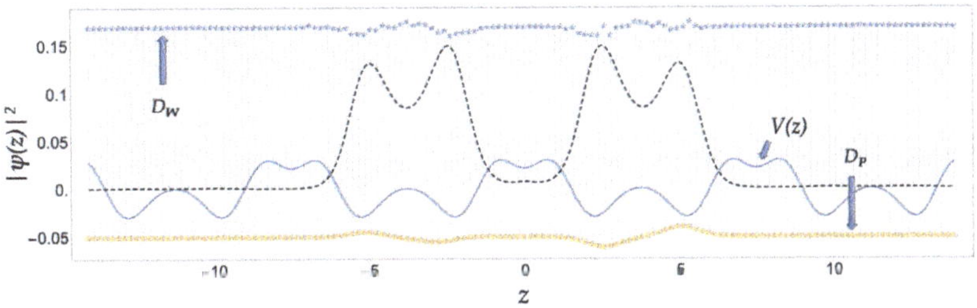

Figure 6. The numerical stability analysis of one of the obtained solutions with $b_1 = 2$ and $b_2 = 1$. The condensate density is depicted by the dotted line, and the trap profile, $V(z)$ (not in scale), is superimposed on it (the solid line). The deviation of the noisy data from their noise-free counterparts is shown for both kinds of the analyses: the dynamical stability, D_W (the upper curve, composed of symbols $*$), and the structural stability, D_P (the lower curve, composed of symbols $\oplus$).

6. Conclusions

In this paper, we reported the exact form of the four-color optical lattice (FOL) trap for the BEC in one dimension, which makes it possible to produce exact solutions for the trapped condensate. A variety of experimentally relevant trap profiles are reported, including one-, two-, three-, and four-color OLs with tunable shapes. It is worthy to stress that there are only two FOL-tuning parameters, b_1 and b_2, instead of four, making the detailed analysis of the exact solutions feasible. For chosen trap parameters, the exact condensate density is illustrated, and its variations after evolving in time are also shown. By means of systematic simulations, we have established dynamical and structural stability of the exact solutions. The stability against structural perturbations is especially important, as the solutions are valid only for the specially designed form of the FOL potentials. This class of FOL trapping potentials offers straightforward potential for use in applications, such as quantum simulation and other quantum technologies [13,28–31,39–41].

Author Contributions: Conceptualization, Data curation, Formal analysis, Investigation, Methodology, Software, Visualization, Validation; Writing—original draft, B.H. and S.G.; Data curation, Formal analysis, Investigation, Software—P.B. and J.B.; Writing—review and editing, B.M. and U.R.; Supervision, B.M. and U.R. All authors have read and agreed to the published version of the manuscript.

Funding: The work of B.A.M. was supported, in part, by the Israel Science Foundation through grant No. 1286/17.

Institutional Review Board Statement: Not applicable.

Informed Consent Statement: Not applicable.

Conflicts of Interest: The authors declare no conflict of interest.

References

1. Bloch, I.; Dalibard, J.; Zwerger, W. Many-body physics with ultracold gases. *Rev. Mod. Phys.* **2008**, *80*, 885. [CrossRef]
2. Greiner, M.; Mandel, O.; Esslinger, T.; Hänsch, T.W.; Bloch, I. Quantum phase transition from a superfluid to a Mott insulator in a gas of ultracold atoms. *Nature* **2002**, *415*, 39–44. [CrossRef] [PubMed]
3. Denschlag, J.H.; Simsarian, J.E.; Häffner, H.; Mckenzie, C.; Browaeys, A.; Cho, D.; Helmerson, K.; Rolston, S.L.; Phillips, W.D. A Bose-Einstein condensate in an optical lattice. *J. Phys. B At. Mol. Opt. Phys.* **2002**, *35*, 3095. [CrossRef]
4. Jaksch, D.; Bruder, C.; Cirac, J.I.; Gardiner, C.W.; Zoller, P. Cold Bosonic Atoms in Optical Lattices. *Phys. Rev. Lett.* **1998**, *81*, 3108. [CrossRef]
5. Brazhnyi, V.A.; Konotop, V.V. Theory of nonlinear matter waves in optical lattices. *Mod. Phys. Lett. B* **2004**, *18*, 627–651. [CrossRef]
6. Morsch, O.; Oberthaler, M. Dynamics of Bose-Einstein condensates in optical lattices. *Rev. Mod. Phys.* **2006**, *78*, 179–212. [CrossRef]
7. Lewenstein, M.; Sanpera, A.; Ahufinger, V. *Ultracold Atoms in Optical Lattices: Simulating Quantum Many-Body Systems*; Oxford University Press: Oxford, UK, 2012.
8. Dutta, O.; Gajda, M.; Hauke, P.; Lewenstein, N.; Luhmann, D.S.; Malomed, B.; Sowinski, T.; Zakrzewski, J. Non-standard Hubbard models in optical lattices: A review. *Rep. Prog. Phys.* **2015**, *78*, 066001. [CrossRef] [PubMed]
9. Hauke, P.; Cucchietti, F.M.; Tagliacozzo, L.; Deutsch, I.; Lewenstein, M. Can one trust quantum simulators? *Rep. Prog. Phys.* **2012**, *75*, 082401. [CrossRef] [PubMed]
10. Gross, C.; Bloch, I. Quantum simulations with ultracold atoms in optical lattices. *Science* **2017**, *357*, 995–1001. [CrossRef]
11. Schäfer, F.; Fukuhara, T.; Sugawa, S.; Takasu, Y.; Takahashi, Y. Tools for quantum simulation with ultracold atoms in optical lattices. *Nat. Rev. Phys.* **2020**, *2*, 411–425. [CrossRef]
12. Brennen, G.K.; Pupillo, G.; Rey, A.M.; Clark, C.W.; Williams, C.J. Scalable register initialization for quantum computing in an optical lattice. *J. Phys. B At. Mol. Opt. Phys.* **2005**, *38*, 1687. [CrossRef]
13. Katori, H. Optical lattice clocks and quantum metrology. *Nat. Photonics* **2011**, *5*, 203–210. [CrossRef]
14. Wang, Z.M.; Wu, L.A.; Modugno, M.; Byrd, M.S.; Yu, T.; You, J.Q. Fault-tolerant breathing pattern in optical lattices as a dynamical quantum memory. *Phys. Rev. A* **2014**, *89*, 042326. [CrossRef]
15. Schulte, T.; Drenkelforth, S.; Kruse, J.; Ertmer, W.; Arlt, J.; Sacha, K.; Zakrzewski, J.; Lewenstein, M. Routes towards Anderson-Like Localization of Bose-Einstein Condensates in Disordered Optical Lattices. *Phys. Rev. Lett.* **2005**, *95*, 170411. [CrossRef] [PubMed]
16. Adhikari, S.K.; Salasnich, L. Localization of a Bose-Einstein condensate in a bichromatic optical lattice. *Phys. Rev. A* **2009**, *80*, 023606. [CrossRef]
17. Nath, A.; Roy, U. Bose–Einstein condensate in a bichromatic optical lattice: An exact analytical model. *Laser Phys. Lett.* **2014**, *11*, 115501. [CrossRef]
18. Yamamoto, D.; Fukuhara, T.; Danshita, I. Frustrated quantum magnetism with Bose gases in triangular optical lattices at negative absolute temperatures. *Nat. Phys.* **2020**, *3*, 56. [CrossRef]
19. Billy, J.; Josse, V.; Zuo, Z.; Bernard, A.; Hambrecht, B.; Lugan, P.; Clement, D.; Sanchez-Palencia, L.; Bouyer, P.; Aspect, A. Direct observation of Anderson localization of matter waves in a controlled disorder. *Nature* **2008**, *453*, 891–894. [CrossRef]
20. Braun, S.; Ronzheimer, J.P.; Schreiber, M.; Hodgman, S.S.; Rom, T.; Bloch, I.; Schneider, U. Negative Absolute Temperature for Motional Degrees of Freedom. *Science* **2013**, *339*, 52–55. [CrossRef]
21. Nath, A.; Bera, J.; Ghosh, S.; Roy, U. Exact Analytical Model for Bose-Einstein Condensate at Negative Temperature. *Sci. Rep.* **2020**, *10*, 9016. [CrossRef]
22. Cheng, Y.; Gong, R.; Li, H. Dynamics of two coupled Bose-Einstein Condensate solitons in an optical lattice. *Opt. Exp.* **2006**, *14*, 3594–3601. [CrossRef]
23. Das, P.; Raju, T.S.; Roy, U.; Panigrahi, P.K. Sinusoidal excitations in two-component Bose-Einstein condensates in a trap. *Phys. Rev. A* **2009**, *79*, 015601. [CrossRef]
24. Nath, A.; Bera, J.; Ghosh, S.; Panigrahi, P.K.; Roy, U. Soliton dynamics for an ingenious trap combination in a Bose-Einstein condensate. *Eur. Phys. J. D* **2020**, *74*, 27. [CrossRef]
25. Sun, Q.; Hu, J.; Wien, L.; Liu, W.M.; Juzeliunas, G.; Ji, A.C. Ground states of a Bose-Einstein Condensate in a one-dimensional laser-assisted optical lattice. *Sci. Rep.* **2016**, *6*, 37679. [CrossRef] [PubMed]
26. Li, J.; Zeng, J. Dark matter-wave gap solitons in dense ultracold atoms trapped by a one-dimensional optical lattice. *Phys. Rev. A* **2021**, *103*, 013320. [CrossRef]
27. Malomed, B.A. *Soliton Management in Periodic Systems*; Springer: Berlin/Heidelberg, Germany, 2006.
28. Qiu, X.; Zou, J.; Qi, X.; Li, X. Precise programmable quantum simulations with optical lattices. *NPJ Quantum Inf.* **2020**, *6*, 87. [CrossRef]
29. Windpassinger, P.; Sengstock, K. Engineering novel optical lattices. *Rep. Prog. Phys.* **2013**, *76*, 086401. [CrossRef]
30. Ghosh, S.; Bera, J.; Panigrahi, P.K.; Roy, U. Sub-fourier quantum metrology through bright solitary trains in Bose–Einstein condensate. *Int. J. Quant. Inf.* **2019**, *17*, 1950019. [CrossRef]
31. Yang, B.; Sun, H.; Huang, C.J.; Wang, H.Y.; Deng, Y.; Dai, H.N.; Yuan, Z.S.; Pan, J.W. Cooling and entangling ultracold atoms in optical lattices. *Science* **2020**, *369*, 550–553. [CrossRef]

32. Bera, J.; Ghosh, S.; Salasnich, L.; Roy, U. Matter-wave fractional revivals in a ring waveguide. *Phys. Rev. A* **2020**, *102*, 063323. [CrossRef]
33. Khaykovich, L.; Schreck, F.; Ferrari, G.; Bourdel, T.; Cubizolles, J.; Carr, L.D.; Castin, Y.; Salomon, C. Formation of a Matter-Wave Bright Soliton. *Science* **2002**, *296*, 1290–1293. [CrossRef]
34. Inouye, S.; Andrews, M.R.; Stenger, J.; Miesner, H.J.; Stamper-Kurn, D.M.; Ketterle, W. Observation of Feshbach resonances in a Bose–Einstein condensate. *Nat. Phys.* **1998**, *392*, 151–154. [CrossRef]
35. Roberts, J.L.; Claussen, N.R.; Cornish, S.L.; Wieman, C.E. Magnetic Field Dependence of Ultracold Inelastic Collisions near a Feshbach Resonance. *Phys. Rev. Lett.* **2000**, *85*, 728. [CrossRef] [PubMed]
36. Kengne, E.; Liu, W.-M.; Malomed, B.A. Spatiotemporal engineering of matter-wave solitons in Bose-Einstein condensates. *Phys. Rep.* **2021**, *899*, 1–62 [CrossRef]
37. Nath, A.; Roy, U. A unified model for an external trap in a cigar-shaped Bose–Einstein condensate. *J. Phys. A Math. Theor.* **2014**, *47*, 415301. [CrossRef]
38. Abramowitz, M.; Stegun, I.A. *Handbook of Mathematical Functions*, 1st ed.; Dover: New York, NY, USA, 1964.
39. Howards, L.A.; Weinhold, T.J.; Shahandeh, F.; Combes, J.; Vanner, M.R.; White, A.G.; Ringbauer, M. Quantum Hypercube States. *Phys. Rev. Lett.* **2019**, *123*, 020402. [CrossRef]
40. Shukla, N.; Akhtar, N.; Sanders, B.C. Quantum tetrachotomous states: Superposition of four coherent states on a line in phase space. *Phys. Rev. A* **2019**, *99*, 063813. [CrossRef]
41. Shukla, N.; Nimmrichter, S.; Sanders, B.C. Squeezed comb states. *Phys. Rev. A* **2021**, *103*, 012408. [CrossRef]
42. Cirac, J.I.; Lewenstein, M.; Molmer, K.; Zoller, P. Quantum superposition states of Bose-Einstein condensates. *Phys. Rev. A* **1998**, *57*, 1208. [CrossRef]
43. Zeng, B.; Zhou, D.L.; Xu, Z.; Sun, C.P.; You, L. Encoding a logical qubit into physical qubits. *Phys. Rev. A* **2005**, *71*, 022309. [CrossRef]
44. Foot, C.J.; Shooter, M.D. Double well potentials and quantum gates. *Am. J. Phys.* **2011**, *79*, 762. [CrossRef]
45. Vo, C.; Riedl, S.; Baur, S.; Rempe, G.; Durr, S. Remote Entanglement between a Single Atom and a Bose-Einstein Condensate. *Phys. Rev. Lett.* **2012**, *109*, 263602. [CrossRef] [PubMed]
46. Gajdacz, M.; Opatrny, T.; Das, K.K. An atomtronics transistor for quantum gates. *Phys. Lett. A* **2014**, *378*, 1919. [CrossRef]

symmetry

MDPI

Article

Zeroth-Order Nucleation Transition under Nanoscale Phase Separation

Vyacheslav I. Yukalov [1,2,*,†] and **Elizaveta P. Yukalova** [3,†]

1 Bogolubov Laboratory of Theoretical Physics, Joint Institute for Nuclear Research, 141980 Dubna, Russia
2 Instituto de Fisica de São Carlos, Universidade de São Paulo, CP 369, 13560-970 São Paulo, Brazil
3 Laboratory of Information Technologies, Joint Institute for Nuclear Research, 141980 Dubna, Russia; yukalova@theor.jinr.ru
* Correspondence: yukalov@theor.jinr.ru; Tel.: +7-496-216-3947
† These authors contributed equally to this work.

Abstract: Materials with nanoscale phase separation are considered. A system representing a heterophase mixture of ferromagnetic and paramagnetic phases is studied. After averaging over phase configurations, a renormalized Hamiltonian is derived describing the coexisting phases. The system is characterized by direct and exchange interactions and an external magnetic field. The properties of the system are studied numerically. The stability conditions define the stable state of the system. At a temperature of zero, the system is in a pure ferromagnetic state. However, at finite temperature, for some interaction parameters, the system can exhibit a zeroth-order nucleation transition between the pure ferromagnetic phase and the mixed state with coexisting ferromagnetic and paramagnetic phases. At the nucleation transition, the finite concentration of the paramagnetic phase appears via a jump.

Keywords: nanoscale phase separation; quasi-equilibrium system; heterophase mixture; zeroth-order transition; nucleation point

Citation: Yukalov, V.I.; Yukalova, E.P. Zeroth-Order Nucleation Transition under Nanoscale Phase Separation. *Symmetry* **2021**, *13*, 2379. https://doi.org/10.3390/sym13122379

Academic Editor: Abraham A. Ungar

Received: 14 October 2021
Accepted: 6 December 2021
Published: 9 December 2021

1. Introduction

Phase transitions are commonly characterized by the appearance of non-analyticities in the system's thermodynamic characteristics. The classification of phase transitions is usually connected with the non-analyticities in the derivatives of thermodynamic potentials. Thus, the non-analyticity in the first-order derivatives implies a first-order phase transition, the non-analyticity in the second-order derivatives of a thermodynamic potential defines a second-order phase transition [1].

Recently, the possible existence of zeroth-order phase transitions has been brought to attention [2]; it is possible when a thermodynamic potential itself exhibits a discontinuity. Zeroth-order phase transitions have been found in the physics of black holes [3–11], holographic superconductors [12–15], and holographic ferromagnets and antiferromagnets [16,17]. The zeroth-order phase transition was also found for some spin models with long-range interactions [18,19]. Note that for the systems with long-range interactions, microcanonical and canonical ensembles are not necessarily equivalent [20].

Metal-insulator phase transitions in some materials, such as V_2O_3, were classified as zeroth-order phase transitions, where the free energy is discontinuous [21,22]. These phase transitions exhibit the phase coexistence and ramified fractal-like nanoscale phase separation in the transition region [21–23].

In this way, the zeroth-order phase transitions can occur when at least one of the features is present: either long-range interactions or nanoscale phase separation. Under this kind of phase separation, the system represents a mixture of nanoscale regions of different phases. The probabilistic weights of the phases are self-consistently defined by the system parameters and thermodynamic variables. Such nanoscale mixtures are also called heterophase or mesoscopic, since the linear size of inclusions of one phase inside the matrix of the other is larger than the interparticle distance but much smaller than the system linear size. The appearance of mesoscopic heterophase mixtures under nanoscale phase separation is a very widespread phenomenon arising around many phase transitions that can be of first or second order. Numerous examples of materials exhibiting the existence of such mixtures are given in the review articles [24–27]. Recently, the possibility of superfluid dislocations inside quantum crystals has been discussed [28–30]. Different types of nanoscale phase separation occur in electrolytes [31–35].

Here we shall concentrate on a heterophase mixture of ferromagnetic and paramagnetic phases. There exist numerous examples of materials exhibiting the coexistence of magnetic (ferromagnetic or antiferromagnetic) and paramagnetic phases. Thus, using the Mössbauer effect, the coexistence of antiferromagnetic and paramagnetic phases is observed in FeF_3 [36], in $CaFe_2O_4$ [37], and in a number of orthoferrites, such as $LaFeO_3$, $PrFeO_3$, $NdFeO_3$, $SmFeO_3$, $EuFeO_3$, $GdFeO_3$, $TbFeO_3$, $DyFeO_3$, $YFeO_3$, $HoFeO_3$, $ErFeO_3$, $TmFeO_3$, and $YbFeO_3$ [38,39]. Ferromagnetic cluster fluctuations, called ferrons or fluctuons, can arise inside a paramagnetic matrix of some semiconductors [40–45]. In some materials, magnetic cluster excitations can occur in the paramagnetic region above T_c or above T_N [46–52], causing the appearance of spin waves in the paramagnetic phase, for instance, in Ni, Fe, EuO, EuS, Pd_3Fe, and Gd [53–58]. The coexistence of ferromagnetic and nonmagnetic phases was also observed in Y_2Co_7, YCo_3, $Co(S_xSe_{1-x})_2$, $Co(Ti_xAl_{1-x})_2$, and $Lu(Co_{1-x}Al_x)_2$ [59,60]. In colossal magnetoresistance materials, such as $La_{1-x}Ca_xMnO_3$ and $La_{1-x}Sr_xCoO_3$, one observes the coexistence of a paramagnetic insulating, or semiconducting, phase and a ferromagnetic metallic phase [61–63], while in $La_{0.67-x}Bi_xCa_{0.33}MnO_3$, paramagnetic and antiferromagnetic phases coexist [64]. Nanoscale phase separation into ferromagnetic and paramagnetic regions has been observed in the colossal magnetoresistence compound, $EuB_{5.99}C_{0.01}$ [65]. Many more examples can be found in the review articles [24–27].

In the present paper, we consider a heterophase system with random phase separation, where the regions of different phases are randomly distributed in space. By averaging the phase configurations, we derive a renormalized, effective Hamiltonian of the mixture. Keeping in mind a spin system, we pass to the quasi-spin representation. Specifically, we consider a mixture of ferromagnetic and paramagnetic phases. Long-range interactions are assumed, such that the mean-field approximation becomes, in the thermodynamic limit, asymptotically exact. The existence of the ferromagnetic–paramagnetic mixture is due to the competition between direct and exchange interactions. We treat the case when the system is placed in an external magnetic field. We show that for some system parameters, there occurs the following situation: at low temperatures, the system is a pure ferromagnet that, when rising in temperature, can transfer into a mixture of ferromagnetic and paramagnetic phases at a nucleation point. For some system parameters, this nucleation transition happens to be a zeroth-order transition.

The plan of the paper is as follows. In Section 2, we recall the Gibbs method of equimolecular surfaces that are used for describing the spatial phase separation. Section 3 explains how the statistical operator of the mixture with phase separation can be defined by minimizing the functional information.The random spatial distribution of competing phases requires the averaging over phase configurations. The results of this averaging are summarized in Section 4. In Section 5, we pass from the field-operator representation to spin representation. Although this conversion is based on the known Bogolubov canonical transformation, it is necessary to recall it in order to elucidate the importance of taking account of direct particle interactions, in addition to exchange interactions. Keeping in mind long-range interactions, in Section 6, we derive the free energy of the mixture. Section 7 formulates the stability conditions that make it straightforward to separate stable states from unstable ones. In Section 8, we present the results of the numerical calculations and accompany them with discussions and conclusions.

2. Spatial Phase Separation

The description of a two-phase system with spatial phase separation starts with the Gibbs method [66] of equimolecular separating surfaces, where the system of volume V and number of particles N is considered to be separated into two parts, with the total volumes V_f and the particle numbers N_f, so that

$$V = V_1 + V_2 \; , \qquad N = N_1 + N_2 \; . \tag{1}$$

The regions $\mathbb{V}_f$ occupied by different phases are assumed to be randomly distributed in space. Their spatial locations are described by the manifold indicator functions

$$\xi_f(\mathbf{r}) = \left\{ \begin{array}{ll} 1 \; , & \mathbf{r} \in \mathbb{V}_f \\ 0 \; , & \mathbf{r} \not\in \mathbb{V}_f \end{array} \right. \; , \tag{2}$$

where

$$V_f \equiv \text{mes} \, \mathbb{V}_f \qquad (f = 1,2) \; .$$

The Hilbert space of microscopic states of the system is the tensor product

$$\mathcal{H} = \mathcal{H}_1 \bigotimes \mathcal{H}_2 \tag{3}$$

of the weighted Hilbert spaces [24–26] corresponding to the phases $f = 1,2$. The algebra of observables in this space is given by the direct sum of the algebra representations on the corresponding subspaces

$$\mathcal{A}(\xi) = \mathcal{A}_1(\xi_1) \bigoplus \mathcal{A}_2(\xi_2) \; . \tag{4}$$

For instance, the system energy Hamiltonian reads as

$$\hat{H}(\xi) = \hat{H}_1(\xi_1) \bigoplus \hat{H}_2(\xi_2) \; , \tag{5}$$

with the general form of the phase replica Hamiltonians

$$\hat{H}_f(\xi_f) = \int \xi_f(\mathbf{r}) \psi_f^\dagger(\mathbf{r}) \left[-\frac{\nabla^2}{2m} + U(\mathbf{r}) \right] \psi_f(\mathbf{r}) \, d\mathbf{r} +$$

$$+ \frac{1}{2} \int \xi_f(\mathbf{r}) \, \xi_f(\mathbf{r}') \, \psi_f^\dagger(\mathbf{r}) \, \psi_f^\dagger(\mathbf{r}') \, \Phi(\mathbf{r} - \mathbf{r}') \, \psi_f(\mathbf{r}') \, \psi_f(\mathbf{r}) \, d\mathbf{r} d\mathbf{r}' \; , \tag{6}$$

where $\Phi(\mathbf{r})$ is an interaction potential, $U(\mathbf{r})$ is an external potential, and the field operators $\psi_f(\mathbf{r})$ are columns with respect to internal degrees of freedom, such as spin. The number-of-particle operator is

$$\hat{N}(\xi) = \hat{N}_1(\xi_1) + \hat{N}_2(\xi_2) \; , \tag{7}$$

with the number-of-particle operators of each phase

$$\hat{N}_f(\xi_f) = \int \xi(\mathbf{r}) \, \psi_f^\dagger(\mathbf{r}) \, \psi_f(\mathbf{r}) \, d\mathbf{r} \; . \tag{8}$$

Here and below, we set the Planck and Boltzmann constants to one.

3. System Statistical Operator

The general procedure of defining the statistical operator for a system is by minimizing the information functional, taking account of the prescribed constraints. The latter is the normalization condition

$$\text{Tr} \int \hat{\rho}(\xi) \, \mathcal{D}\xi = 1 \; , \tag{9}$$

the definition of the system energy

$$\text{Tr} \int \hat{\rho}(\xi) \, \hat{H}(\xi) \, \mathcal{D}\xi = E \; , \tag{10}$$

and of the total number of particles in the system

$$\text{Tr} \int \hat{\rho}(\xi)\, \hat{N}(\xi)\, \mathcal{D}\xi = N\,. \tag{11}$$

Here and in what follows, the trace operation is taken over the whole Hilbert space (3), and $\mathcal{D}\xi$ implies the averaging over phase configurations describing the random locations and shapes of separated phases.

The information functional in the Kullback–Leibler form [67,68] reads as

$$I[\,\hat{\rho}\,] = \text{Tr} \int \hat{\rho}(\xi)\, \ln \frac{\hat{\rho}(\xi)}{\hat{\rho}_0(\xi)}\, \mathcal{D}\xi \, + \, \alpha\left[\text{Tr} \int \hat{\rho}(\xi)\, \mathcal{D}\xi - 1 \right] +$$

$$+ \, \beta\left[\text{Tr} \int \hat{\rho}(\xi)\, \hat{H}(\xi)\, \mathcal{D}\xi - E \right] \, + \, \gamma\left[\text{Tr} \int \hat{\rho}(\xi)\, \hat{N}(\xi)\, \mathcal{D}\xi - N \right], \tag{12}$$

with the Lagrange multipliers α, $\beta = 1/T$, and $\gamma = -\beta\mu$, and with a trial statistical operator $\hat{\rho}_0(\xi)$ characterizing some a priori information if any. If no a priori information is available, $\hat{\rho}_0(\xi)$ is a constant. Then minimizing the information functional over $\hat{\rho}(\xi)$ yields the statistical operator

$$\hat{\rho}(\xi) = \frac{1}{Z}\, \exp\{-\beta H(\xi)\}\,, \tag{13}$$

with the grand Hamiltonian

$$H(\xi) = \hat{H}(\xi) - \mu \hat{N}(\xi) \tag{14}$$

and the partition function

$$Z = \text{Tr} \int \exp\{-\beta H(\xi)\}\, \mathcal{D}\xi\,. \tag{15}$$

Introducing the effective renormalized Hamiltonian by the relation

$$\exp\{-\beta \widetilde{H}\} = \int \exp\{-\beta H(\xi)\}\, \mathcal{D}\xi \tag{16}$$

gives the partition function

$$Z = \text{Tr}\, \exp\{-\beta \widetilde{H}\}\,. \tag{17}$$

Then we get the grand thermodynamic potential

$$\Omega = -T \ln Z\,. \tag{18}$$

This picture describes a heterophase system where the phase-separated regions are random in the sense that they are randomly located in space and can move and change their shapes. In that sense, strictly speaking, the system is in quasi-equilibrium. However, the averaging over phase configurations reduces the consideration to an effective system equilibrium on average [24–26].

4. Averaging over Phase Configurations

In order to explicitly accomplish the averaging over phase configurations, it is necessary to define the functional integration over the manifold indicator functions (2). This functional integration has been defined and explicitly realized in papers [24,69–73]. Here we formulate the main results of this functional integration over the manifold indicator functions with the differential measure $\mathcal{D}\xi$, which realizes the averaging over phase configurations.

Theorem 1. *Let us consider the functional*

$$A_f(\xi_f) = \sum_{n=0}^{\infty} \int \xi_f(\mathbf{r}_1)\, \xi_f(\mathbf{r}_2) \ldots \xi_f(\mathbf{r}_n)\, A_f(\mathbf{r}_1, \mathbf{r}_2, \ldots, \mathbf{r}_n)\, d\mathbf{r}_1 d\mathbf{r}_2 \ldots d\mathbf{r}_n \; . \tag{19}$$

The integration of this function over the manifold indicator functions gives

$$\int A_f(\xi_f)\, \mathcal{D}\xi = A_f(w_f) \; , \tag{20}$$

where

$$A_f(w_f) = \sum_{n=0}^{\infty} w_f^n \int A_f(\mathbf{r}_1, \mathbf{r}_2, \ldots, \mathbf{r}_n)\, d\mathbf{r}_1 d\mathbf{r}_2 \ldots d\mathbf{r}_n \; , \tag{21}$$

while

$$w_f = \frac{1}{V} \int \xi_f(\mathbf{r})\, d\mathbf{r} = \frac{V_f}{V} \tag{22}$$

defines the geometric probability of an f-th phase.

Theorem 2. *The thermodynamic potential*

$$\Omega = -T \ln \text{ Tr} \int \exp\{-\beta H(\xi)\}\, \mathcal{D}\xi \; , \tag{23}$$

after the averaging over phase configurations, becomes

$$\Omega = -T \ln \text{ Tr}\, \{-\beta \widetilde{H}\} = \sum_f \Omega_f \equiv \Omega(w) \; , \tag{24}$$

where

$$\Omega_f = -T \ln \text{ Tr}_{\mathcal{H}_f}\, \{-\beta H_f(w_f)\} \equiv \Omega_f(w_f) \; , \tag{25}$$

and the renormalized Hamiltonian is

$$\widetilde{H} = \bigoplus_f H_f(w_f) \equiv \widetilde{H}(w) \; , \tag{26}$$

with the phase probabilities w_f being the minimizers of the thermodynamic potential,

$$\Omega = \text{abs} \min_{\{w_f\}} \Omega(w) \; , \tag{27}$$

under the normalization condition

$$\sum_f w_f = 1 \; , \qquad 0 \leq w_f \leq 1 \; . \tag{28}$$

Theorem 3. *The observable quantities, given by the averages*

$$\langle\, \hat{A}\, \rangle = \text{Tr} \int \hat{\rho}(\xi)\, \hat{A}(\xi)\, \mathcal{D}\xi \tag{29}$$

of the operators from the algebra of observables (4),

$$\hat{A}(\xi) = \bigoplus_f \hat{A}_f(\xi_f) \; , \tag{30}$$

with $\hat{A}_f(\xi_f)$ defined as in (19), after the averaging over phase configurations, reduce to the form

$$\langle\, \hat{A}\, \rangle = \text{Tr}\, \hat{\rho}(w)\, \hat{A}(w) \; , \tag{31}$$

where the renormalized operator of an observable is

$$\hat{A}(w) = \bigoplus_f \hat{A}_f(w_f) \,, \tag{32}$$

with $\hat{A}_f(w_f)$ *defined as in (21), and the renormalized statistical operator is*

$$\hat{\rho}(w) = \frac{1}{Z} \exp\{-\beta\widetilde{H}(w)\} \,, \tag{33}$$

with the partition function (17).

The proofs of these theorems are given in the papers [24,69–73].

5. Hamiltonian in Spin Representation

Since we aim to study the magnetic properties of a system with phase separation, it is useful to transform Hamiltonian (6) into spin representation. For this purpose, we assume that the system is periodic over a lattice with the lattice sites $\mathbf{r}_j$, where $j = 1, 2, \ldots, N$, and we expand the field operators over Wannier functions:

$$\psi_f(\mathbf{r}) = \sum_j c_{jf}\varphi_f(\mathbf{r} - \mathbf{r}_j) \,. \tag{34}$$

Keeping in mind well-localized Wannier functions [74], we retain in the Hamiltonian only the terms expressed through the matrix elements over Wannier functions containing not more than two lattice sites, since the overlap of Wannier functions located at three or four different lattice sites is negligibly small.

The remaining matrix elements are: the tunneling term

$$T_{ijf} = -\int \varphi_f^*(\mathbf{r} - \mathbf{r}_i) \left[-\frac{\nabla^2}{2m} + U(\mathbf{r}) \right] \varphi_f(\mathbf{r} - \mathbf{r}_j) \, d\mathbf{r} \,, \tag{35}$$

the term of direct interactions

$$\Phi_{ijf} = \int |\, \varphi_f(\mathbf{r} - \mathbf{r}_i) \,|^2 \, \Phi(\mathbf{r} - \mathbf{r}') \,|\, \varphi_f(\mathbf{r}' - \mathbf{r}_j) \,|^2 \, d\mathbf{r} d\mathbf{r}' \,, \tag{36}$$

and the term of exchange interactions

$$J_{ijf} = -\int \varphi_f^*(\mathbf{r} - \mathbf{r}_i) \, \varphi_f^*(\mathbf{r}' - \mathbf{r}_j) \, \Phi(\mathbf{r} - \mathbf{r}') \, \varphi_f(\mathbf{r}' - \mathbf{r}_i) \, \varphi_f(\mathbf{r} - \mathbf{r}_j) \, d\mathbf{r} d\mathbf{r}' \,. \tag{37}$$

Then the Hamiltonian (6) transforms into the form

$$H_f = -w_f \sum_{ij} (T_{ijf} + \mu\delta_{ij}) c_{if}^\dagger \, c_{jf} \, +$$

$$+ \, \frac{1}{2} \, w_f^2 \sum_{ij} \left(\Phi_{ijf} \, c_{if}^\dagger \, c_{jf}^\dagger \, c_{jf} \, c_{if} - J_{ijf} \, c_{if}^\dagger \, c_{jf}^\dagger \, c_{if} \, c_{jf} \right) . \tag{38}$$

To exclude self-interactions, one sets

$$\Phi_{jjf} \equiv J_{jjf} = 0 \,. \tag{39}$$

Then we introduce spin operators following the method of canonical transformations [75–77], generalized in the case of heterophase systems [78–80]. Keeping in mind the particles with spin one-half, the operators c_{jf} are to be treated as spinors

$$c_{jf} = \begin{bmatrix} c_{jf}(\uparrow) \\ c_{jf}(\downarrow) \end{bmatrix} \tag{40}$$

of two components, one with spin up and the other with spin down. When each lattice site is occupied by a single particle, the unipolarity condition is valid

$$c_{jf}^\dagger(\uparrow)\, c_{jf}(\uparrow) + c_{jf}^\dagger(\downarrow)\, c_{jf}(\downarrow) = 1\,. \tag{41}$$

The canonical transformations introducing spin operators $\mathbf{S}_{jf}$, acting on the space $\mathcal{H}_f$, read as

$$c_{jf}^\dagger(\uparrow)\, c_{jf}(\uparrow) = \frac{1}{2} + S_{jf}^z\,, \qquad c_{jf}^\dagger(\uparrow)\, c_{jf}(\downarrow) = S_{jf}^x + i\, S_{jf}^y$$

$$c_{jf}^\dagger(\downarrow)\, c_{jf}(\downarrow) = \frac{1}{2} - S_{jf}^z\,. \tag{42}$$

Employing these canonical transformations and wishing to write the Hamiltonian in a compact form, we define the average direct interactions

$$\Phi_f \equiv \frac{1}{N} \sum_{i \neq j} \Phi_{ijf}\,, \tag{43}$$

the average exchange interactions

$$J_f \equiv \frac{1}{N} \sum_{i \neq j} J_{ijf}\,, \tag{44}$$

and the effective chemical potentials

$$\mu_f \equiv \mu + \frac{1}{N} \sum_{ij} T_{ijf}\,. \tag{45}$$

Then Hamiltonian (38) becomes

$$H_f = \frac{1}{2}\, w_f^2 U_f N - w_f^2 \sum_{i \neq j} J_{ijf} \mathbf{S}_{if} \cdot \mathbf{S}_{jf} - w_f \mu_f N\,, \tag{46}$$

where

$$U_f \equiv \Phi_f - \frac{1}{2}\, J_f\,. \tag{47}$$

For localized particles, the tunneling term T_{ijf} is small and can be neglected. Hence, as is seen from expression (45), $\mu_f = \mu$. Then the last term in Hamiltonian (46) becomes $-w_f \mu N$. Such linear scalar terms in w_f can be omitted since they enter the Hamiltonian (26) as $w_1\mu + w_2\mu = \mu$, which is as a constant shift. The value (47) characterizes an average potential acting on each particle in the system and is mainly due to direct interactions that are usually much larger than the exchange interactions. It is reasonable to assume that this average potential does not depend on the kind of magnetic phases, so that $U_f = U$. For generality, it is also necessary to take into account an external magnetic field $\mathbf{B}_0$. As a result, we come to the Hamiltonian

$$H_f = \frac{1}{2}\, w_f^2 U N - w_f^2 \sum_{i \neq j} J_{ijf} \mathbf{S}_{if} \cdot \mathbf{S}_{jf} - w_f \sum_j \mu_0 \mathbf{B}_0 \cdot \mathbf{S}_{jf}\,. \tag{48}$$

The main feature of the paramagnetic phase is, clearly, the absence of long-range order. The direct way of taking this into account on the microscopic level is to notice that the term of exchange interactions (37) essentially depends on the localization of Wanier functions. From expression (37), it is evident that the better Wannier functions are localized, the smaller the exchange term. Therefore, accepting that the paramagnetic exchange term is very small, automatically degrades the long-range order. Keeping this in mind, we set to zero the paramagnetic exchange interactions, $J_{ij2} = 0$. Then the Hamiltonian (48) yields for the ferromagnetic phase

$$H_1 = \frac{1}{2}\, w_1^2 UN - w_1^2 \sum_{i \neq j} J_{ij1} \mathbf{S}_{i1} \cdot \mathbf{S}_{j1} - w_1 \sum_j \mu_0 \mathbf{B}_0 \cdot \mathbf{S}_{j1} \; , \tag{49}$$

and for the paramagnetic phase

$$H_2 = \frac{1}{2}\, w_2^2 UN - w_2 \sum_j \mu_0 \mathbf{B}_0 \cdot \mathbf{S}_{j2} \; . \tag{50}$$

The external magnetic field is assumed to be directed along the z-axis

$$\mathbf{B}_0 = B_0 \mathbf{e}_z \qquad (B_0 \geq 0) \; . \tag{51}$$

Recall that the total system Hamiltonian, according to (26), reads as

$$\widetilde{H} = H_1 \bigoplus H_2 \; . \tag{52}$$

The order parameters can be defined by the averages

$$s_f \equiv \left\langle \frac{2}{N} \sum_j S_{jf}^z \right\rangle , \tag{53}$$

which lie in the interval

$$0 \leq s_f \leq 1 \qquad (f = 1, 2) \; . \tag{54}$$

For the ferromagnetic phase, there exist such low temperatures where

$$\lim_{B_0 \to 0} s_1 > 0 \qquad (T \to 0) \; , \tag{55}$$

while for the paramagnetic phase at all temperatures, one has

$$\lim_{B_0 \to 0} s_2 = 0 \; . \tag{56}$$

Accepting that interparticle interactions are of a long-range order, Hamiltonian (49) can be simplified by resorting to the mean-field approximation

$$S_{i1}^z S_{j1}^z = S_{i1}^z \langle S_{j1}^z \rangle + \langle S_{i1}^z \rangle S_{j1}^z - \langle S_{i1}^z \rangle \langle S_{j1}^z \rangle \; , \tag{57}$$

where $i \neq j$. This reduces that Hamiltonian to the form

$$H_1 = \frac{1}{2}\, w_1^2 \left(U + \frac{1}{2}\, J s_1^2 \right) N - \left(w_1^2 J s_1 + w_1 \mu_0 B_0 \right) \sum_j S_{j1}^z \; . \tag{58}$$

6. Free Energy of Mixture

Defining the reduced free-energy in the standard way

$$F = -\, \frac{T}{N} \ln \operatorname{Tr} e^{-\beta \widetilde{H}} \; , \tag{59}$$

introducing the dimensionless parameters

$$u \equiv \frac{U}{J} , \qquad h \equiv \frac{\mu_0 B_0}{J} , \tag{60}$$

where

$$J \equiv J_1 = \frac{1}{N} \sum_{i \neq j} J_{ij1} , \tag{61}$$

and measuring temperature in units of J, we come to the mixture free energy

$$F = F_1 + F_2 . \tag{62}$$

Here the free energy of the magnetic component is

$$F_1 = \frac{1}{2}\, w_1^2 \left(u + \frac{1}{2}\, s_1^2 \right) - T \, \ln \left[\, 2 \cosh \left(\frac{w_1 h + w_1^2 s_1}{2T} \right) \right] , \tag{63}$$

with the order parameter

$$s_1 = \tanh \left(\frac{w_1 h + w_1^2 s_1}{2T} \right) , \tag{64}$$

and the free energy of the paramagnetic component is

$$F_2 = \frac{1}{2}\, w_2^2 \, u - T \, \ln \left[\, 2 \cosh \left(\frac{w_2 h}{2T} \right) \right] , \tag{65}$$

with the order parameter

$$s_2 = \tanh \left(\frac{w_2 h}{2T} \right) . \tag{66}$$

Studying the properties of the free energy, it is convenient to represent it in the form symmetric with respect to both phase components, introducing the quantity

$$g_f \equiv \frac{1}{N} \sum_{i \neq j} \frac{J_{ijf}}{J} . \tag{67}$$

By definition, $g_1 = 1$, while $g_2 \to 0$. Then the partial free energy

$$F_f = - \frac{T}{N} \, \ln \, \mathrm{Tr} \, e^{-\beta H_f} \tag{68}$$

becomes

$$F_f = \frac{1}{2}\, w_f^2 \left(u + \frac{1}{2}\, g_f s_f^2 \right) - T \, \ln \left[\, 2 \cosh \left(\frac{w_f h + w_f^2 g_f s_f}{2T} \right) \right] , \tag{69}$$

with the order parameter

$$s_f = \tanh \left(\frac{w_f h + w_f^2 g_f s_f}{2T} \right) . \tag{70}$$

7. Stability Conditions

The statistical system is stable when it is in the state of the absolute minimum of the thermodynamic potential, which in the present case is the free energy. The system is in the mixed state, provided the free energy (62) corresponds to a minimum with respect to the variables w_1, s_1, and s_2. The variable w_2 is expressed through the relation $w_2 = 1 - w_1$. For convenience, it is possible to use the notation

$$w_1 \equiv w , \qquad w_2 = 1 - w \tag{71}$$

and consider only the variable w, instead of w_1 and w_2 connected by the normalization condition. The conditions of the extremum are

$$\frac{\partial F}{\partial s_1} = 0 \, , \qquad \frac{\partial F}{\partial s_2} = 0 \, , \qquad \frac{\partial F}{\partial w} = 0 \, . \tag{72}$$

The first and second conditions give the expressions (64) and (66) for the order parameters s_1 and s_2. The third equation, due to the normalization condition, can be written as

$$\frac{\partial F}{\partial w_1} = \frac{\partial F}{\partial w_2} \, .$$

Using the derivative

$$\frac{\partial F}{\partial w_f} = w_f \left(u - \frac{1}{2} \, g_f s_f^2 \right) - \frac{1}{2} \, h s_f$$

results in the probability of the ferromagnetic component

$$w = \frac{2u + h(s_1 - s_2)}{4u - s_1^2} \, . \tag{73}$$

The extremum is a minimum provided the principal minors of the Hessian matrix are positive. The Hessian matrix is expressed through the second derivatives

$$\frac{\partial^2 F_f}{\partial w_f^2} = u - \frac{1}{2} \, g_f s_f^2 - \frac{1 - s_f^2}{4T} \left(h + 2 w_f g_f s_f \right)^2 \, ,$$

$$\frac{\partial^2 F_f}{\partial w_f \partial s_f} = - \, \frac{1 - s_f^2}{4T} \, w_f^2 g_f \left(h + 2 w_f g_f s_f \right) \, , \qquad \frac{\partial^2 F_f}{\partial s_f^2} = \frac{1}{2} \, w_f^2 g_f \left(1 - \frac{1 - s_f^2}{4T} \, w_f^2 g_f \right) \, .$$

For the considered system, we have

$$\frac{\partial^2 F}{\partial w^2} = 2u - \frac{1}{2} \, s_1^2 - \frac{1 - s_1^2}{4T} \, (h + 2 w s_1)^2 - \frac{1 - s_2^2}{4T} \, h^2 \, ,$$

$$\frac{\partial^2 F}{\partial w \partial s_1} = - \, \frac{1 - s_1^2}{4T} \, w^2 \, (h + 2 w s_1) \, ,$$

$$\frac{\partial^2 F}{\partial w \partial s_2} = \frac{\partial^2 F}{\partial s_1 \partial s_2} = \frac{\partial^2 F}{\partial s_2^2} = 0 \, , \qquad \frac{\partial^2 F}{\partial s_1^2} = \frac{1}{2} \, w^2 \left(1 - \frac{1 - s_1^2}{2T} \, w^2 \right) \, .$$

The minimum of the free energy implies the stability conditions that for the present case become

$$\frac{\partial^2 F}{\partial w^2} > 0 \, , \qquad \frac{\partial^2 F}{\partial s_1^2} > 0 \, ,$$

$$\frac{\partial^2 F}{\partial w^2} \, \frac{\partial^2 F}{\partial s_1^2} - \left(\frac{\partial^2 F}{\partial w \partial s_1} \right)^2 > 0 \, . \tag{74}$$

We need to solve the system of equations for the order parameter s_1, given in (64) and satisfying condition (55), for the order parameter s_2, given in (66) and satisfying condition (56), and for the probability of ferromagnetic phase w, defined in (73) and satisfying conditions (28). If there occur several solutions, it is necessary to choose the solution that corresponds to the minimal free energy and satisfies the stability conditions (74).

Furthermore, it is necessary to choose the state with the minimal free energy between the free energy F of the mixture, free energy F_{fer} of the pure ferromagnetic phase

$$F_{fer} = \frac{1}{2}\left(u + \frac{1}{2}\, s_{fer}^2\right) - T\, \ln\left[2\cosh\left(\frac{h + s_{fer}}{2T}\right)\right], \tag{75}$$

with the order parameter

$$s_{fer} = \tanh\left(\frac{h + s_{fer}}{2T}\right), \tag{76}$$

and the free energy of the pure paramagnetic phase

$$F_{par} = \frac{1}{2}\, u - T\ln\left[2\cosh\left(\frac{h}{2T}\right)\right], \tag{77}$$

with the order parameter

$$s_{par} = \tanh\left(\frac{h}{2T}\right). \tag{78}$$

8. Results and Discussion

We have derived the model of a mixed system describing the coexistence of different phases when at least one of the phases represents nanoscale regions of a competing phase inside a host phase. The spatial distribution of the phases is random. This picture is often termed nanoscale phase separation. As a concrete example, we have studied the mixture of ferromagnetic and paramagnetic phases, modeling a ferromagnet with paramagnetic fluctuations. The choice of this example is dictated by the fact that spin models serve as typical illustrations of phase transitions of different nature.

After averaging over phase configurations, we obtain a renormalized Hamiltonian, taking into account the coexistence of mesoscopic phases. In the resulting effective picture, thermodynamic potentials are represented as the sums of replicas characterizing different phases. This, however, is not a simple sum of the terms corresponding to pure phases, as in the case of the Gibbs macroscopic mixture, where, for instance, free energy is a linear combination, in our case

$$F_G = w_1 F_{fer} + w_2 F_{par}. \tag{79}$$

The separation of phases is connected with the existence of surface free energy. The latter is not a microscopic notion and is not defined at the level of operators and microscopic states. The surface free energy is a thermodynamic notion defined by the difference between the actual free energy of the system and the free energy of the Gibbs macroscopic mixture [81–83]. That is, the surface free energy is defined by the difference

$$F_{sur} = F - F_G. \tag{80}$$

In our case, this is

$$F_{sur} = F_1 + F_2 - w_1 F_{fer} - w_2 F_{par}. \tag{81}$$

Contrary to a pure phase needing a one-order parameter (that can be a vector or a tensor), the mixed state requires, for its correct description, a larger number of parameters. Thus, compared to the pure ferromagnetic phase, described by a single order parameter s_{fer}, the mixed ferromagnetic–paramagnetic state needs three parameters: the order parameter (reduced magnetization) of the ferromagnetic component, s_1, the order parameter (reduced magnetization) of the paramagnetic component, s_2, and the probability of one of the phases, say w, with the probability of the other phase given by $1 - w$.

In Figures 1–8, we present the results of the numerical investigation for different parameters u and h. Only stable solutions are shown. The absence of F in a figure implies that F is unstable. Depending on the values of the parameters, there can exist two types of behavior.

(i) At low temperatures, the system is a pure ferromagnet described by the free energy F_{fer} and the order parameter $s_{fer} \equiv s_1$, with $w \equiv 1$. When increasing temperature, F_{fer} gradually approaches F_{par} corresponding to a paramagnet. The order parameter $s_{fer} \equiv s_1$ has the form typical of the ferromagnetic magnetization. This behavior, for instance, happens for $u < 0.25$ and all $h > 0$.

(ii) For $u > 0.25$, at low temperatures, below the nucleation temperature T_n, the system is a pure ferromagnet, with the free energy F_{fer}, the order parameter $s_{fer} \equiv s_1$, and $w \equiv 1$. At the nucleation temperature T_n, there appears a solution for the mixed state with the free energy F and the order parameters s_1 and s_2. The free energy F is lower than F_{fer}, but does not intersect it so that the nucleation is to be classified as a zeroth-order transition.

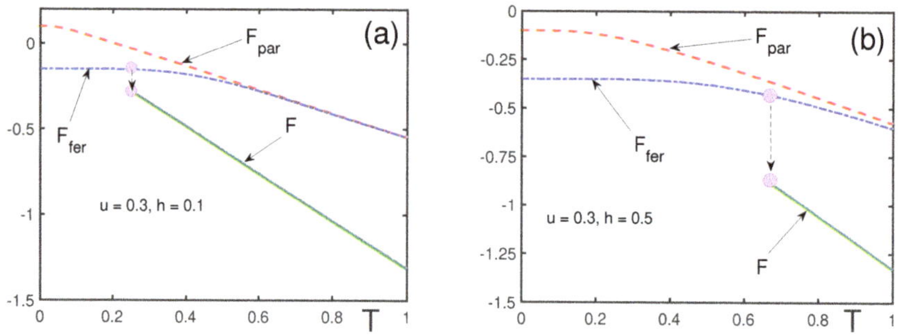

Figure 1. Free energies of the mixed state, F (solid line), ferromagnetic state, F_{fer} (dash–dotted line), and of the paramagnetic state, F_{par} (dashed line), for $u = 0.3$ and different magnetic fields: (**a**) $h = 0.1$; (**b**) $h = 0.5$.

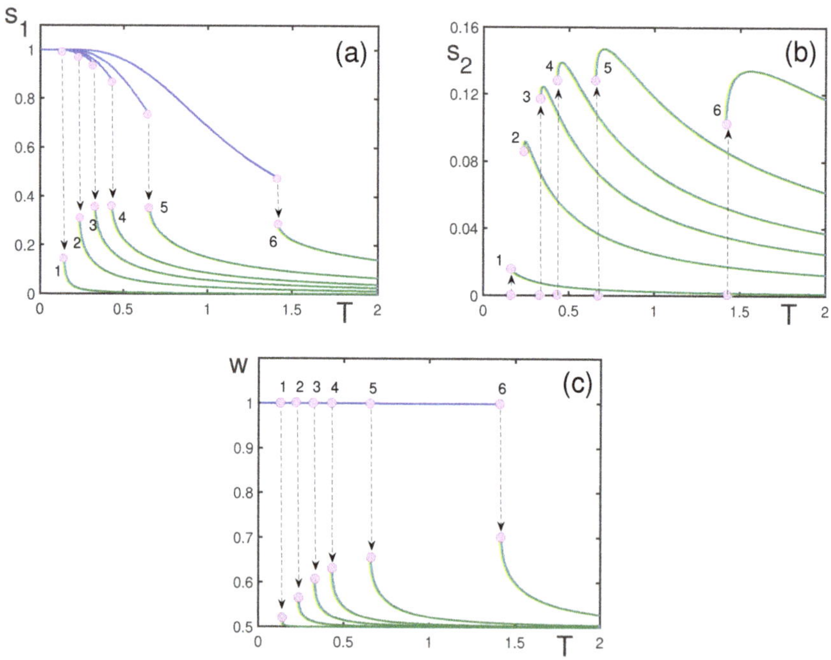

Figure 2. Order parameters s_1 (**a**), s_2 (**b**), and w as functions of dimensionless temperature T (**c**), for $u = 0.3$ and different fields: (1) $h = 0.01$; (2) $h = 0.1$; (3) $h = 0.2$; (4) $h = 0.3$; (5) $h = 0.5$; (6) $h = 1$. The corresponding nucleation temperatures are: (1) $T_n = 0.15$; (2) $T_n = 0.24$; (3) $T_n = 0.34$; (4) $T_n = 0.44$; (5) $T_n = 0.66$; (6) $T_n = 1.42$.

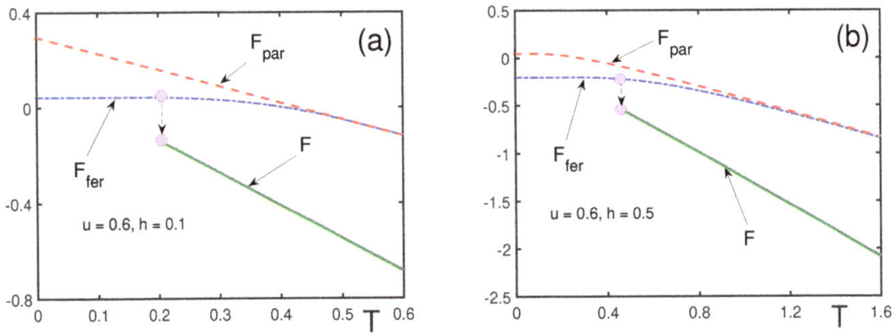

Figure 3. Free energies of the mixed state, F (solid line), ferromagnetic state, F_{fer} (dash-dotted line), and of the paramagnetic state, F_{par} (dashed line), for $u = 0.6$ and different magnetic fields: (**a**) $h = 0.1$; (**b**) $h = 0.5$.

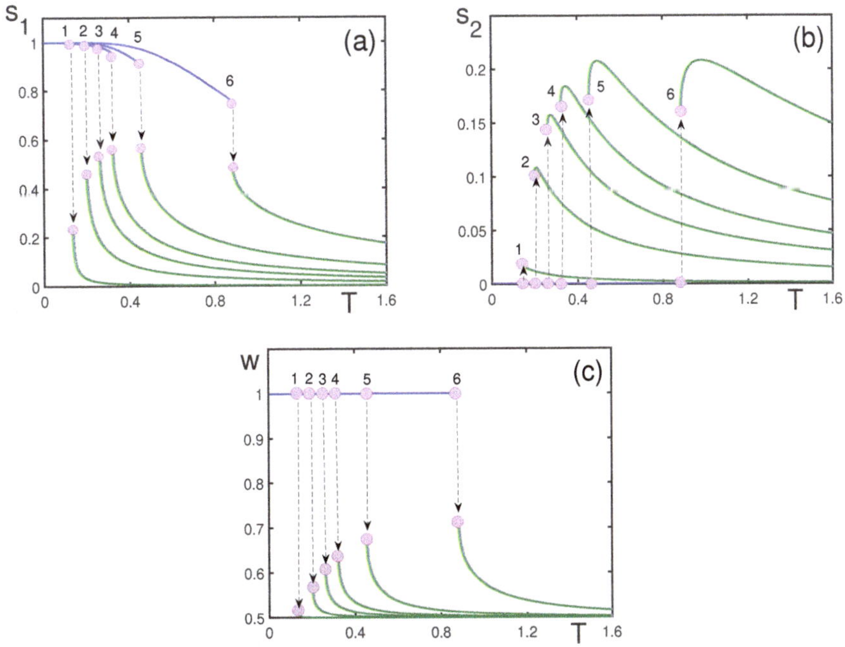

Figure 4. Order parameters s_1 (**a**), s_2 (**b**), and w as functions of dimensionless temperature T (**c**), for $u = 0.6$ and different fields: (1) $h = 0.01$; (2) $h = 0.1$; (3) $h = 0.2$; (4) $h = 0.3$; (5) $h = 0.5$; (6) $h = 1$. The corresponding nucleation temperatures are: (1) $T_n = 0.14$; (2) $T_n = 0.20$; (3) $T_n = 0.26$; (4) $T_n = 0.33$; (5) $T_n = 0.46$; (6) $T_n = 0.89$.

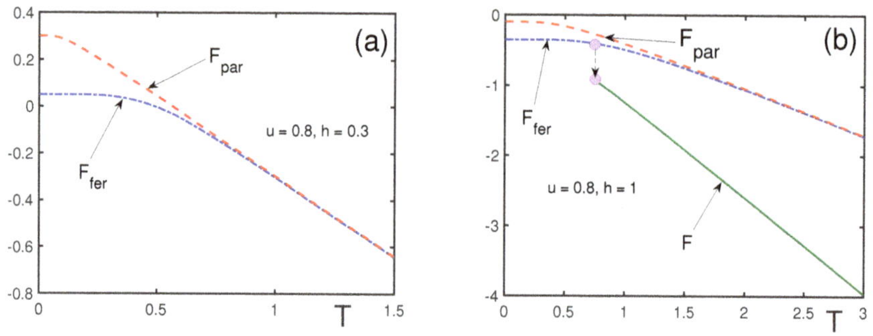

Figure 5. Free energies of the mixed state, F (solid line), ferromagnetic state, F_{fer} (dash–dotted line), and of the paramagnetic state, F_{par} (dashed line), for $u = 0.8$ and different magnetic fields: (**a**) $h = 0.3$; (**b**) $h = 1$. For $h = 0.3$, the mixed state is not stable. For $h = 1$, the zeroth-order nucleation transition occurs at the nucleation temperature $T_n = 0.72$.

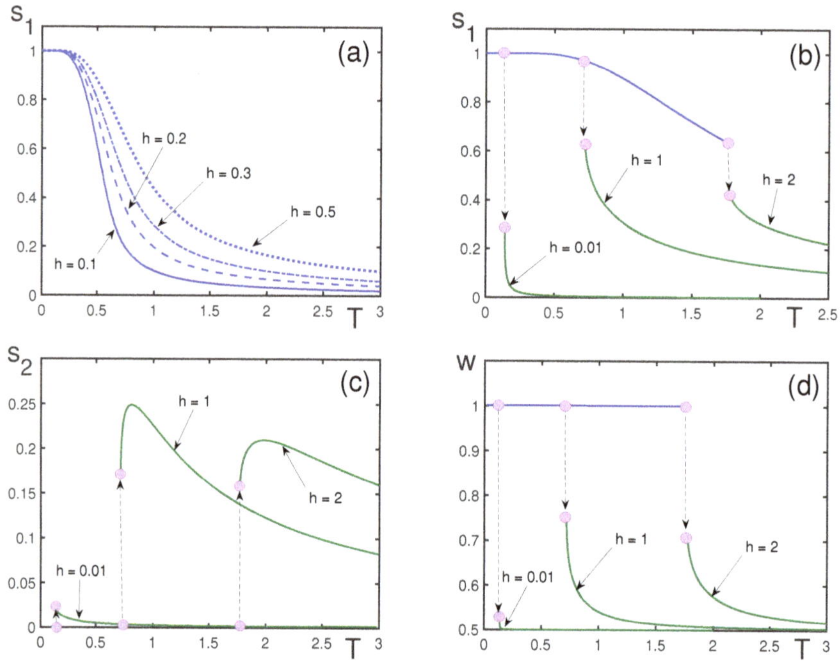

Figure 6. Order parameters s_1, s_2, and w as functions of dimensionless temperature T, for $u = 0.8$ and different fields: (**a**) $h = 0.1$ (solid line), $h = 0.2$ (dashed line); $h = 0.3$ (dash–dotted line); $h = 0.5$ (dotted line); (**b**) $h = 0.01$; $h = 1$; $h = 2$. The corresponding nucleation temperatures are $T_n = 0.14$, $T_n = 0.72$, and $T_n = 1.77$; (**c**) $h = 0.01$; $h = 1$; $h = 2$; (**d**) $h = 0.01$; $h = 1$; $h = 2$.

In this way, the nucleation transition is the transition of a system from a pure phase into a mixed phase. In the considered case, this is the transition between the pure ferromagnetic phase and a mixed state, where ferromagnetic regions start coexisting with paramagnetic fluctuations.

As follows from the figures, the zeroth-order nucleation transition is accompanied by the abrupt appearance inside the ferromagnetic phase of a finite concentration of nanoscale paramagnetic regions. Hence, when the concentration of the paramagnetic admixture does not continuously grow from zero but increases by a jump, this suggests the possible occurrence of a zeroth-order nucleation transition. The appearance of paramagnetic regions can be noticed by means of Mössbauer experiments.

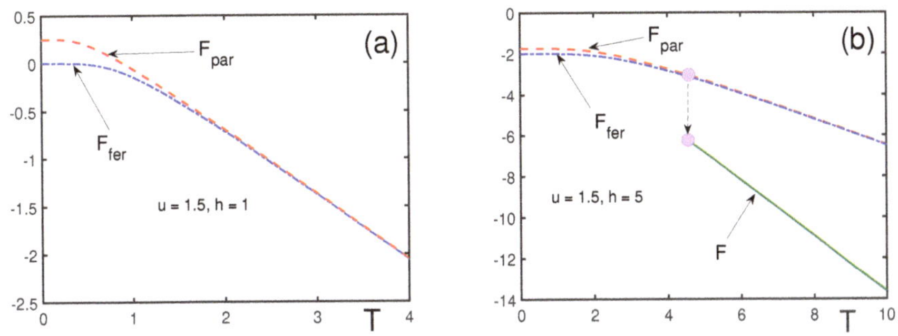

Figure 7. Free energies of the mixed state, F (solid line), ferromagnetic state, F_{fer} (dash–dotted line), and of the paramagnetic state, F_{par} (dashed line), for $u = 1.5$ and different magnetic fields: (**a**) $h = 1$; (**b**) $h = 5$. For $h = 1$, the mixed state is not stable. For $h = 5$, the zeroth-order nucleation transition occurs at the nucleation temperature $T_n = 4.55$.

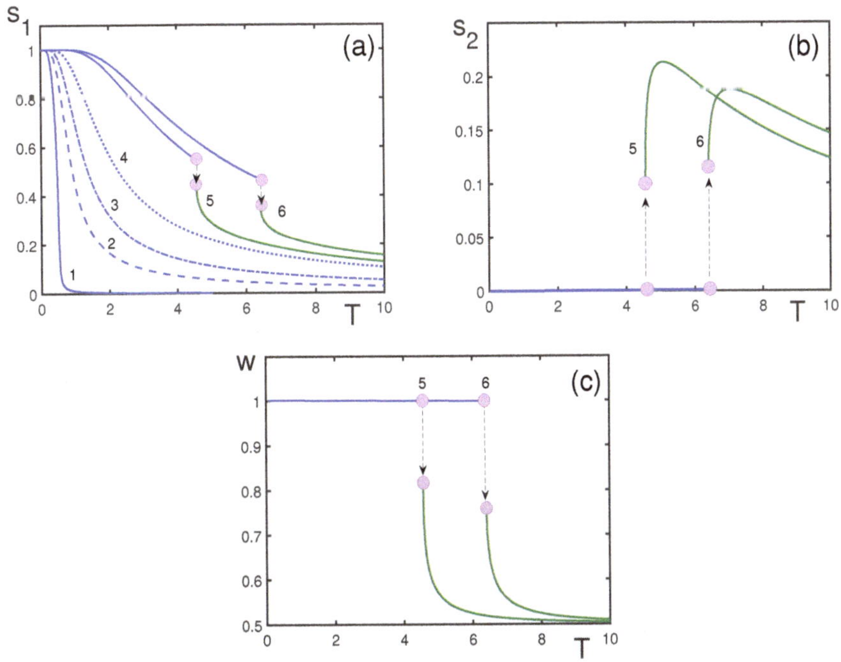

Figure 8. Order parameters s_1 (**a**), s_2 (**b**), and w as functions of dimensionless temperature T (**c**), for $u = 1.5$ and different fields: (1) $h = 0.01$; (2) $h = 0.5$; (3) $h = 1$; (4) $h = 2$; (5) $h = 5$; (6) $h = 6$. The nucleation temperatures are $T_n = 4.55$ for $h = 5$ and $T_n = 6.4$ for $h = 6$.

We show, numerically, that the nucleation transition can be of zeroth order. From one side, this could be a consequence of approximations involved in the process of calculations. From the other side, strictly speaking, nucleation is not a typical phase transition, because of which, it is not compulsorily required to be classified as either first or second order. Although, as is discussed in the Introduction, there are works showing that even classical phase transitions could be of zeroth order. Even more so is allowed for such a non-classical transition as a nucleation transition.

Author Contributions: V.I.Y. and E.P.Y. equally contributed to this review. All authors have read and agreed to the published version of the manuscript.

Funding: This research received no external funding.

Institutional Review Board Statement: Not applicable.

Informed Consent Statement: Not applicable.

Data Availability Statement: Not applicable.

Conflicts of Interest: The authors declare no conflict of interest.

References

1. Kubo, R. *Thermodynamics;* North-Holland: Amsterdam, The Netherlands, 1968.
2. Maslov, V.P. Zeroth-order phase transitions. *Math. Notes* **2004**, *76*, 697–710. [CrossRef]
3. Gunasekaran, S.; Kubiznak, D.; Mann, R.B. Extended phase space thermodynamics for charged and rotating black holes and Born-Infeld vacuum polarization. *J. High Energy Phys.* **2012**, *2012*, 110. [CrossRef]
4. Altamirano, N.; Kubiznak, D.; Mann, R.B. Reentrant phase transitions in rotating anti-de Sitter black holes. *Phys. Rev. D* **2013**, *88*, 101502. [CrossRef]
5. Altamirano, N.; Kubiznak, D.; Mann, R.B.; Sherkatghanad, Z. Kerr-AdS analogue of triple point and solid/liquid/gas phase transition. *Class. Quantum Grav.* **2014**, *31*, 042001. [CrossRef]
6. Frassino, A.M.; Kubiznak, D.; Mann, R.B.; Simovic, F. Multiple reentrant phase transitions and triple points in Lovelock thermodynamics. *J. High Energy Phys.* **2014**, *2014*, 1–47. [CrossRef]
7. Hennigar, R.A.; Brenna, W.G.; Mann, R.B. $P-V$ criticality in quasitopological gravity. *J. High Energy Phys.* **2015**, *2015*, 77. [CrossRef]
8. Hennigar R.A.; Mann, R.B. Reentrant phase transitions and van der Waals behaviour for hairy black holes. *Entropy* **2015**, *17*, 8056. [CrossRef]
9. Kubiznak, D.; Simovic, F. Thermodynamics of horizons: De Sitter black holes and reentrant phase transitions. *Class. Quantum Grav.* **2016**, *33*, 245001. [CrossRef]
10. Dehyadegari, A.; Sheykhi, A.; Montakhab, A. Novel phase transition in charged dilaton black holes. *Phys. Rev. D* **2017**, *96*, 084012. [CrossRef]
11. Dehyadegari A.; Sheykhi, A. Reentrant phase transition of Born-Infeld-AdS black holes. *Phys. Rev. D* **2018**, *98*, 024011. [CrossRef]
12. Zeng, H.B. Possible Anderson localization in a holographic superconductor. *Phys. Rev. D* **2013**, *88*, 126004. [CrossRef]
13. Cai, R.G.; Li, L.; Li, L.F. A holographic *P*-wave superconductor model. *J. High. Energy Phys.* **2014**, *1401*, 032. [CrossRef]
14. Cai, R.G.; Li, L.; Li, L.F.; Yang, R.Q. Towards complete phase diagrams of a holographic *P*-wave superconductor model. *J. High. Energy Phys.* **2014**, *1404*, 016. [CrossRef]
15. Zeng, H.B.; Zhang, H.Q. Zeroth order phase transition in a holographic superconductor with single impurity. *Nucl. Phys. B* **2015**, *897*, 276–288. [CrossRef]
16. Cai, R.G.; Yang, R.Q. Paramagnetism-ferromagnetism phase transition in a dyonic black hole. *Phys. Rev. D* **2014**, *90*, 081901. [CrossRef]
17. Cai, R.G.; Yang, R.Q. Holographic model for the paramagnetism-antiferromagnetism phase transition. *Phys. Rev. D* **2015**, *91*, 086001. [CrossRef]
18. Hou, J.X. Zeroth order phase transition induced by ergodicity breaking in a mean-field system. *Eur. Phys. J. B* **2021**, *94*, 6. [CrossRef]
19. Hou, J.X. Zeroth-order phase transition in the Blume-Emery-Griffiths model without bilinear exchange coupling. *Phys. Rev. E* **2021**, *104*, 024114. [CrossRef]
20. Touchette, H. Equivalence and nonequivalence of ensembles: Thermodynamic, macrostate, and measure levels. *J. Stat. Phys.* **2015**, *159*, 987–1016. [CrossRef]
21. Bar, T.; Choudhary, S.K.; Ashraf, M.A.; Sujith, K.S.; Puri, S.; Raj, S.; Bansal, B. Kinetic spinodal instabilities in the Mott transition in V_2O_3: Evidence from hysteresis scaling and dissipative phase ordering. *Phys. Rev. Lett.* **2018**, *121*, 045701. [CrossRef]
22. Kundu, S.; Bar, T.; Nayak, R.K.; Bansal, B. Critical slowing down at the abrupt Mott transition: When the first-order phase transition becomes zeroth order and looks like second order. *Phys. Rev. Lett.* **2020**, *124*, 095703. [CrossRef] [PubMed]
23. Liu, S.; Phillabaum, B.; Carlson, E.W.; Dahmen, K.A.; Vidhyadhiraja, N.S.; Qazilbash, M.M.; Basov, D.N. Random field driven spatial complexity at the Mott transition in VO_2. *Phys. Rev. Lett.* **2016**, *116*, 036401. [CrossRef]
24. Yukalov, V.I. Phase transitions and heterophase fluctuations. *Phys. Rep.* **1991**, *208*, 395–489. [CrossRef]
25. Yukalov, V.I. Mesoscopic phase fluctuations: General phenomenon in condensed matter. *Int. J. Mod. Phys. B* **2003**, *17*, 2333–2358. [CrossRef]

26. Yukalov, V.I. Systems with symmetry breaking and restoration. *Symmetry* **2010**, *2*, 40–68. [CrossRef]
27. Kagan, M.Y.; Kugel, K.I.; Rakhmanov, A.L. Electronic phase separation: Recent progress in the old problem. *arXiv* **2021**, arXiv:2109.12684.
28. Kuklov, A.B.; Prokof'ev, N.V.; Svistunov, B.V. How solid is supersolid? *Physics* **2011**, *4*, 109. [CrossRef]
29. Boninsegni, M.; Prokof'ev, N.V. Supersolids: What and where are they? *Rev. Mod. Phys.* **2012**, *84*, 759–776. [CrossRef]
30. Yukalov, V.I. Saga of superfluid solids. *Physics* **2020**, *2*, 6. [CrossRef]
31. Buggy, N.C.; Du, Y.; Kuo, M.C.; Seifert, S.; Gasvoda, R.J.; Agarwal, S.; Coughlin, E.B.; Herring, A.M. Designing anion-exchange ionomers with oriented nanoscale phase separation at a Silver interface. *J. Phys. Chem. C* **2021**, *125*, 20592–20605. [CrossRef]
32. Batys, P.; Fedorov, D.; Mohammadi, P.; Lemetti, L.; Linder, M.B.; Sammalkorpi, M. Self-assembly of silk-like protein into nanoscale bicontinuous networks under phase-separation conditions. *Biomacromolecules* **2021**, *22*, 690–700. [CrossRef]
33. Park, J.; Zhao, H.; Kang, S.D. Fictitious phase separation in Li layered oxides driven by electro-autocatalysis. *Nat. Mater.* **2021**, *20*, 991–999. [CrossRef] [PubMed]
34. Martin, E.W.; Harmon, T.S.; Hopkins, J.B. A multi-step nucleation process determines the kinetics of prion-like domain phase separation. *Nat. Commun.* **2021**, *12*, 4513. [CrossRef] [PubMed]
35. Kyndiah, A.; Checa, M.; Leonardi, F.; Millan-Solsona, R.; Di, M.; Tanwar, S.; Fumagalli, L.; Mas-Torrent, M.; Gomila, G. Nanoscale mapping of the conductivity and interfacial capacitance of an electrolyte-gated organic field-effect transistor under operation. *Funct. Mater.* **2021**, *31*, 2008032. [CrossRef]
36. Bertelsen, U.; Knudsen, J.; Krogh, H. Mössbauer effect in FeF_3. *Phys. Status Solidi* **1967**, *22*, 59–64. [CrossRef]
37. Yamamoto, H.; Osaka, T.; Watanabe, H.; Fukase, M. Mössbauer effect study of spin relaxation in $CaFe_2O_4$. *J. Phys. Soc. Jpn.* **1968**, *24*, 275–279. [CrossRef]
38. Eibschütz, M.; Shtrikman, S.; Treves, D. Mössbauer studies of Fe^{57} in orthoferrites. *Phys. Rev.* **1967**, *156*, 562–577. [CrossRef]
39. Levinson, L.; Luban, M.; Shtrikman, S. Mössbauer studies of Fe^{57} near the Curie temperature. *Phys. Rev.* **1969**, *177*, 864–870. [CrossRef]
40. Krivoglaz, M.A. Fluctuon states of electrons. *Phys. Usp.* **1974**, *16*, 856–877. [CrossRef]
41. Nagaev, E.L. Ferromagnetic and antiferromagnetic semiconductors. *Phys. Usp.* **1975**, *18*, 863–892. [CrossRef]
42. Belov, K.P. *Rare-Earth Magnets and Their Applications*; Nauka: Moscow, Russia, 1980.
43. Belov, K.P.; Tretyakov, Y.D.; Gordeev, I.V.; Koroleva, L.I.; Kesler Y.A. *Magnetic Semiconductors–Chalcogenide Spinels*; Moscow State University: Moscow, Russia, 1981.
44. Reissner, M.; Steiner, W.; Kappler, J.; Bauer, P.; Besnus, M. Magnetic behaviour of $Y(Fe_xAl_{1-x})_2$ alloys. *J. Phys. F* **1984**, *14*, 1249–1260. [CrossRef]
45. Kumeishin, V.F.; Ivanov, O.A. Investigation of relaxation processes in nickel near the Curie temperature by means of nuclear gamma-resonance. *Phys. Met. Metallogr.* **1975**, *40*, 1295–1299.
46. Baryakhtar, V.G.; Vitebsky, I.M.; Yablonsky, D.A. Theory of creation of nuclei at magnetic first order phase tarnsition between paramagnetic and magneto-ordered phases. *Phys. Solid State.* **1977**, *19*, 347–352.
47. Goldman, M. Nuclear dipolar magnetic ordering. *Phys. Rep.* **1977**, *32*, 1–47. [CrossRef]
48. Reimann, H.; Hagen, H.; Waldner, F.; Arend, H. Observation of excitation of the antiferromagnetic mode in the paramagnetic state of $(C_2H_5NH_3)_2CuCl_4$. *Phys. Rev. Lett.* **1978**, *40*, 1344–1346. [CrossRef]
49. Bhargava, S.; Zeeman, N. Mössbauer study of $Ni_{0.25}Zn_{0.75}Fe_2O_4$: Spin fluctuations. *Phys. Rev. B* **1980**, *21*, 1717–1724. [CrossRef]
50. Uen, T.M.; Tseng, P.K. Mössbauer-effect studies on the magnetic properties of the Ni-Zn-ferrite system. *Phys. Rev. B* **1982**, *25*, 1848–1859. [CrossRef]
51. Srivastava, J.; Muraleedharan, K.; Vijayaraghavan, R. On anomalous Mössbauer spectra in spinel ferrites. *Phys. Lett. A* **1984**, *104*, 482–486. [CrossRef]
52. Hälg, B.; Furrer, A.; Vogt, O. Coexistence of different short-range-ordered spin fluctuations in $Ce_{1-x}(LaY)_xSb$. *Phys. Rev. Lett.* **1985**, *54*, 1388–1391. [CrossRef] [PubMed]
53. Lynn, J.W. Temperature dependence of the magnetic excitations in iron. *Phys. Rev. B* **1975**, *11*, 2624–2637. [CrossRef]
54. Liu, S.H. Magnetic excitations above the critical temperature. *Phys. Rev. B* **1976**, *13*, 2979–2985. [CrossRef]
55. Lynn, J.W.; Mook, H.A. Temperature dependence of the dynamic susceptibiity of nickel. *Phys. Rev. B* **1981**, *23*, 198–206. [CrossRef]
56. Cable, J.W.; Nicklow, R.M.; Wakabayashi, N. Temperature dependence of the magnetic excitations in gadolinium. *Phys. Rev. B* **1986**, *32*, 1710–1719. [CrossRef] [PubMed]
57. Lynn, J.W.; Mook, H.A. Nature of the magnetic excitations above T_c in Ni anf Fe. *J. Magn. Magn. Mater.* **1986**, *54*, 1169–1170. [CrossRef]
58. Cable, J.W.; Nicklow, R.M. Spin dynamics of Gd at high temperatures. *Phys. Rev. B* **1989**, *39*, 11732–11741. [CrossRef]
59. Goto, T.; Sakakabara, T.; Yamaguchi, M. Coexistence of nonmagnetic and ferromagnetic Co in Y_2Co_7 and YCo_3 hydrides. *J. Magn. Magn. Mater.* **1986**, *54*, 1085–1086. [CrossRef]
60. Shinogi, A.; Saito, T.; Endo, K. Coexistence of nonmagnetic and magnetic Co in cubic Laves phase compounds $Lu(Co_{1-x}Al_x)_2$. *J. Phys. Soc. Jpn.* **1987**, *56*, 2633–2636. [CrossRef]
61. Jaime, M.; Lin, P.; Chun, S.H.; Salamon, M.B.; Dorsey, P.; Rubinshtein, M. Coexistence of localized and itinerant carriers near T_C in calcium-doped manganites. *Phys. Rev. B* **1999**, *60*, 1028–1032. [CrossRef]

62. Merithew, R.D.; Weissman, M.B.; Hess, F.M.; Spradling, P.; Nowak, E.R.; O'Donnell, J.; Ekstein, J.M.; Tokura, Y.; Tomioka, Y. Mesoscopic thermodynamics of an inhomogeneous colossal-magnetoresistive phase. *Phys. Rev. Lett.* **2000**, *84*, 3442–3445. [CrossRef]
63. Baio, J.; Barucca, G.; Caciuffo, R.; Rinaldi, D.; Mira, J.; Rivas, J.; Senaris-Rodriguez, M.A.; Fiorani, D. Phase separation, thermal history and magnetic behavior of Sr dopped $LaCoO_3$. *J. Phys. Condens. Matter* **2000**, *12*, 9761–9770. [CrossRef]
64. Sun, J.R.; Shen, B.G.; Gao, J.; Fei, Y.; Nie, Y.P. Presence of a paramagnetic phase well below the ferromagnetic onset in $La_{0.67-x}Bi_xCa_{0.33}MnO_3$. *Eur. Phys. Lett.* **2003**, *62*, 732–738. [CrossRef]
65. Batko, I.; Batkova, M.; Tran, V.H.; Keiderling, U.; Filipov, V.B. Evidence for magnetic phase separation in colossal magnetoresistence compound $EuB_{5.99}C_{0.01}$. *Solid State Commun.* **2014**, *190*, 23–27. [CrossRef]
66. Gibbs, J.W. *Collected Works*; Longmans: New York, NY, USA, 1928.
67. Kullback, S.; Leibler, R.A. On information and sufficiency. *Ann. Math. Stat.* **1951**, *22*, 79–86. [CrossRef]
68. Kullback, S. *Information Theory and Statistics*; Wiley: New York, NY, USA, 1959.
69. Yukalov, V.I. Theory of melting and crystallization. *Phys. Rev. B* **1985**, *32*, 436–446. [CrossRef]
70. Yukalov, V.I. Effective Hamiltonians for systems with mixed symmetry. *Physica A* **1986**, *136*, 575–587. [CrossRef]
71. Yukalov, V.I. Renormalization of quasi-Hamiltonians under heterophase averaging. *Phys. Lett. A* **1987**, *125*, 95–100. [CrossRef]
72. Yukalov, V.I. Procedure of quasiaveraging for heterophase mixtures. *Physica A* **1987**, *141*, 352–374. [CrossRef]
73. Yukalov, V.I. Lattice mixtures of fluctuating phases. *Physica A* **1987**, *144*, 369–389. [CrossRef]
74. Marzari, N.; Mostofi, A.A.; Yates, J.R.; Souza, I.; Vanderbilt, D. Maximally localized Wannier functions: Theory and applications. *Rev. Mod. Phys.* **2012**, *84*, 1419–1475. [CrossRef]
75. Bogolubov, N.N. *Lectures on Quantum Statistics*; Gordon and Breach: New York, NY, USA, 1967; Volume 1.
76. Bogolubov, N.N. *Lectures on Quantum Statistics*; Gordon and Breach: New York, NY, USA, 1970; Volume 2.
77. Bogolubov, N.N. *Quantum Statistical Mechanics*; World Scientific: Singapore, 2015.
78. Yukalov, V.I. Microscopic theory of spin reorientations: General analysis. *Physica A* **1990**, *167*, 861–876. [CrossRef]
79. Yukalov, V.I.; Yukalova, E.P. Statistics of multiscale fluctuations in macromolecular systems. *J. Phys. Chem. B* **2012**, *116*, 8435–8448. [CrossRef] [PubMed]
80. Yukalov, V.I.; Yukalova, E.P. Mesoscopic disorder in double well optical lattices. *Laser Phys.* **2011**, *21*, 1448–1458. [CrossRef]
81. Ono, S.; Kondo, S. *Molecular Theory of Surface Tension in Liquids*; Springer: Berlin, Germany, 1960.
82. Rusanov, A.I. Problems of surface thermodynamics. *Pure Appl. Chem.* **1992**, *64*, 111–124. [CrossRef]
83. Kjelstrup, S.; Bedeaux, D. *Non-Equilibrium Thermodynamics of Heterogeneous Systems*; World Scientific: Singapore, 2008.

symmetry

MDPI

Article

Small-Angle Scattering from Fractional Brownian Surfaces

Eugen Mircea Anitas [1,2]

[1] Joint Institute for Nuclear Research, 141980 Dubna, Russia; anitas@theor.jinr.ru
[2] Horia Hulubei, National Institute of Physics and Nuclear Engineering, Bucharest-Magurele, RO-077125 Bucharest, Romania

Abstract: Recent developments in nanotechnology have allowed the fabrication of a new generation of advanced materials with various fractal-like geometries. Fractional Brownian surfaces (fBs) are often used as models to simulate and characterize these complex geometries, such as the surface of particles in dilute particulate systems (e.g., colloids) or the interfaces in non-particulate two-phase systems (e.g., semicrystalline polymers with crystalline and amorphous phases). However, for such systems, a realistic simulation involves parameters averaged over a macroscopic volume. Here, a method based on small-angle scattering technique is proposed to extract the main structural parameters of surfaces/interfaces from experimental data. It involves the analysis of scattering intensities and the corresponding pair distance distribution functions. This allows the extraction of information with respect to the overall size, fractal dimension, Hurst and spectral exponents. The method is applied to several classes of fBs, and it is shown that the obtained numerical values of the structural parameters are in very good agreement with theoretical ones.

Keywords: small-angle scattering; fractional Brownian surfaces; fractal dimension; Hurst exponent; spectral exponent

Citation: Anitas, E.M. Small-Angle Scattering from Fractional Brownian Surfaces. *Symmetry* **2021**, *13*, 2042. https://doi.org/10.3390/sym13112042

Academic Editors: V. I. Yukalov, V. S. Bagnato, Rashid G. Nazmitdinov and Raffaele Barretta

Received: 15 September 2021
Accepted: 28 October 2021
Published: 30 October 2021

1. Introduction

In recent years, various advanced techniques, such as printing [1,2], gas-phase (aerosol) synthesis [3] or powder compact foaming [4], have been developed for large-scale preparation of fractal materials at nano and micro scales. An important feature for many applications is the possibility to control the surface properties of such materials and of its components as well as their symmetry and dimensionality, since it allows incorporating advanced functionalities at a design stage.

For artificially created structures, interface roughness affects coherent dynamical processes in quantum dots [5] while for solar cells, it increases their power conversion efficiency [6]. For natural rough surfaces arising in materials science, chemistry, biology or geology, the roughness is often useful for tuning superhydrophobicity [7], biocompatibility [8,9] or flexibility [7,10]. In addition to roughness, the third dimension has been proved to be important for the interpretation of experimental data on singlet-triplet transitions in the ground states of the two-electron quantum dots under a perpendicular magnetic field [11,12].

Theoretically, for both artificial and natural surfaces and interfaces, a frequently employed realistic model that aims to relate the observed physical/chemical/biological properties with the roughness is based on the concept of fractional Brownian surface (fBs) [13]. This has been successfully used in describing various rough structures, including the contact zone between two distinct materials in layered composites [14], substrates subjected to plasma-chemical etching [15] or soil structures [16,17].

A fBs is defined in terms of the Hurst exponent, and it is related to the fractal dimension [18] of the surface. This is one of the most fundamental parameters characterizing a surface since it does not depend on the sampling length or on the instrument resolution. Therefore, various methods for practical estimations of fractal dimension are commonly used, such as the wavelet based multifractal analysis [19], the root mean square method [20], the variogram method [21], the structure function method [22] or the variation method [23].

However, for macroscopic volumes consisting of a large number of randomly distributed surfaces/interfaces, such real-space methods are not appropriate since the underlying microscopy techniques may introduce artefacts in sample preparation and can be used to provide only information for small surface areas. To overcome this issue, one can describe the average of the correlations among atoms positions by using an appropriate statistical or ensemble average of the electron density distribution within the particle [24].

In this paper, small-angle scattering (SAS) technique is used to estimate the overall shape, size and fractal dimension of 2D disordered fBs. This involves an analysis of scattering intensity $I(q)$ and its Fourier transform and the pair distance distribution function (pddf; $p(r)$). Depending on the parameters sought, either $I(q)$ or $p(r)$ may be more convenient for detailed analysis [25]. While symmetry and self-similarity characteristics provide more pronounced effects in the reciprocal space, the determination of the shape and size is more intuitive by using $p(r)$. As such, in this work, the advantages provided by both analyses are exploited.

In order to illustrate the general applicability of employing SAS technique in revealing structural properties of naturally occuring (i.e., statistically self-similar/affine) fractal surfaces/interfaces, in this work, several classes of fBs observed on a regular grid are investigated. It is shown that the obtained fractal dimensions provide values for the Hurst and spectral exponents in very good agreement with theoretical ones. The main steps on how $I(q)$ (and $p(r)$) can be employed to differentiate between various fBs are described in detail, and similarities with SAS from exact self-similar surface fractals [26–28] are highlighted.

2. Theoretical Background

SASs of X-rays (SAXS) or neutrons (SANS) are experimental techniques used for the investigation of structures with dimensions from 1 nm up to several hundreds of nanometers [29]. In the case of SAXS, the incoming wave induce dipole oscillations in the atoms, and the electrons are excited due to the high energy of X-rays. In turn, the accelerated charges generate secondary waves, which then add up at large distances and provide the scattering amplitude. This is related to the electron density distribution of the scattering object by a Fourier transform. However, in a scattering experiment, due to the high frequency, only the square of the amplitudes (scattering intensities) are recorded as a function of the scattering angle [30].

In SANS, neutrons interact with the nuclei of the atoms and with unpaired electrons, and they are sensitive to the isotopic composition of the sample. Neutrons can be used as a magnetic probe (since they posses a magnetic moment), and this allows us to investigate bulk properties of matter (due to their weak interaction with matter) [30]. In contrast to SAXS, where scattering amplitudes increase regularly with atomic number, in SANS, neutron-coherent amplitudes vary irregularly and are related by a Fourier transform to the scattering length density distribution [31].

Therefore, SAXS and SANS have their own advantages depending on the sample investigated. In particular, SANS is often used in combination with contrast-variation to probe the structure of multicomponent macromolecular complexes. In the following, the theoretical background is focused on SAXS but it applies to SANS as well when electron density is replaced by a scattering length density distribution.

2.1. Small-Angle Scattering Technique

In SAS, the differential elastic cross-section per unit angle, i.e., the scattering intensity, is obtained as the product between the scattering amplitude $A(\boldsymbol{q})$ of the irradiated volume and its complex conjugate $A(\boldsymbol{q})^*$. Here, $\boldsymbol{q}$ is the scattering vector with length $q = 4\pi\lambda^{-1}\sin\theta$, λ is the radiation wavelength and 2θ is the scattering angle. Furthermore, one considers a scattering process that involves a two-phase system consisting of a large number of disordered particles with rough surfaces described by fBs (see below) embedded in a homogeneous matrix/solution.

Let us denote $\rho(\boldsymbol{r})$ as the electron density for a particle of volume V in a fixed orientation, i.e., the number of electrons per unit volume at position $\boldsymbol{r}$. Then, the scattered amplitude can be written as follows: $A(\boldsymbol{q}) = \int\int\int \rho(\boldsymbol{r}) \exp(-\mathrm{i}\boldsymbol{q} \cdot \boldsymbol{r}) \mathrm{d}V$, where $\mathrm{d}V$ is a small volume element situated at position $\boldsymbol{r}$ and which contains $\rho(\boldsymbol{r})\mathrm{d}V$ electrons. Therefore, the scattering intensity becomes the following [30]:

$$I(\boldsymbol{q}) \equiv A(\boldsymbol{q})A(\boldsymbol{q})^* = \int\int\int \widetilde{\rho}^2(\boldsymbol{r}) \exp(-\mathrm{i}\boldsymbol{q} \cdot \boldsymbol{r}) \mathrm{d}V, \quad (1)$$

where $\widetilde{\rho}^2(\boldsymbol{r}) = \int\int\int \rho(\boldsymbol{r}_1)\rho(\boldsymbol{r}_1 - \boldsymbol{r}) \mathrm{d}V$ is the convolution square [30,32].

In order to take into account the contribution of the matrix/solvent, the electron density in Equation (1) shall be replaced by the difference between electron densities of the particle and that of the matrix/solvent (ρ_0), i.e., by $\Delta\rho = \rho - \rho_0$. This is also known in the literature as the *contrast*. The random orientations of the particles over orientations are taken into account such that $\langle \exp(-\mathrm{i}\boldsymbol{q}\boldsymbol{r}) \rangle = \sin qr/qr$. This results in the following [30]:

$$I(q) = 4\pi \int_0^\infty p(r) \frac{\sin qr}{qr} \mathrm{d}r, \quad (2)$$

where $p(r) = r^2 \Delta\widetilde{\rho}^2(r)$ is the pair distance distribution function (pddf) and provides the number of different electron pairs found in the range $(r, r + \mathrm{d}r)$ within the particle. Geometrically, $p(r)$ is the distance histogram of the particle and has the property that $p(r) = 0$ at $r = 0$ and at $r > D_{\mathrm{max}}$, where D_{max} is the maximum dimension of the particle. For a finite number N of point-like scatterers, Equation (2) can be approximated by the following [33]:

$$I(q) = N + 2 \sum_i^{N_{\mathrm{bin}}} p(r_i) \frac{\sin qr_i}{qr_i}, \quad (3)$$

where N_{bin} is the number of bins, and $p(r_i)$ is the population at pair distance r_i. This approach brings an important computational advantage since it can handle systems consisting of a large number of scatterers in reasonable timescales [33].

2.2. Small-Angle Scattering from Fractal Surfaces

Within the class of fractal surfaces, one distinguishes three main subclasses of fractals. Figure 1 provides a schematic illustration for 2D case for each subclass. The first subclass (Figure 1 left) consists of a dense object with a fractal surface. The corresponding fractal dimension of the mass is $D_{\mathrm{m}} = 2$, the fractal dimension of the surface is $1 < D_{\mathrm{s}} < 2$ and the fractal dimension of the pores (i.e., the surrounding) is $D_{\mathrm{p}} = 2$ as for the mass. When $D_{\mathrm{s}} \to 1$, the surface is perfectly smooth, while for $D_{\mathrm{s}} \to 2$, the surface is so folded that it almost completely fills the plane. They are known in the literature as surface fractals. Such surfaces are specific to erosion surfaces (materials or mountains), chemically dissolved surfaces, thin films, corrosion surfaces, fractures, etc. [34].

The second subclass (Figure 1 middle) is a fractal resembling a branched cluster or network and for which its surface is also a fractal. For this configuration, we have $D_{\mathrm{m}} = D_{\mathrm{s}} < 2$ and $D_{\mathrm{p}} = 2$. The higher the value of D_{m}, the more close the structure is, while for $D_{\mathrm{m}} \to 1$, the object becomes a line. They are known as mass fractals, and they are specific to polymer chains or various types of aggregates (carbon, soils, etc.).

The third class (Figure 1 right) is also a dense object but within which there exists a distribution of pores or holes with a fractal structure. This is called a pore fractal, and it has the properties that $D_{\mathrm{m}} = 2$ and $D_{\mathrm{s}} = D_{\mathrm{p}} < 2$. The higher the value of D_{p}, the more porous the structure becomes. As the name implies, they are specific to various porous structures, such as carbon nanopores, bituminous coals, etc.

Figure 1. Schematic representation of the three main classes of fractals that can be described in a SAS experiment. (**Left**) Surface fractal ($D_m = D_p = d$ and $D_s < d$). (**Middle**) Mass fractal ($D_s = D_m < d$ and $D_p = d$). (**Right**) Pore/volume fractal ($D_s = D_p < d$ and $D_m = d$). Here, $d = 2$, and it represents the Euclidean dimension of the embedding space. See main text for details.

The fractal dimension of the fractal surface is related to the power-law behaviour of the scattering intensity (Equations (1)–(3)) by [35,36].

$$I(q) \propto q^{-\alpha}, \tag{4}$$

Here, α is the scattering exponent and carries out information about the fractal dimension: $\alpha = D_m$ for mass fractals [37], $\alpha = 4 - D_s$ [38] for surface fractals and $\alpha = D_p$ for pore fractals [39]. These relations allow inferring the type of fractal from SAS data: if the measured scattering exponent of $I(q)$ vs. q is smaller than two, then we deal with a mass fractal, while if it is higher than two (but smaller than three), we have scattering from a surface fractal [35,36].

2.3. Fractional Brownian Surfaces

Let us consider a two dimensional Euclidean space. A fBs V_H on $\mathbb{R}^2$ is a function for which its increments have a Gaussian distribution with the variance of the following [40]:

$$\left\langle |V_H(\boldsymbol{x}) - V_H(\boldsymbol{y})|^2 \right\rangle \propto |\boldsymbol{x} - \boldsymbol{y}|^{2H}, \tag{5}$$

where $\langle \cdots \rangle$ denotes an ensemble average over many samples of V_H, $0 < H < 1$ is a parameter known as a Hurst exponent and $\boldsymbol{x}, \boldsymbol{y} \in \mathbb{R}^2$. The parameter H controls the roughness of the surface: the larger its value, the smoother the surface. It is related, together with the spectral exponent β ($1 < \beta < 3$), to the fractal dimension of the surface by the following [40].

$$D = 3 - H = 2 + \frac{3 - \beta}{2}. \tag{6}$$

Here, β is useful for the determination of the spectral density or the two point autocorrelation function of V_H, which provides information about the correlations in the surface in turn.

3. Methodology for Generating the Fractional Brownian Surfaces and for Calculating the Pair Distance Distribution Function

In the present paper, fractional Brownian surfaces (fBss) are generated at different values of H (and implicitly of β) based on Equation (5) and by using the Fourier filtering method suggested in Reference [41] (Figure 2). The obtained surfaces are discretized on a rectangular grid and are recorded as elevation data relative to a plane at $z = 0$ (Figure 3). The smallest distance between grid points is denoted by l_{min}. The length of the surface in either x or y direction is denoted by a (Figure 3). In this approach, the resulting structure is a point-like distribution similar to the one shown in the middle of Figure 1 but with "branches" not limited to a single plane. Instead, they are confined into a limited range $z_{min} < z < z_{max}$, where z_{min} is the lowest elevation point, and z_{max} is the highest one (see Figure 3, middle). Therefore, the model considers the space between the elevation planes (through the heights of each point relative to the $z = 0$ plane), and fBs divides this space into two non-fractal regions. This is similar to the division of the plane into two non-fractal regions by the boundary of the disk in Figure 1 (Left). This separation gives rise to surface fractals, and the parameter D in Equation (6) is related the surface fractal dimension D_s described in Section 2.2. Note that although the resulting structure is embedded in the 3D Euclidean space, the self-similarity properties are manifested only along two directions (x and y).

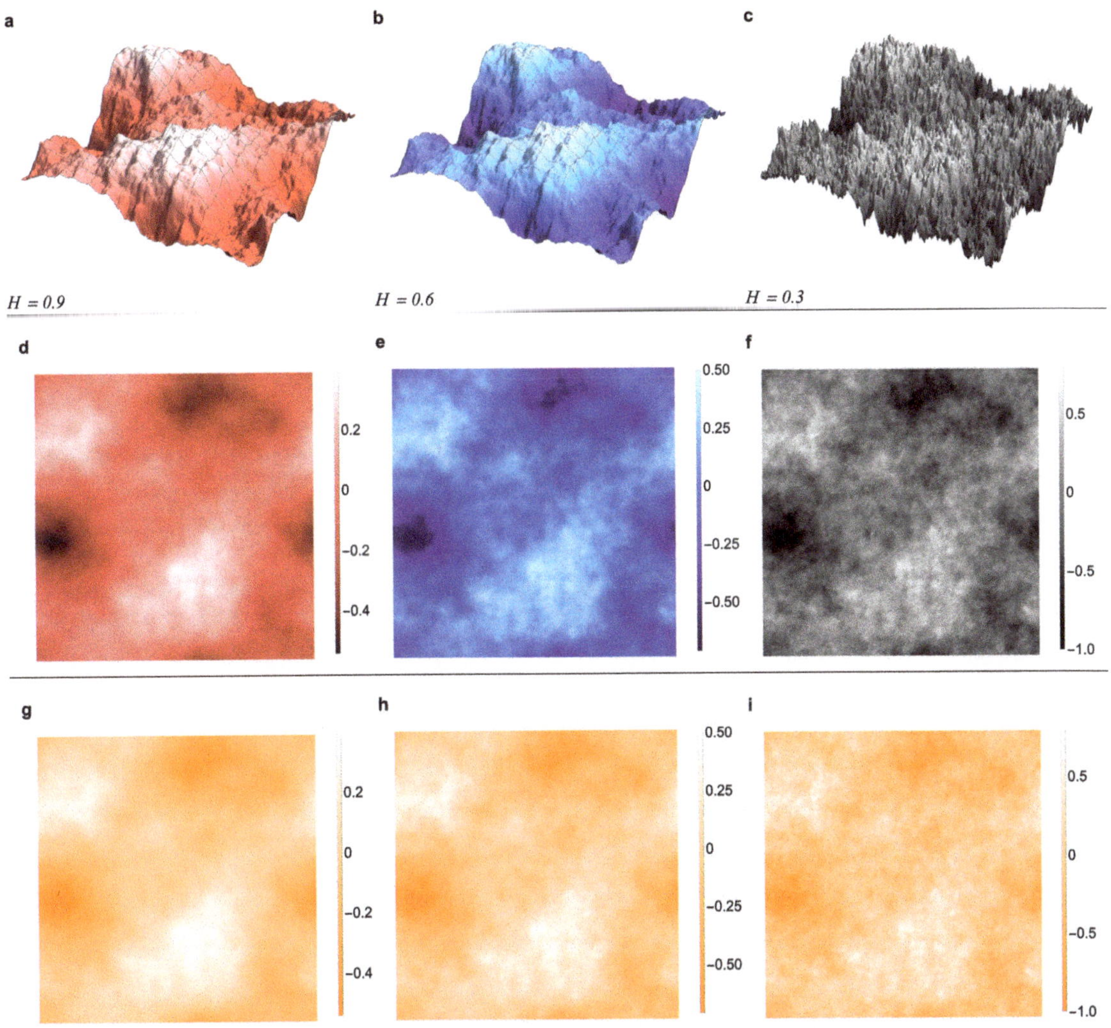

Figure 2. Fractional Brownian surfaces on a square grid with dimensions $x = y$ at various values of Hurst exponent H. (**a–c**) 3D representation. (**d–f**) Density plot. (**a**,**d**,**g**) $H = 0.9$. (**b**,**e**,**h**) $H = 0.6$. (**c**,**f**,**i**) $H = 0.3$. The peaks and bottoms are represented by light and dark regions along Oz-axis, respectively. (**g–i**) are the same as (**d–f**) but are represented in a single color for better visualization of the variation of density plot roughness. (**a**) $x \vee y = 2.31z$. (**b**) $x \vee y = 1.78z$. (**c**) $x \vee y = 1.19z$.

The geometry of the fBss, investigated here, correspond to Hurst exponents $H = 0.9$, $H = 0.6$ and, respectively, at $H = 0.3$, as shown in Figure 2. In this figure, the lighter the region, the higher the surface and vice versa. For each H, the same random seed generator was used in order to compare fBss of the same global features. As expected, the roughness of fBss increases with decreasing H (Figure 2a–c), while large scale features are preserved, i.e., the positions of maxima and minima are unchanged with H. This can be observed more clearly in the density plots in Figure 2d–f. In addition, they provide another type of method visualizing the variation of fBss roughness: the higher the value of H, the better the local variations become in terms of visibility. An equivalent representation for illustrating the differences in the local structural differences in fBss is shown in Figure 2g–i where a single color is used for all H.

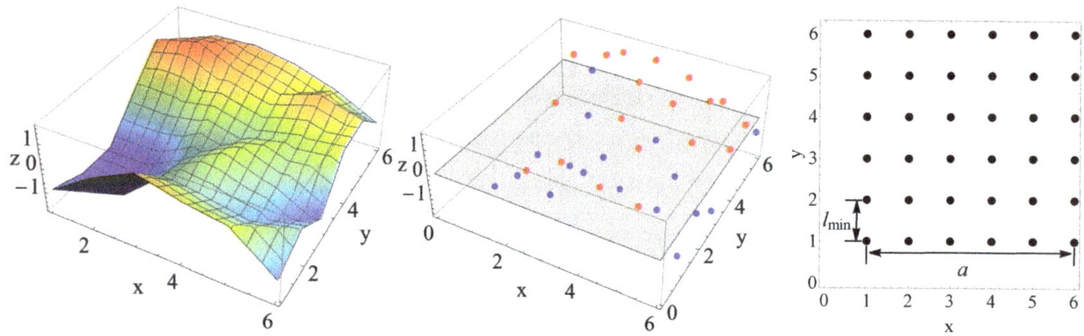

Figure 3. Schematic representation of fBS and the associated grid (6 × 6 points) at $H = 0.9$. (**Left**) fBs. Red—the highest points (at 1.5); Blue—the lowest ones (at −1.5). (**Middle**) The corresponding grid used. The gray plane is at $z = 0$ and stands as the reference level for the heights of the grid points. Red points are above the plane, and blue ones are below it. (**Right**) Projection of the grid on the 2D xy plane. l_{min} is the minimum distance between the points in the grid, and a is the length of the grid in either x or y direction.

Depending on their extension along x, y and z dimensions, fBss are simulated for three main cases:

- Class I fBss (CI): distances between points are kept unchanged; thus, x, y and z are of the same orders of magnitude. This corresponds to the classical structure of fBss, as shown in Figure 1, with a globular-like shape.
- Class II fBss (CII): distances between points are stretched by the same amount along x and y directions by a factor of b; thus, $x = y \ll z$. This gives rise to fBss with rectangular, planar-like shapes.
- Class III fBss (CIII): distances between points are stretched along a single direction by a factor of b; thus, x or $y \ll z$. This gives rise to fBss with rod-like shapes.

Therefore, in terms of the fractal dimension D in Equation (6) and on the classes considered above, one should expect a behaviour of scattering intensities characterized by different successions of power-law decays reflecting both the spatial and self-similarity symmetries of fBss. In particular, for the power-law decays arising from the self-similarity symmetry, one should expect a behaviour of the type $I(q) \propto q^{-D}$, where $D < 2$. Note that a random surface fractal can be built based on fBss by assigning a volume/area to each point such that their sizes follow a continuous power-law distribution similar to the case of deterministic surface fractals and where the scattering units within the fractal have a discrete power-law distribution of sizes [26,42].

The pddf $p(r)$ is calculated by using the distance histogram approach suggested in Reference [33]. This involves discretization of fBs (Figure 3, Middle), recording the position of each point and calculating all the distances between them. To this aim, the dimensions of the grid are equal to the maximum dimensions of fBs along its length, width and height.

Finally, the pair distances are discretized in a histogram of a bin size commensurate with the resolution of the data, and the scattering intensity is calculated by Equation (3) [33].

4. Results and Discussion

4.1. Pair-Distance Distribution Functions

The pair distance distribution functions (pddfs) at $H = 0.9$, $H = 0.6$ and $H = 0.3$ are calculated as described in Section 2.1 and are presented in Figure 4. The image size used for each class is 200×200 pixels, which corresponds to 4×10^4 point scatterers. For classes CII and CIII, the stretching factor is $b = 10$. Although the pddfs are different within each class, they, however, have a common feature that allows distinguishing fBss belonging to different classes. For class CI and CII (Figure 4a,b), the pddfs have a symmetric bell-like shape specific to globular or flat-like structures [30]. However, for class CIII (Figure 4c), the right side of the bell becomes completely linear, which is specific to elongated structures [30].

Within class CI, the pddfs of fBss show that the maximum diameter $D_{\max}$ occurs at $r/a \simeq 2.33$ for $H = 0.3$, $r/a \simeq 1.48$ for $H = 0.6$ and at $r/a \simeq 1.12$ at $H = 0.9$ (Figure 4a). This decrease in maximum dimension with increasing H arises from the contribution of elevation along the z-axis, since in all cases the dimensions along x and y axes are kept fixed ($a = 200 l_{\min}$). Such behaviour is in line with elevation data of fBss models as observed in the legends of Figure 2d–i. Here, the smallest difference in height occurs at $H = 0.9$, while the largest one occurs at $H = 0.3$. A second important feature of pddfs is that the position of maximum shifts to the left and increases in height with increasing H. It shows that the *value* of most common distances within fBss decreases as a consequence of decreasing elevation of points along z-axis. However, the height increases of pddf reflects an increase in the *number* of most common distances with H. Thus, the decrease in surface roughness gives rise to a larger number of point-distances with similar values. Note that the globular type of class CI fBss can be inferred also from the end region of pddf, which shows pronounced decay followed by a flat region.

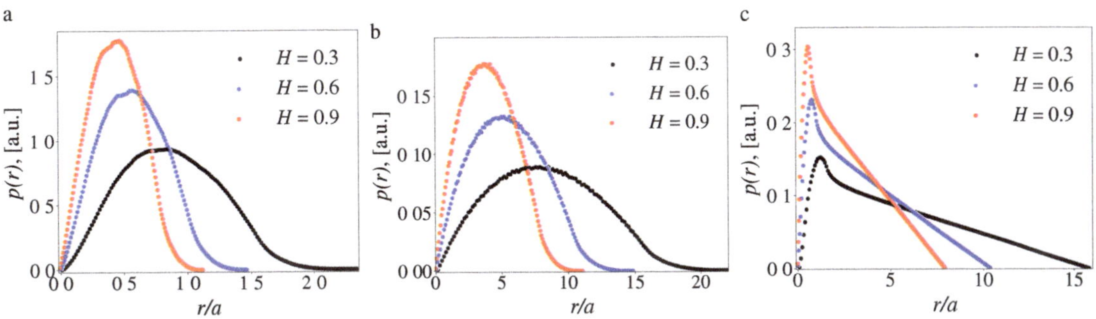

Figure 4. Pddfs from fBss at various geometries. (**a**) Class CI: symmetric bell-like curves reveal the globular-like shape. (**b**) Class CII with stretching factor $b = 10$: symmetric bell-like curves reveal planar-like structures, since one dimension is kept fixed while the other two are stretched, by a factor of $b = 10$. (**c**) Class CIII with stretching factor $b = 10$: curves with long linear domains reveal elongated structures.

The pddfs of fBss of class CII are shown in Figure 4b, and their behaviour is quite similar to fBss of class CI. However, the position and height of maxima are now related to the cross-sectional area, since they now provide the most common distances within the surface. In addition, due to the increased length sizes along x and y directions, the values of r/a at which the maximum diameter is attained also increased by a factor of b. Similarly, the number of distances decreases by a factor of b relative to fBss of class CI as a consequence of stretching the surface along the x and y directions. Note that the overall shape of the pddfs resemble quite closely those of the structures strictly confined to a plane, such as 2D DLA or surface fractals [25].

The pddfs of class CIII fBss are shown in Figure 4c and are characterized by a linear region with different slopes. One can relate it with the surface roughness of fBss: The smoother the surface, the steeper is the slope. The variation of maxima position and of its height is quite similar to class CI fBss. However, the maxima correspond to the cross section here since one can find the most common distances within them. Moreover, the curves are characterized by inflection points at $r/a \simeq 1.25$ for $H = 0.9$, $r/a \simeq 1.30$ for $H = 0.6$ and $r/a \simeq 1.35$ for $H = 0.3$, and this reveals an increase in cross-sectional area. This is also in agreement with models shown in Figure 1: The lower the H, the rougher the surface and, thus, the higher the surface area. Another particular feature for this class is that the maximum size of the surfaces significantly varies with H, i.e., $D_{\max} \simeq 16$ at $H = 0.3$, $D_{\max} \simeq 10.7$ at $H = 0.6$ and $D_{\max} \simeq 7.8$ at $H = 0.9$. One reason for this is the variation of heights along the z direction on length scales comparable with only *one* other direction (x or y). In particular, the length of z-range at $H = 0.9$ is about half of that for $H = 0.3$ (see Figure 2d,f), which is reflected in the value of their maximum sizes.

4.2. Scattering Intensities

The scattering intensities from fBss demonstrated in Figure 1 are calculated according to Equation (3) for the same classes and parameters H used for pddfs in Figure 4. The results are presented in Figure 5 on a double logarithmic scale, and they show that, within the calculated q-range, the scattering curves are characterized by the presence of a Guinier region (i.e., a region where $I(q) \propto q^0$) at $q \lesssim 2\pi/a$, followed by one or more power-law decays of the type described by Equation (4) at higher values. The scattering exponent depends on the values of H; thus, it reveals the surface roughness, while a particular succession of power-law decays or the presence of a single power-law decay is specific to the class the fBss belong to.

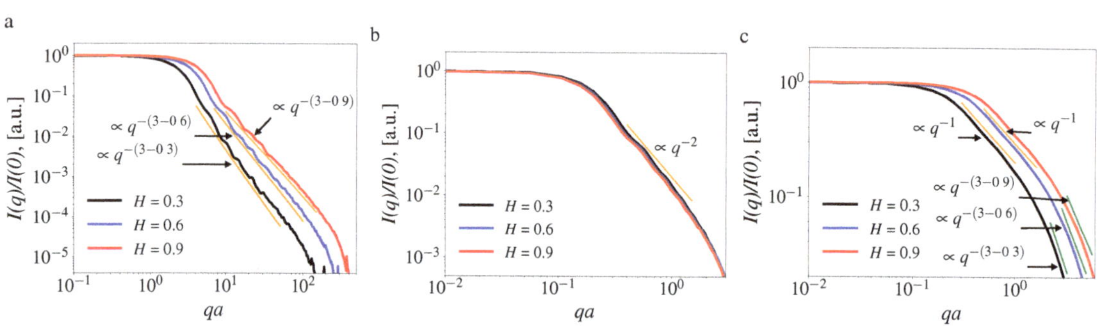

Figure 5. SAS from fBs at various grid geometries. (**a**) Three-dimensional. Fractal regions follow immediately the Guinier region (i.e., the region where $I(q) \propto q^0$) and, thus, are completely visible. (**b**) Two-dimensional. Fractal regions are expected to follow the region where $I(q) \propto q^{-2}$. (**c**) One-Dimensional. Fractal regions follow the region where $I(q) \propto q^{-1}$ and are partially visible.

The SAS intensity curves for class CI fBss are shown in Figure 5a. The results show that the length of the Guinier region increases with H and reflects the increase in the size of fBss, which is in line with the corresponding pddfs shown in Figure 4a. Then, at each H, the Guinier region is followed by a power-law decay. Here, the number of points is chosen in such a manner that the length of the power-law regimes spans at least one order of magnitude, as indicated in Reference [37]. The main feature of these power-law decays is the dependence of the scattering exponent α on H in the form $\alpha = 3 - H (\equiv D)$, and it reflects the decrease in surface roughness by decreasing the fractal dimension (see discussion in Section 2.2). This behaviour shows that the slope of scattering intensity following a power-law decay can be used to differentiate between fBss of different roughness.

For the class CII fBss, within the calculated q-range, the proposed approach show a Guinier region followed by a power-law decay $I(q) \propto q^{-2}$ for all H in the range $2\pi \lesssim qa \lesssim$

$2\pi a/d$, where $d = \sqrt{2}|z_{\text{max}} + z_{\text{min}}|/b$ is a measure of the size of the cross-sectional area. Here, z_{max} and z_{min} are given in the legends of Figure 1. However, since the dominant contribution comes from *equal* dimensions x and y in this configuration, the overall size changes insignificantly. Therefore, the corresponding scattering intensities in Guinier and $I(q) \propto q^{-2}$ power-law decay are very similar. Since $2\pi/d \simeq 35.9$ here is well beyond the investigated q-range, one expects that to observe more pronounced differences for $qa \gtrsim 35.9$ (see Figure 5c), which would allow revealing the fractal dimension of each fBs. This would require more extensive hardware resources than only a desktop computer.

For class CIII fBss, the Guinier region is followed first by a power-law decay $I(q) \propto q^{-1}$ in the range $2\pi \lesssim qa \lesssim 2\pi a/d$, where $d = \sqrt{2}|z_{\text{max}} + z_{\text{min}}|$ is a measure of the cross-section size. This is subsequently followed by a second power-law decay, which is similar to those from class I fBss. The succession of these types of power-law decays is a signature of an elongated structure with rough surfaces. Similar to the classes CI fBss, here the roughness also increases with decreasing H and is reflected in the value of the scattering exponent.

Note that for fractal surfaces with exact self-similarity, the power-law decay corresponding to the fractal region has an exponent equal to the fractal dimension of the surface. However, the simple power-law decay observed here is replaced by a succession of maxima and minima superimposed on a simple power-law decay. For such systems, the periodicity and number of these minima can be used to extract additional structural information such as the fractal iteration number or the value of the scaling factor [26–28].

5. Conclusions

The main structural properties of fBss at nano-scales and micro-scales are studied in both real and reciprocal space by exploiting the behaviour of pddfs and, respectively, of the associated small-angle scattering intensities.

The proposed approach allows us to reveal the dependence of fractal dimension and the overall size and shape of fBss on the Hurst (and implicitly, spectral) exponent. The obtained values of these structural parameters (obtained from analysis of data in Figures 4 and 5) are in a good agreement with the simulated ones (obtained from analysis of Figure 1). In particular, the simulated fractal dimensions resulting from the slope of SAS intensity at high q regions (see Figure 5) are in very good agreement with theoretical ones (given by Equation (6)).

It is shown how the SAS technique can distinguish between fBss embedded in Euclidean dimensions of different dimensionalities. The distinction is based on the presence (for flat-like and rod-like fBss) or absence (for globular-like fBss) of a succession of two power law-decays with different scattering exponents. The scattering exponent α of the first power-law is an integer reflecting the Euclidean dimensionality ($\alpha = 1$ for 1D and $\alpha = 2$ for 2D), while the scattering exponent for the second power-law is $\alpha = 3 - H$, reflecting the fractal dimension. For both power-law regions, their lower bounds allow us to determine the overall size and, respectively the cross-section size of fBs, as described in Section 4.2 *Scattering intensities*. Therefore, such a succession allows a structural characterization of fBss at various scales.

The results shown here could be a starting point for a multi-scale analysis of more complex structures involving fBss, such as mass fractals (see Figure 1 Middle) in which the branches themselves are rod-like fBss. In this case, one should expect that the first power-law decay will be replaced by a decay of the type $I(q) \propto q^{-D_{\text{m}}}$, where D_{m} is the fractal dimension of the mass fractal. Other complex geometries can be modeled by considering that the fBss form a closed surface over a domain with a given shape (i.e., ball, ellipsoid, torus, etc.).

Funding: This research received no external funding.

Institutional Review Board Statement: Not applicable.

Informed Consent Statement: Not applicable.

Data Availability Statement: Not applicable.

Conflicts of Interest: The author declares no conflict of interest.

References

1. Zeng, P.; Tian, B.; Tian, Q.; Yao, W.; Li, M.; Wang, H.; Feng, Y.; Liu, L.; Wu, W. Screen-Printed, Low-Cost, and Patterned Flexible Heater Based on Ag Fractal Dendrites for Human Wearable Application. *Adv. Mater. Technol.* **2019**, *4*, 1800453. [CrossRef]
2. Reinhardt, H.; Kroll, M.; Karstens, S.L.; Schlabach, S.; Hampp, N.A.; Tallarek, U. Nanoscaled Fractal Superstructures via Laser Patterning—A Versatile Route to Metallic Hierarchical Porous Materials. *Adv. Mater. Interfaces* **2021**, *8*, 2000253. [CrossRef]
3. Kelesidis, G.A.; Pratsinis, S.E. A perspective on gas-phase synthesis of nanomaterials: Process design, impact and outlook. *Chem. Eng. J.* **2021**, *421*, 129884. [CrossRef]
4. Liu, T.; Liu, P.; Guo, X.; Zhang, J.; Huang, Q.; Luo, Z.; Zhou, X.; Yang, Q.; Tang, Y.; Lu, A. Preparation, characterization and discussion of glass ceramic foam material: Analysis of glass phase, fractal dimension and self-foaming mechanism. *Mater. Chem. Phys.* **2020**, *243*, 122614. [CrossRef]
5. Culcer, D.; Hu, X.; Das Sarma, S. Interface roughness, valley-orbit coupling, and valley manipulation in quantum dots. *Phys. Rev. B* **2010**, *82*, 205315. [CrossRef]
6. Bonnín-Ripoll, F.; Martynov, Y.B.; Cardona, G.; Nazmitdinov, R.G.; Pujol-Nadal, R. Synergy of the ray tracing+carrier transport approach: On efficiency of perovskite solar cells with a back reflector. *Sol. Energy Mater. Sol. Cells* **2019**, *200*, 110050. [CrossRef]
7. Jiang, G.; Hu, J.; Chen, L. Preparation of a Flexible Superhydrophobic Surface and Its Wetting Mechanism Based on Fractal Theory. *Langmuir* **2020**, *36*, 8435–8443. [CrossRef]
8. Gonzalez-Torres, M.; Ramirez-Mata, A.; Melgarejo-Ramirez, Y.; Alvarez-Perez, M.A.; Jose Montesinos, J.; Leyva-Gomez, G.; Sanchez-Sanchez, R.; Eugenia-Baca, B.; Velasquillo, C. Assessment of biocompatibility and surface topography of poly(ester urethane)–silica nanocomposites reveals multifunctional properties. *Mater. Lett.* **2020**, *276*, 128269. [CrossRef]
9. Wang, Y.; Zhang, J.; Li, K.; Hu, J. Surface characterization and biocompatibility of isotropic microstructure prepared by UV laser. *J. Mater. Sci. Technol.* **2021**, *94*, 136–146. [CrossRef]
10. Vipul, S.; Anastasia, K.; Kyriacos, Y.; Kimmo, L.; Veikko, S. Flexible biodegradable transparent heaters based on fractal-like leaf skeletons. *NPJ Flex. Electron.* **2020**, *4*, 27. [CrossRef]
11. Nazmitdinov, R.G.; Simonović, N.S.; Rost, J.M. Semiclassical analysis of a two-electron quantum dot in a magnetic field: Dimensional phenomena. *Phys. Rev. B* **2002**, *65*, 155307. [CrossRef]
12. Nazmitdinov, R.G.; Simonović, N.S. Finite-thickness effects in ground-state transitions of two-electron quantum dots. *Phys. Rev. B* **2007**, *76*, 193306. [CrossRef]
13. Mandelbrot, B.B.; Van Ness, J.W. Fractional Brownian Motions, Fractional Noises and Applications. *SIAM Rev.* **1968**, *10*, 422–437. [CrossRef]
14. Calame, J.P.; Garven, M. Dielectric permittivity simulations of layered composites with rough interfacial surfaces. *IEEE Trans. Dielectr. Electr. Insul.* **2007**, *14*, 287–295. [CrossRef]
15. Glyanko, M.S.; Volkov, A.V.; Fomchenkov, S.A. Assessment of surface roughness of substrates subjected to plasma-chemical etching. *J. Phys. Conf. Ser.* **2014**, *541*, 012100. [CrossRef]
16. Zribi, M.; Ciarletti, V.; Taconet, O.; Paillé, J.; Boissard, P. Characterisation of the Soil Structure and Microwave Backscattering Based on Numerical Three-Dimensional Surface Representation: Analysis with a Fractional Brownian Model. *Remote Sens. Environ.* **2000**, *72*, 159–169. [CrossRef]
17. Clivati-McIntyre, A.A.; McCoy, E.L. Fractional Brownian description of aggregate surfaces within undisturbed soil samples using penetration resistance measurements. *Soil Tillage Res.* **2006**, *88*, 144–152. [CrossRef]
18. Mandelbrot, B.B. *The Fractal Geometry of Nature*; W.H. Freeman: San Francisco, CA, USA, 1982; p. 460.
19. Arrault, J.; Arnéodo, A.; Davis, A.; Marshak, A. Wavelet Based Multifractal Analysis of Rough Surfaces: Application to Cloud Models and Satellite Data. *Phys. Rev. Lett.* **1997**, *79*, 75–78. [CrossRef]
20. Majumdar, A.; Tien, C. Fractal characterization and simulation of rough surfaces. *Wear* **1990**, *136*, 313–327. [CrossRef]
21. Liang, X.; Lin, B.; Han, X.; Chen, S. Fractal analysis of engineering ceramics ground surface. *Appl. Surf. Sci.* **2012**, *258*, 6406–6415. [CrossRef]
22. Liu, Y.; Wang, Y.; Chen, X.; Zhang, C.; Tan, Y. Two-stage method for fractal dimension calculation of the mechanical equipment rough surface profile based on fractal theory. *Chaos Solitons Fractals* **2017**, *104*, 495–502. [CrossRef]
23. Dubuc, B.; Zucker, S.W.; Tricot, C.; Quiniou, J.F.; Wehbi, D.; Berry, M.V. Evaluating the fractal dimension of surfaces. *Proc. R. Soc. Lond. A Math. Phys. Sci.* **1989**, *425*, 113–127. [CrossRef]
24. Schaefer, D.W.; Justice, R.S. How Nano Are Nanocomposites? *Macromolecules* **2007**, *40*, 8501–8517. [CrossRef]
25. Anitas, E.M. Small-Angle Scattering from Fractals: Differentiating between Various Types of Structures. *Symmetry* **2020**, *12*, 65. [CrossRef]
26. Cherny, A.Y.; Anitas, E.M.; Osipov, V.A.; Kuklin, A.I. Small-angle scattering from the Cantor surface fractal on the plane and the Koch snowflake. *Phys. Chem. Chem. Phys.* **2017**, *19*, 2261–2268. [CrossRef]
27. Anitas, E.M.; Slyamov, A.; Szakacs, S. Microstructural characterization of surface fractals using small-angle scattering. *Rom. J. Phys.* **2018**, *63*, 104.

28. Cherny, A.Y.; Anitas, E.M.; Osipov, V.A.; Kuklin, A.I. The structure of deterministic mass and surface fractals: Theory and methods of analyzing small-angle scattering data. *Phys. Chem. Chem. Phys.* **2019**, *21*, 12748–12762. [CrossRef] [PubMed]
29. Feigin, L.A.; Svergun, D.I. *Structure Analysis by Small-Angle X-ray and Neutron Scattering*; Springer: Boston, MA, USA, 1987; p. 335. [CrossRef]
30. Glatter, O.; May, R. Small-Angle Techniques. In *International Tables for Crystallography Volume C: Mathematical, Physical and Chemical Tables*; Prince, E., Ed.; Springer: Dordrecht, The Netherland, 2004; pp. 89–112.
31. Bacon, G.E. *Neutron Diffraction*, 2nd ed.; Oxford University Press: London, UK, 1962; p. 438.
32. Bracewell, R. *The Fourier Transform and Its Applications*, 3rd ed.; Mcgraw-Hill College: Singapore, 2000; p. 486.
33. Pantos, E.; Bordas, J. Supercomputer simulation of small angle X-ray scattering, electron micrographs and X-ray diffraction patterns of macromolecular structures. *Pure Appl. Chem.* **1994**, *66*, 77–82. [CrossRef]
34. Russ, J.C. *Fractal Surfaces*; Springer Science + Business Media: Raleigh, NC, USA, 1994; p. 309. [CrossRef]
35. Martin, J.E.; Hurd, A.J. Scattering from fractals. *J. Appl. Crystallogr.* **1987**, *20*, 61–78. [CrossRef]
36. Schmidt, P.W. Small-angle scattering studies of disordered, porous and fractal systems. *J. Appl. Crystallogr.* **1991**, *24*, 414–435. [CrossRef]
37. Teixeira, J. Small-angle scattering by fractal systems. *J. Appl. Crystallogr.* **1988**, *21*, 781–785. [CrossRef]
38. Bale, H.D.; Schmidt, P.W. Small-Angle X-Ray-Scattering Investigation of Submicroscopic Porosity with Fractal Properties. *Phys. Rev. Lett.* **1984**, *53*, 596–599. [CrossRef]
39. Pfeifer, P.; Ehrburger-Dolle, F.; Rieker, T.P.; González, M.T.; Hoffman, W.P.; Molina-Sabio, M.; Rodríguez-Reinoso, F.; Schmidt, P.W.; Voss, D.J. Nearly Space-Filling Fractal Networks of Carbon Nanopores. *Phys. Rev. Lett.* **2002**, *88*, 115502. [CrossRef] [PubMed]
40. Stein, M.L. Fast and Exact Simulation of Fractional Brownian Surfaces. *J. Comput. Graph. Stat.* **2002**, *11*, 587–599. [CrossRef]
41. Saupe, D. Algorithms for random fractals. In *The Science of Fractal Images*; Peitgen, H.O., Saupe, D., Eds.; Springer: New York, NY, USA, 1988; pp. 71–113.
42. Cherny, A.Y.; Anitas, E.M.; Osipov, V.A.; Kuklin, A.I. Scattering from surface fractals in terms of composing mass fractals. *J. Appl. Crystallogr.* **2017**, *50*, 919–931. [CrossRef]

symmetry

MDPI

Article

Characteristic Length Scale during the Time Evolution of a Turbulent Bose-Einstein Condensate

Lucas Madeira [1,*], Arnol D. García-Orozco [1], Michelle A. Moreno-Armijos [1], Francisco Ednilson Alves dos Santos [2] and Vanderlei S. Bagnato [1,3]

1 Instituto de Física de São Carlos, Universidade de São Paulo, CP 369, São Carlos 13560-970, Brazil; arnolgarcia@ifsc.usp.br (A.D.G.-O.); michelle.moreno@ifsc.usp.br (M.A.M.-A.); vander@ifsc.usp.br (V.S.B.)
2 Departamento de Física, Universidade Federal de São Carlos, São Carlos 13565-905, Brazil; santos@ufscar.br
3 Hagler Fellow, Hagler Institute for Advanced Study, Texas A&M University, College Station, TX 77843, USA
* Correspondence: madeira@ifsc.usp.br

Abstract: Quantum turbulence is characterized by many degrees of freedom interacting non-linearly to produce disordered states, both in space and in time. In this work, we investigate the decaying regime of quantum turbulence in a trapped Bose-Einstein condensate. We present an alternative way of exploring this phenomenon by defining and computing a characteristic length scale, which possesses relevant characteristics to study the establishment of the quantum turbulent regime. We reconstruct the three-dimensional momentum distributions with the inverse Abel transform, as we have done successfully in other works. We present our analysis with both the two- and three-dimensional momentum distributions, discussing their similarities and differences. We argue that the characteristic length allows us to intuitively visualize the time evolution of the turbulent state.

Keywords: quantum turbulence; Bose-Einstein condensate; out-of-equilibrium

Citation: Madeira, L.; García-Orozco, A.D.; Moreno-Armijos, M.A.; dos Santos, F.E.A.; Bagnato, V.S. Characteristic Length Scale during the Time Evolution of a Turbulent Bose-Einstein Condensate. *Symmetry* **2021**, *13*, 1865. https://doi.org/10.3390/sym13101865

Academic Editor: Luca Salasnich

Received: 20 August 2021
Accepted: 29 September 2021
Published: 3 October 2021

Publisher's Note: MDPI stays neutral with regard to jurisdictional claims in published maps and institutional affiliations.

1. Introduction

Turbulence is a process that occurs in many types of fluids and a broad range of length scales, and it is characterized by chaotic changes in the flow velocity and pressure. The field of quantum turbulence (QT) investigates turbulence in quantum fluids, mainly liquid helium and trapped Bose-Einstein condensates (BECs) [1,2]. Many features of classical turbulence are not entirely explained, to the extent that Feynman deemed it the most important unsolved problem in classical physics [3]. Consequently, dealing with QT may seem a challenging task [3]. However, turbulence in quantum fluids might be more manageable than its classical equivalent because the vortex circulation is quantized in the former and continuous in the latter. Furthermore, from a technical point of view, the advances in cooling and tuning the interparticle interactions in trapped atomic BECs make them attractive candidates for investigating quantum turbulence and connecting it to related fields [4].

The first observation of turbulence in a trapped BEC, and its self-similar expansion, dates to 2009 [5]. Since then, considerable progress has been made in identifying and characterizing QT. A significant breakthrough in the area was the observation of a particle cascade, which appears as a power law in the momentum distribution [6,7],

$$n(k) \propto k^{-\delta}, \tag{1}$$

where δ is a positive constant, and its value depends on the mechanism behind the generation of the turbulent state.

There are some intrinsic obstacles in determining the momentum range, where the power law is observed and its characteristic exponent. The range of momentum scales present in trapped BEC systems is small compared to superfluid helium, for example. The exponent of the power law depends on the mechanism behind the turbulence, and different theoretical models predict exponents to be close together. Unfortunately, experiments do not have the necessary precision to distinguish between them. Hence, strategies other than the power law identification have been employed to identify and characterize QT. Energy and particle fluxes have been used in simulations [8,9] and experiments [10,11] to overcome some of these difficulties. Since turbulence and disorder are intimately related, an approach based on the entropy of turbulent BECs has also been successfully applied [12] as an alternative method to investigate and characterize quantum turbulence.

A typical scale used to study turbulence in liquid helium is the vortex line density. Its time dependence provides evidence of the mechanism behind the turbulent regime [13]. In some ^{4}He experiments, where visualization techniques are well-developed, the geometry and interactions of vortices can be directly observed [14]. In trapped BECs, where the range of length scales available is much smaller, the visualization techniques have not yet reached the same level of detail.

In this work, we employed a length scale associated with the momentum distribution $n(k)$ to study the onset of turbulence and the turbulent state. It is inspired by the integral length scale, which is a quantity commonly used in classical turbulence. If we assume isotropic flow, then the integral length scale L_E can be written in terms of the incompressible kinetic energy spectrum $E(k)$ [13,15,16],

$$L_E(t) \propto \frac{\int_0^\infty dk\, E(k,t)/k}{\int_0^\infty dk\, E(k,t)}. \quad (2)$$

Casting Equation (2) in this form also illustrates that it is the length scale that contains most of the energy of the system.

It is known from numerical simulations that in some cases turbulence is mainly in the form of waves, and in sother cases mainly in the form of vortices, depending on the excitation protocol and boundary conditions of the system. Since moving vortices radiate waves and strong waves can create vortices, the relative proportion of waves and vortices depends on the particular experiment. In numerical simulations, one has access to the phase of the wave function [17]. Thus, the circulation can be computed to distinguish vortices from waves. Moreover, in the simulations, one can formally identify the compressible kinetic energy, which comes from waves, and the incompressible kinetic energy related to vortices. Unlike in the simulations, in the experiments, we cannot separate waves and vortices so easily.

However, with current experimental techniques, we can measure the momentum distribution $n(k)$ independently of its origin: waves, vortices, or a combination of both. Hence, with Equation (2) in mind, we define the following length scale,

$$L(t) = \frac{\int_0^\infty dk\, n(k,t)/k}{\int_0^\infty dk\, n(k,t)}. \quad (3)$$

Intuitively, L is associated with the scale where most of the particles reside. In this work, we investigated the behavior of this quantity, and we showed that it is possible to use it to study a turbulent BEC.

Besides the difficulties mentioned above, there is also an experimental challenge when studying QT in trapped BECs. The momentum distribution of the cloud is obtained using a two-dimensional (2D) projection of the three-dimensional (3D) condensate. We employed the inverse Abel transform, an integral transform that connects the 2D projection of an axially or spherically symmetric function to its 3D value, to reconstruct the momentum distribution of the three-dimensional cloud. We showed that the results for the characteristic length scale are qualitatively the same if calculated using the two-dimensional projection. This indicates that it is possible to study some aspects of the turbulent states using the experimental data directly, without reconstructing the three-dimensional cloud.

This work is structured as follows. First, we provide the experimental details of how the BECs are produced and excited. Then, we present the momentum distributions in both two and three dimensions. These are used to compute the characteristic length scale and other quantities related to it. We discuss both the implications of our findings regarding the length scale and the projection of the cloud. In the Appendix A, we provide the Abel transforms of momentum distributions relevant to our system.

2. Experimental Procedure

The first step is the production of a Bose-Einstein condensate in equilibrium. A typical BEC contains $\approx 4 \times 10^5$ rubidium-87 atoms in the hyperfine state $|F, m_F\rangle = |2,2\rangle$, confined in a Quadrupole–Ioffe configuration (QUIC) magnetic trap of frequencies $\omega_r/2\pi = 130.7(8)$ Hz and $\omega_x/2\pi = 21.8(2)$ Hz. Hereafter, we adopted the convention of reporting the uncertainties as one standard deviation between parenthesis. Before any excitation is applied, the BEC in equilibrium has a condensate fraction of 70(5)%. The chemical potential at the center of the cloud is $\mu_0/k_B = 124(5)$ nK, and the healing length is $\xi = 0.15(2)$ µm.

In Figure 1a, we present schematically the protocol we employed to drive the BEC out of equilibrium. Following the condensate production, an oscillating magnetic field is applied while it is still in the trap. The field is produced by a pair of coils, placed in an anti-Helmholtz configuration, with their axis tilted by a small angle of approximately 5° with respect to the axis of the trap. The excitation potential is given by $V_{\text{exc}}(\mathbf{r}, t) = A[1 - \cos(\Omega t)]x'/R_x$, where x' is the coordinate in the rotated frame and $R_x = 42$ µm is the in-trap extent of the BEC along the x-axis of the trap. Since the perturbation is not aligned with the axes of the trap, the oscillations generate deformations, displacements, and rotations. Several excitation parameters can be varied, such as the amplitude A, total excitation time, and perturbation frequency. In this work, we performed the parametric excitation of fixed frequency $\Omega/(2\pi) = 132.8$ Hz, close to the radial trapping frequency, $\omega_r/2\pi = 130.7(8)$ Hz.

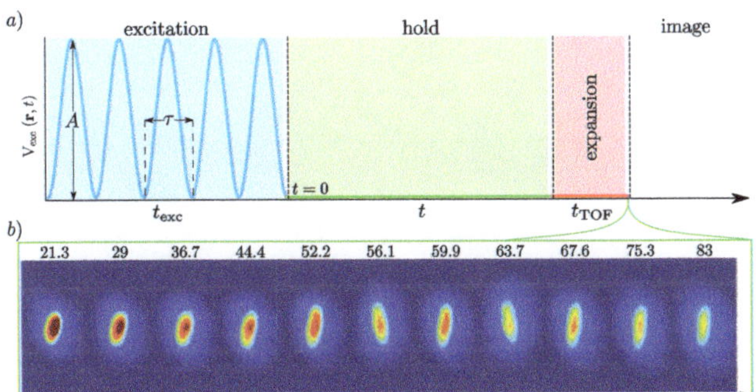

Figure 1. (**a**) Schematic representation of the excitation protocol. The experiment begins with the production of an unperturbed BEC in the trap. Then, a sinusoidal potential of amplitude A and period τ is applied during t_{exc}. The system evolves during a time t, after which the trap is released, and an absorption image is taken after a time-of-flight t_{ToF}. (**b**) Absorption images for an excitation amplitude of $A = 1.8\,\mu_0$ as a function of the holding time.

We increased the excitation amplitude until reaching a value where the momentum distribution corresponds to an out-of-equilibrium state, and we look for turbulent characteristics. The energy input to the condensate is related to both the total excitation time and the amplitude of the perturbation. Larger amplitudes need less time to reach similar conditions than it would take for smaller amplitudes. The range of amplitudes and excitation times to obtain a turbulent state was the topic of investigation in previous works [18]. In this work, we chose to apply the excitation protocol during a time $t_{\mathrm{exc}} = 5\tau$, where $\tau = 2\pi/\Omega$. The amplitude A is varied, ranging from 0 (no perturbation) to 2.2 μ_0.

After the excitation is turned off, we hold the cloud for a time t inside the trap, often called holding time, which is varied from 20 to 90 ms. During this time, the temporal evolution of the momentum distribution $n(k,t)$, which we are interested in, occurs. Next, we turn off the trap potential and measure $n(k,t)$ using absorption images taken from the ballistic expansion of the cloud after a time of flight (ToF) of $t_{\mathrm{ToF}} = 30\,\mathrm{ms}$. In Figure 1b, we show typical absorption images corresponding to an excitation amplitude of $A = 1.8\,\mu_0$ for different holding times t.

For each excitation amplitude and holding time, we perform several realizations of the experiment and then average the results. In Figure 2, we show a typical two-dimensional momentum distribution obtained from the absorption images. The distance that an atom has traveled from the center of the cloud, after a time t_{ToF}, is given by $r = \hbar t_{\mathrm{ToF}} k/m$, where m is the mass of a rubidium-87 atom. Thus, in practice, the ToF technique corresponds to a Fourier transform of the spatial distribution, which yields the momentum distribution, $n(r) \propto n(\hbar t_{\mathrm{ToF}} k/m)$. There are known shortcomings of the ToF technique, which do not significantly impact the measurement of our momentum distributions, mainly because the turbulent state is kinetically dominated [19]. For a more detailed discussion, the reader is referred to Reference [12] and references therein. This technique has been used successfully in the past to obtain the momentum distribution of turbulent trapped BECs [6,7].

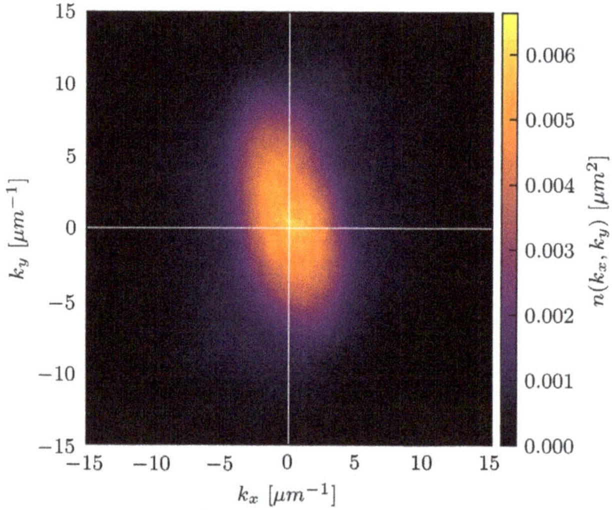

Figure 2. Momentum distribution $n(k_x, k_y)$ obtained from the absorption images of the cloud for an excitation of amplitude $A = 1.8\,\mu_0$ and $t = 36.7\,\mathrm{ms}$.

3. Momentum Distributions

We performed angular averages on the momentum distributions $n(k_x, k_y, t)$ obtained from the absorption images, such that the resulting profiles depend only on $k = (k_x^2 + k_y^2)^{1/2}$. The two-dimensional momentum distributions $n_{\mathrm{2D}}(k,t)$ are normalized according to

$$2\pi \int dk\, k\, n_{\mathrm{2D}}(k,t) = 1. \tag{4}$$

As discussed above, an experimental challenge when studying momentum distributions of trapped BECs is that the absorption images correspond to a projection of the cloud. We overcome this difficulty by considering the symmetry of the trapped BEC in momentum space. The inverse Abel transform [20–22] has been successfully used in the literature [6,7] to obtain the 3D momentum distribution from its two-dimensional projection. It is an integral transform given by

$$n_{3D}(k,t) = -\frac{1}{\pi}\int_k^\infty \frac{dn_{2D}(k',t)}{dk'}\frac{dk'}{\sqrt{k'^2-k^2}}. \quad (5)$$

We normalized the distributions according to

$$4\pi \int dk\, k^2\, n_{3D}(k,t) = 1. \quad (6)$$

The signature of a particle cascade is a power law, $n(k) \propto k^{-\delta}$. Equations (4) and (6), together with dimensional analysis, suggest that if we observe a power law both in the two-dimensional momentum distribution, $n_{2D}(k) \propto k^{-\delta_{2D}}$, and in the three-dimensional one, $n_{3D}(k) \propto k^{-\delta_{3D}}$, then their exponents differ by one, $\delta_{3D} - \delta_{2D} = 1$. To go beyond simple dimension analysis, in Appendix A, we derive this relation analytically for the case where a power law is present over the whole k-range. Although this toy-model is nonphysical, it sheds light on how we can reconstruct the three-dimensional momentum distribution based on symmetry arguments.

In our system, the low-momenta region of $n(k)$ is dominated by the presence of the condensate, which corresponds to a Gaussian distribution. We show in Appendix A that the inverse Abel transform of a Gaussian function is also a Gaussian with the same width. This symmetry is extremely useful because we can work with the two-dimensional projections for quantities related to the Gaussian shape without the need for 3D reconstruction. This is the case of the temperature, which is related to the width of the Gaussian.

All the arguments presented above indicate that the power-law exponents in an ideal situation would be related through $\delta_{3D} - \delta_{2D} \approx 1$. However, the fact that we have the power-law behavior superimposed with the condensate at the low-momenta region of the momentum distribution alters this relation. Hence, we need to verify the exponents with the experimental data.

A power law is simply a line in a log–log plot of the momentum distribution as a function of k. We then look for a time window when, in a certain k-range, $n(k)$ is proportional to $k^{-\delta}$, the particle cascade characteristic of a turbulent cloud. We performed the experiment described in Section 2 employing six different excitation amplitudes. For only the three highest ones, A = 1.8, 2.0, and 2.2 μ_0, we observed the appearance of a power-law, around $t \approx 35$ ms and in the region $10\,\mu\text{m}^{-1} \leqslant k \leqslant 15\,\mu\text{m}^{-1}$. We found the exponents $\delta_{2D} = 3.1(1)$ and $\delta_{3D} = 3.8(2)$, which lead to $\delta_{3D} - \delta_{2D} = 0.7(3)$. It is interesting to see that even in our finite-sized non-homogeneous system inside an anisotropic trap, we still have $\delta_{3D} - \delta_{2D}$ close to one.

In Figure 3, we present the time evolution of both the two- and three-dimensional momentum distributions for an excitation amplitude of A = 1.8 μ_0 (which is qualitatively the same for A = 2.0 and 2.2 μ_0). As we wait after the external excitation has been turned off, the distribution evolves, promoting the population from low to high momentum values, as can be seen in Figure 3.

Some words regarding the values of the exponents we found are in order. To the best of our knowledge, there is no theoretical work that describes all aspects of the experiments we performed, mainly for two reasons. First, our condensate is produced in an anisotropic trap. From the theoretical perspective, it is much easier to implement periodic boundary conditions and describe, or simulate, bulk systems. Second, the route we take is the inverse of most experiments. We begin with a BEC in equilibrium and then excite it, while, for example, quench experiments usually start with a thermal gas and produce a condensate [23,24]. However, after these considerations, we can compare the momentum distributions and exponent we obtained with other works that share similarities with our experiment.

One of the first predictions for the time evolution of the momentum distribution describing Bose-Einstein condensation in a far-from-equilibrium system is given in Reference [25]. The authors find a plateau in the lower momenta region and a power law at higher momenta, akin to what we observe in our experiments; see Figure 3. Since then, much progress has been made in characterizing turbulent flows in BECs. The advances and state of the art concerning this topic can be found in Reference [26], for example. Here, we will discuss two references that capture the essential physical aspects of our experiment.

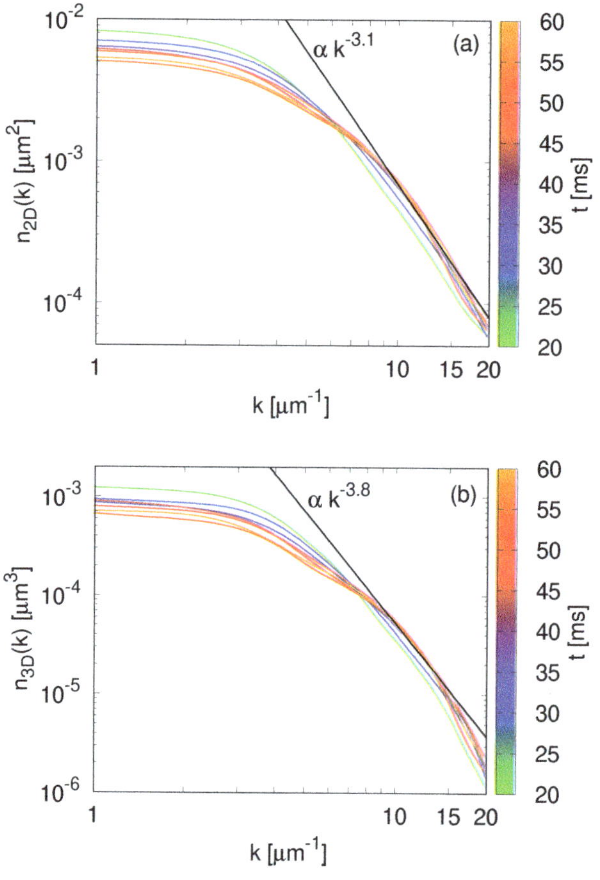

Figure 3. Time evolution of the momentum distributions for an excitation amplitude of $A = 1.8\,\mu_0$. We present the results obtained with both (**a**) the angular average of the absorption image and (**b**) the three-dimensional reconstruction of the cloud using the inverse Abel transform. In both plots, we include the curve corresponding to the power-law behavior characteristic of the turbulent states as a guide to the eye.

The authors of [27] address the topic of turbulence in ultracold Bose gases under the light of the so-called non-thermal fixed points [28]. They consider a variety of scenarios and analyze each region of the momentum distributions. For strong turbulence and a freely (without external energy input or dissipation) decaying initial state, scaling arguments lead to $\delta_{3D} = 5$ prediction, which was confirmed by accompanying numerical simulations. We did not expect quantitative agreement with their results, since we have dissipation in our system, and there are no indications that we are in the strong turbulent regime. Nonetheless, this is one of the few references that address the decay of turbulence.

Another interesting study that allows comparison, to some extent, to our work is presented in Reference [29]. The authors performed numerical simulations employing the forced-dissipated Gross–Pitaevskii equation to study Bose-Einstein condensation under non-equilibrium conditions. They observed that the momentum distribution for the late time dynamics, that is, after the kinetic stage is over, displays an approximately constant value at low-momenta, in accordance with our findings, and a power law in a higher momentum region with the exponent $\delta_{3D} = 4.4$ (for some values of the nonlinear term). This value is relatively close to what we observed in this work, $\delta_{3D} = 3.8(2)$. We should remark that the boundary conditions in their simulations and our experiments are different and that we are exciting a BEC while they are looking at the inverse process. However, the agreement warrants further investigations.

4. The Characteristic Length Scale

We computed the characteristic length scale given by Equation (3) using the experimental data available,

$$\begin{aligned} L_{2D}(t) &= \frac{\int_{k_D}^{k_d} dk\, n_{2D}(k,t)}{\int_{k_D}^{k_d} dk\, k\, n_{2D}(k,t)} = 2\pi \int_{k_D}^{k_d} dk\, n_{2D}(k,t), \\ L_{3D}(t) &= \frac{\int_{k_D}^{k_d} dk\, k\, n_{3D}(k,t)}{\int_{k_D}^{k_d} dk\, k^2\, n_{3D}(k,t)} = 4\pi \int_{k_D}^{k_d} dk\, k\, n_{3D}(k,t), \end{aligned} \quad (7)$$

where $k_D \sim 0.05\ \mu\text{m}^{-1}$ and $k_d \sim 30\ \mu\text{m}^{-1}$ are the smallest and largest wave vectors we can measure, respectively. It is worth noting that although our definition relies on $n(k)/k$, the singular behavior when $k \to 0$ will never be reached. The lower limit of the integrals, k_D, is inversely proportional to the largest length scale of the system, which can be large concerning other scales, but always finite.

Using the momentum distributions obtained with different excitation amplitudes and holding times, we can study the time evolution of the BEC, ranging from a slightly perturbed cloud up to a turbulent state. In Figure 4, we present our results for the characteristic length scale computed with both the two-dimensional projection of the cloud and its three-dimensional reconstruction. Although they differ quantitatively, their qualitative behavior is remarkably the same.

The $L(t)$ value can be interpreted as the evolution of the length scale where most of the particles are located. If we think of $1/k$ as being a weight in Equation (3), then $L(t)$ is related to the inverse of the momentum value for which $n(k)/k$ is peaked. Nucleation of excitations occurs during the excitation, whether in the form of vortices or waves. Then the interaction of these excitations takes place, leading to different stages of deviation from equilibrium.

For small excitation amplitudes, $A = 0.8\ \mu_0$, the system is only slightly disturbed and removed from equilibrium, but it does not have enough energy to reach what is considered a disordered state. In this case, it evolves differently from the others, and the value of L is approximately constant with time.

For intermediate perturbations, $A = 1.4$ and $1.6\ \mu_0$, there is a separation between these results and the smallest amplitude, besides a clear dependence with time. For these excitation amplitudes, we are in a regime best characterized as the onset of turbulence. The characteristic length scale decreases on time, indicating the particle transfer to higher-momenta, but slower than the higher excitation amplitudes.

In this work and previous investigations [12], we identified the highest excitation amplitudes with turbulent clouds, $A = 1.8, 2.0,$ and $2.2\ \mu_0$. The value of $L(t)$ at the end of the processes seems to depend on the amplitude and, more importantly, if we deal with a moderate perturbation, the onset of turbulence, or a state with turbulent characteristics. It is interesting to observe that the turbulent states quickly reach lengths comparable to the healing length ($\xi = 0.15(2)\ \mu\text{m}$), where dissipation processes are expected to occur.

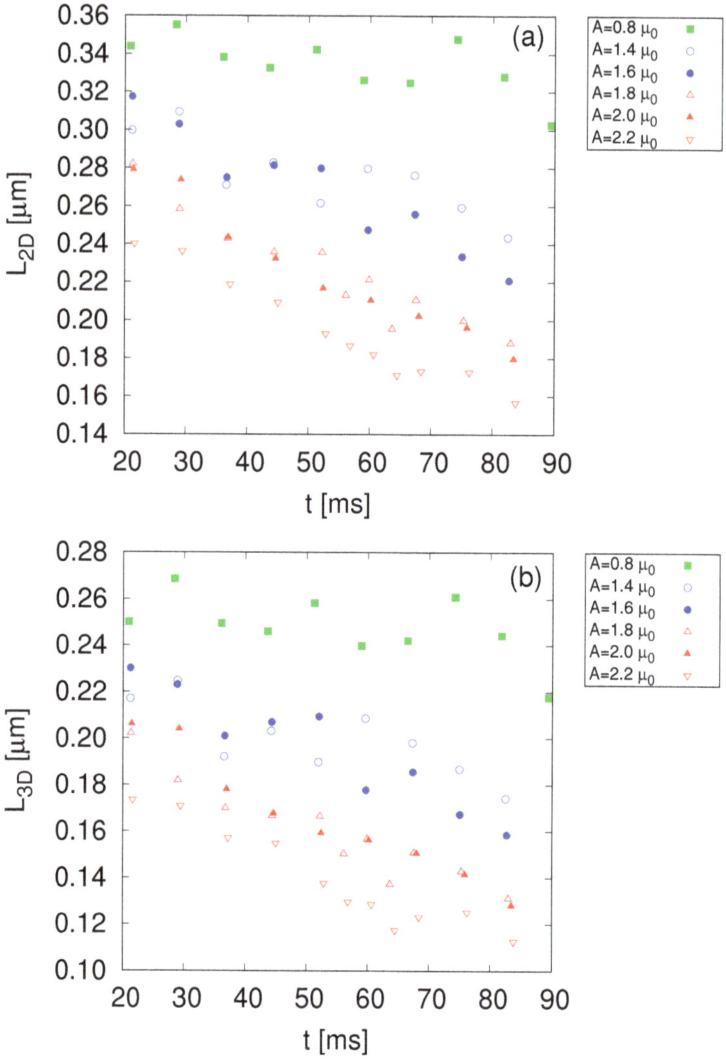

Figure 4. Time evolution of the characteristic length scale for different excitation amplitudes computed with (**a**) the two-dimensional projection of the cloud and (**b**) its three-dimensional reconstruction using the inverse Abel transform. Although the values computed with two-dimensional profiles are higher, their qualitative behavior is the same.

The behavior of the characteristic length scale can be described by an exponential decay,

$$L(t) = L_0 \exp(-t/t_0), \tag{8}$$

where L_0 is the extrapolation of the characteristic length scale to the instant when the excitation was introduced, and t_0 is its characteristic time.

In Figure 5, we present the values of L_0 and t_0 fitted to the functional form of Equation (8) for different excitation amplitudes. We did not include the results for $A = 0.8\,\mu_0$, since we obtain a value of $t_0 \approx 750$ ms, of the same order as the lifetime of the condensate.

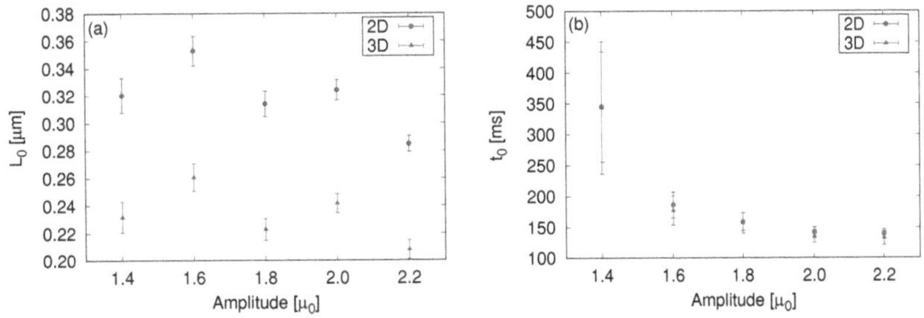

Figure 5. (**a**) Extrapolation of the characteristic length scale to the instant when the excitation is turned on, L_0, and (**b**) the characteristic time of the particle transfer, t_0, as a function of the excitation amplitude. The results were obtained fitting the data to the functional form of Equation (8). Although the analysis employing two- or three-dimensional momentum distributions produces different values of L_0, both approaches yield the same values for the characteristic time.

Figure 5a shows that the length containing most particles of the system is approximately the same with respect to the excitation amplitudes. However, if we perform the analysis employing the two- or three-dimensional momentum distributions, we arrive at different values for L_0. A possible reason for this is that the inverse Abel transform slightly shifts the $n(k)$ profiles toward higher-momena—see Figure 3—which then implies a smaller value of L_0.

In Figure 5b, we present the characteristic time that the particle transfer takes as a function of the excitation amplitude. It is observed that t_0 decreases with amplitude, as expected, since larger amplitudes lead to a faster formation of excitations and, therefore, speed up the decay. It is possible to see that the results obtained in 2D and 3D are in remarkable agreement. This opens the possibility of studying dynamical processes in 2D without the need to reconstruct the three-dimensional cloud, depending on the quantity of interest.

5. Discussion and Final Remarks

In this work, we defined and computed a characteristic length scale related to the momentum distribution of a trapped BEC, which allowed us to identify distinct out-of-equilibrium stages: a slightly perturbed cloud, the onset of turbulence, and the turbulent state. This quantity complements the formal analysis of identifying a power-law in the momentum distribution as a hallmark of turbulence.

We also focused our efforts on calculating this characteristic length scale using both two- and three-dimensional momentum distributions. The former is obtained straightforwardly from the experimental data, and the latter is reconstructed based on the symmetry of the cloud. From a technical point of view, it is preferable to work only with the two-dimensional distributions, since no assumptions about the symmetry of the cloud have to be made. Although the length scales are affected by the inverse Abel transform, which shifts the momentum distributions to higher momenta, the qualitative behavior calculated in 2D and 3D is remarkably similar. The excellent agreement in the characteristic time of the particle transfer indicates that the two-dimensional analysis may be appropriate to investigate dynamical aspects of these systems.

One important remark is that isotropy is assumed through Equation (3) only for the kinetically dominated regions in Fourier space. Such an assumption can be made even for inhomogeneous cigar-shaped clouds. This is because such large-scale inhomogeneities affect only regions in Fourier space up to the order of $k \sim 2\pi/L_{\mathrm{min}}$, where L_{min} is the smaller linear size of the cigar-shaped cloud. In a previous work [11], we studied the impact of anisotropy in the energy transfer during the evolution of turbulence in a trapped BEC. Like the integral length scale, the energy flux can also be computed from the kinetic energy spectrum. We found that the turbulent state can be identified and characterized in terms of the energy flux regardless of whether we employ the whole cloud or just a region close to the major axis of the expanded cloud. We should note that the axial trapping frequencies of this work and of Reference [11] are very close; however, in this work, we employ a radial trapping frequency that is ≈ 0.7 smaller than the one used in Reference [11]. Thus, the BECs in this work are much less elongated and closer to a spherical shape than the ones in Reference [11]. Therefore the range of validity for Equation (3) is even larger than in previous works.

In future works, we intend to vary the excitation protocol to investigate the changes in the characteristic length. Since there is a compromise between the excitation amplitude and time [18], it may prove insightful to investigate situations where the same amount of energy is introduced in the system but varying the time it takes to be injected from very slow inputs up to abrupt changes.

Author Contributions: Conceptualization, F.E.A.d.S. and V.S.B.; methodology, L.M.; software, L.M.; validation, L.M., A.D.G.-O. and M.A.M.-A.; formal analysis, L.M.; investigation, L.M., A.D.G.-O. and M.A.M.-A.; resources, V.S.B.; data curation, L.M., A.D.G.-O. and M.A.M.-A.; writing—original draft preparation, L.M.; writing—review and editing, F.E.A.d.S. and V.S.B.; visualization, L.M.; supervision, V.S.B.; project administration, L.M.; funding acquisition, V.S.B. All authors have read and agreed to the published version of the manuscript.

Funding: This work was supported by the São Paulo Research Foundation (FAPESP) under the grants 2013/07276-1, 2014/50857-8, and 2018/09191-7, and by the National Council for Scientific and Technological Development (CNPq) under the grant 465360/2014-9. FEAS thanks CNPq (Conselho Nacional de Desenvolvimento Científico e Tecnológico, National Council for Scientific and Technological Development) for support through Bolsa de produtividade em Pesquisa Grant No. 305586/2017-3.

Institutional Review Board Statement: Not applicable.

Informed Consent Statement: Not applicable.

Data Availability Statement: The data presented in this study are available on request from the corresponding author.

Acknowledgments: We thank L. Galantucci and C. Barenghi for the useful discussions.

Conflicts of Interest: The authors declare no conflict of interest.

Abbreviations

The following abbreviations are used in this manuscript:

QT	Quantum turbulence
BEC	Bose-Einstein condensate
QUIC	Quadrupole–Ioffe configuration
ToF	Time-of-flight

Appendix A. The Abel Transform

We used the inverse Abel transform to reconstruct a three-dimensional momentum distribution from its two-dimensional projection in the main text. It is insightful to take the inverse route to see what the two-dimensional projection is of a known three-dimensional $n_{3D}(k)$. In this appendix, we considered two relevant cases for our physical system that possess analytical solutions.

The Abel transform is given by

$$n_{2D}(k) = 2\int_{k}^{\infty} dk' \frac{n_{3D}(k')k'}{\sqrt{k'^2 - k^2}}. \tag{A1}$$

The first case we considered is a Gaussian normalized according to Equation (6),

$$n_{\mathrm{G;3D}}(k) = \frac{\sqrt{2}}{\sqrt{\pi\sigma^3}}e^{-k^2/(2\sigma^2)}. \tag{A2}$$

Using Equation (A1), the normalized two-dimensional projection is

$$n_{\mathrm{G;2D}}(k) = \frac{1}{\sigma^2}e^{-k^2/(2\sigma^2)}. \tag{A3}$$

Hence, the Abel transform of a Gaussian is a Gaussian of the same width. This is very convenient in the case of the temperature of the cloud, for example, since it is estimated through the width of the Gaussian profile of the momentum distribution.

The second case we considered is a power-law with exponent δ_{3D},

$$n_{\mathrm{P;3D}}(k) = Ak^{-\delta_{\mathrm{3D}}}, \tag{A4}$$

with A constant. The standard normalization procedure, Equation (6), is going to fail because this momentum distribution is not valid in the entire domain. In reality, the power-law would be observed over a certain k-range, $k_i \leqslant k \leqslant k_f$ with $k_i \neq 0$. However, this simplified example will have an interesting result as we will see. The Abel transformation yields

$$n_{\mathrm{P;2D}}(k) = \frac{\sqrt{\pi}\Gamma\left(\frac{\delta_{\mathrm{3D}}-1}{2}\right)}{\Gamma\left(\frac{\delta_{\mathrm{3D}}}{2}\right)}Ak^{-(\delta_{\mathrm{3D}}-1)} = A'k^{-\delta_{\mathrm{2D}}}, \tag{A5}$$

where Γ is the gamma function, A' is another constant, and $\delta_{\mathrm{2D}} = \delta_{\mathrm{3D}} - 1$. The conclusion is that the bidimensional projection of a power-law with an exponent of δ_{3D} in three dimensions is also a power-law, but with the exponent increased by one.

Clearly, the momentum distributions we presented in the main text cannot be fully described by these two simple examples. However, they provide indications of the expected behavior of the projection procedure.

References

1. Vinen, W.F.; Niemela, J.J. Quantum Turbulence. *J. Low Temp. Phys.* **2002**, *128*, 167–231. [CrossRef]
2. Barenghi, C.F.; Skrbek, L.; Sreenivasan, K.R. Introduction to quantum turbulence. *Proc. Natl. Acad. Sci. USA* **2014**, *111*, 4647–4652. [CrossRef] [PubMed]
3. Madeira, L.; Caracanhas, M.; dos Santos, F.; Bagnato, V. Quantum Turbulence in Quantum Gases. *Annu. Rev. Condens. Matter Phys.* **2020**, *11*, 37–56. [CrossRef]
4. Dalfovo, F.; Giorgini, S.; Pitaevskii, L.P.; Stringari, S. Theory of Bose-Einstein condensation in trapped gases. *Rev. Mod. Phys.* **1999**, *71*, 463–512. [CrossRef]
5. Henn, E.A.L.; Seman, J.A.; Roati, G.; Magalhães, K.M.F.; Bagnato, V.S. Emergence of Turbulence in an Oscillating Bose-Einstein Condensate. *Phys. Rev. Lett.* **2009**, *103*, 045301. [CrossRef]
6. Thompson, K.J.; Bagnato, G.G.; Telles, G.D.; Caracanhas, M.A.; dos Santos, F.E.A.; Bagnato, V.S. Evidence of power law behavior in the momentum distribution of a turbulent trapped Bose–Einstein condensate. *Laser Phys. Lett.* **2014**, *11*, 015501. [CrossRef]
7. Navon, N.; Gaunt, A.L.; Smith, R.P.; Hadzibabic, Z. Emergence of a turbulent cascade in a quantum gas. *Nature* **2016**, *539*, 72–75. [CrossRef]
8. Baggaley, A.W.; Barenghi, C.F.; Sergeev, Y.A. Three-dimensional inverse energy transfer induced by vortex reconnections. *Phys. Rev. E-Stat. Nonlinear Soft Matter Phys.* **2014**, *89*, 013002. [CrossRef] [PubMed]
9. Marino, Á.V.M.; Madeira, L.; Cidrim, A.; dos Santos, F.E.A.; Bagnato, V.S. Momentum distribution of Vinen turbulence in trapped atomic Bose–Einstein condensates. *Eur. Phys. J. Spec. Top.* **2021**, *230*, 809–812. [CrossRef]
10. Navon, N.; Eigen, C.; Zhang, J.; Lopes, R.; Gaunt, A.L.; Fujimoto, K.; Tsubota, M.; Smith, R.P.; Hadzibabic, Z. Synthetic dissipation and cascade fluxes in a turbulent quantum gas. *Science* **2019**, *366*, 382–385. [CrossRef]
11. Daniel García-Orozco, A.; Madeira, L.; Galantucci, L.; Barenghi, C.F.; Bagnato, V.S. Intra-scales energy transfer during the evolution of turbulence in a trapped Bose-Einstein condensate. *EPL Europhys. Lett.* **2020**, *130*, 46001. [CrossRef]
12. Madeira, L.; García-Orozco, A.D.; dos Santos, F.E.A.; Bagnato, V.S. Entropy of a Turbulent Bose-Einstein Condensate. *Entropy* **2020**, *22*, 956. [CrossRef] [PubMed]

13. Stagg, G.W.; Parker, N.G.; Barenghi, C.F. Ultraquantum turbulence in a quenched homogeneous Bose gas. *Phys. Rev. A* **2016**, *94*, 053632. [CrossRef]
14. Bewley, G.P.; Lathrop, D.P.; Sreenivasan, K.R. Superfluid helium: Visualization of quantized vortices. *Nature* **2006**, *441*, 588. [CrossRef] [PubMed]
15. Batchelor, G.K.; Press, C.U. *The Theory of Homogeneous Turbulence*; Cambridge Science Classics; Cambridge University Press: Cambridge, UK, 1953.
16. WANG, H.; GEORGE, W.K. The integral scale in homogeneous isotropic turbulence. *J. Fluid Mech.* **2002**, *459*, 429–443. [CrossRef]
17. Tsubota, M.; Fujimoto, K.; Yui, S. Numerical Studies of Quantum Turbulence. *J. Low Temp. Phys.* **2017**, *188*, 119–189. [CrossRef]
18. Seman, J.A.; Henn, E.A.; Shiozaki, R.F.; Roati, G.; Poveda-Cuevas, F.J.; Magalhães, K.M.; Yukalov, V.I.; Tsubota, M.; Kobayashi, M.; Kasamatsu, K.; et al. Route to turbulence in a trapped Bose-Einstein condensate. *Laser Phys. Lett.* **2011**, *8*, 691–696. [CrossRef]
19. Pitaevskii, L.; Stringari, S. *Bose-Einstein Condensation and Superfluidity*; Oxford University Press: Oxford, UK, 2016. [CrossRef]
20. Bracewell, R.N.; Bracewell, R.N. *The Fourier Transform and Its Applications*; McGraw-Hill Series in Electrical Engineering; McGraw-Hill: New York, NY, USA, 1986.
21. Polyanin, P.; Manzhirov, A.V. *Handbook of Integral Equations*, 2nd ed.; Handbooks of Mathematical Equations; CRC Press: Boca Raton, FL, USA, 2008.
22. Hickstein, D.D.; Gibson, S.T.; Yurchak, R.; Das, D.D.; Ryazanov, M. A direct comparison of high-speed methods for the numerical Abel transform. *Rev. Sci. Instrum.* **2019**, *90*, 065115. [CrossRef]
23. Proukakis, N.P.; Snoke, D.W.; Littlewood, P.B. (Eds.) *Universal Themes of Bose-Einstein Condensation*; Cambridge University Press: Cambridge, UK, 2017. [CrossRef]
24. Glidden, J.A.P.; Eigen, C.; Dogra, L.H.; Hilker, T.A.; Smith, R.P.; Hadzibabic, Z. Bidirectional dynamic scaling in an isolated Bose gas far from equilibrium. *Nat. Phys.* **2021**, *17*, 457–461. [CrossRef]
25. Semikoz, D.V.; Tkachev, I.I. Kinetics of Bose Condensation. *Phys. Rev. Lett.* **1995**, *74*, 3093–3097. [CrossRef]
26. Semisalov, B.; Grebenev, V.; Medvedev, S.; Nazarenko, S. Numerical analysis of a self-similar turbulent flow in Bose–Einstein condensates. *Commun. Nonlinear Sci. Numer. Simul.* **2021**, *102*, 105903. [CrossRef]
27. Nowak, B.; Schole, J.; Sexty, D.; Gasenzer, T. Nonthermal fixed points, vortex statistics, and superfluid turbulence in an ultracold Bose gas. *Phys. Rev. A* **2012**, *85*, 043627. [CrossRef]
28. Schmied, C.M.; Mikheev, A.N.; Gasenzer, T. Non-thermal fixed points: Universal dynamics far from equilibrium. *Int. J. Mod. Phys. A* **2019**, *34*, 1941006. [CrossRef]
29. Shukla, V.; Nazarenko, S. Non-equilibrium Bose-Einstein Condensation. *arXiv* **2021**, arXiv:2105.07274.

symmetry MDPI

Article

Probing Many-Body Systems near Spectral Degeneracies

Klaus Ziegler

Institut für Physik, Universität Augsburg, D-86135 Augsburg, Germany; klaus.ziegler@physik.uni-augsburg.de

Abstract: The diagonal elements of the time correlation matrix are used to probe closed quantum systems that are measured at random times. This enables us to extract two distinct parts of the quantum evolution, a recurrent part and an exponentially decaying part. This separation is strongly affected when spectral degeneracies occur, for instance, in the presence of spontaneous symmetry breaking. Moreover, the slowest decay rate is determined by the smallest energy level spacing, and this decay rate diverges at the spectral degeneracies. Probing the quantum evolution with the diagonal elements of the time correlation matrix is discussed as a general concept and tested in the case of a bosonic Josephson junction. It reveals for the latter characteristic properties at the transition to Hilbert-space localization.

Keywords: dynamics of closed quantum systems; random probing; separation of time scales; Hilbert-space localization

Citation: Ziegler, K. Probing Many-Body Systems near Spectral Degeneracies. *Symmetry* **2021**, *13*, 1796. https://doi.org/10.3390/sym13101796

Academic Editors: V. I. Yukalov, V. S. Bagnato and Rashid G. Nazmitdinov

Received: 30 August 2021
Accepted: 23 September 2021
Published: 26 September 2021

1. Introduction

Symmetries play a central role in classical and in quantum many-body systems. They determine the macroscopic behavior of these systems. Moreover, symmetries of macroscopic states reflect symmetries and spontaneous symmetry breaking of the underlying system. For instance, the Hamiltonian of the Ising model has a global Z_2 symmetry, and with ferromagnetic nearest-neighbor spin–spin coupling its ground state is two-fold degenerate with $|\uparrow,\ldots,\uparrow\rangle$ and $|\downarrow,\ldots,\downarrow\rangle$. Spin flip dynamics will create excited states that are generically not Z_2 symmetric. Moreover, we can prepare the initial state of the Ising system in one of the ground states, e.g., in the state $|\uparrow,\ldots,\uparrow\rangle$ that breaks the Z_2 symmetry. Then the dynamics prefer the vicinity of the initial state because it would cost too much energy to overcome the barrier to the other ground state $|\downarrow,\ldots,\downarrow\rangle$, although the Hamiltonian and the spin-flip operator obey the Z_2 symmetry. As a result, the system experiences dynamical symmetry breaking by which it prefers to remain in the vicinity of a symmetry-broken state.

In the following we consider the unitary evolution of closed quantum many-body systems. It is based on the idea that the extraction of information about the quantum system in an experiment is limited. In other words, not all properties or degrees of freedom of the quantum model are accessible by an experiment. Typical exceptions are the return and transition probabilities for quantum states. Moreover, quantum systems have complex dynamics. Although the evolution is deterministic, even for a few particles it can looks erratic, similarly to a classical random walk. Such behavior suggests a statistical approach to extract generic information about the quantum evolution, using averaged quantities. A statistical approach is also supported by the fact that large sets of experimental data are available whose properties can be treated statistically. This idea is not new and found very successful realization in random matrix theory (RMT). It has been applied to many physical systems, such as nuclei, atoms and mesoscopic systems [1–7]. The motivation for RMT is that there is no way of knowing the Hamiltonian of even a relatively small many-body quantum system, such as an atomic nucleus. On the other hand, the spectra of these systems, complex though they are, have some characteristic features, such as level repulsion. Thus, instead of guessing a specific Hamiltonian, a random ensemble of Hamiltonians is chosen, which describes the generic features of a class of quantum systems. The class is characterized by the invariance of the random ensemble with respect to symmetry transformations. These are typically orthogonal, unitary or symplectic transformations. Another application of RMT has been recently proposed for the description of random measurements. It is based on Dyson's circular matrix ensemble [3,6,8,9]), which represents random unitary matrices and has been used as a tool to determine the trace of powers of the density matrix and the related Rényi entropy [10–14].

In contrast to these RMT approaches, we consider in the following a dynamical approach in which only the time of a measurement is random, whereas the energy levels $\{E_j\}$ of the Hamiltonian H and the overlaps $\langle E_j|\Psi_0\rangle$, $\langle E_j|\Psi\rangle$ of the energy eigenstates $\{|E_j\rangle\}$ with a given initial state $|\Psi_0\rangle$ and a measured state $|\Psi\rangle$ are not random. This leads to the time correlation matrix (TCM) as the central tool for the definition of the statistical model, instead of the random ensemble of Hamiltonians in RMT. We employ this approach, which was previously described in [15] to analyze the evolution of the return and transition probabilities. In more concrete terms, for a given time t_k we evaluate (in a calculation or in a real experiment) the probability p_k that the system is in a certain state. Then we evaluate the probabilities $\{p_1, p_2, ...\}$ at different discrete and randomly chosen times $\{t_1, t_2, ...\}$. This can be translated into practical observations, for which it was assumed that each experiment was prepared in the same initial state and all measurements were performed for the same final state of the evolution at different times. These experiments provide an ensemble of probabilities $\{p_1, p_2, ...\}$ with the corresponding times $\{t_1, t_2, ...\}$.

For given overlaps $\langle E_j|\Psi_0\rangle$, $\langle E_j|\Psi\rangle$ we can immediately predict some restrictions for the evolution of the probability p_k in the N-dimensional Hilbert space. When the overlaps vanish for some of the eigenstates $|E_n\rangle$, the evolution cannot reach those states and the accessible Hilbert space is restricted to the states $|E_j\rangle$ with $j \neq n$. This reduction of the Hilbert space can be interpreted as Hilbert-space localization [16] or Hilbert-space fragmentation [17]. This effect can be associated with spontaneous symmetry breaking, induced by the choice of the initial and measured states. In the case that the overlaps with some states are not strictly zero but very small, the access to those states may be negligible and can be ignored. This corresponds to complex dynamical behavior and requires a careful analysis. It is addressed briefly for the example of a bosonic Josephson junction in Section 3, where the mirror symmetry of the junction is spontaneously broken.

This paper is organized as follows. After the definition of the TCM in Section 2 we focus on the properties of its diagonal elements (Section 2.1). Then the effect of spectral degeneracies on the diagonal TCM elements are discussed in Section 2.2. In Section 3, we analyze the diagonal TCM elements in the specific example of a bosonic Josephson junction.

2. Time Correlation Matrix

We consider the transition amplitude from $|\Psi_0\rangle$ to $|\Psi\rangle$

$$u_k = \langle\Psi|e^{-iHt_k}|\Psi_0\rangle \, , \tag{1}$$

which is based on the unitary evolution with the Hamiltonian H from the initial state $|\Psi_0\rangle$. The probability of measuring the state $|\psi\rangle$ at time t_k is given by $|\langle\Psi|e^{-iHt_k}|\Psi_0\rangle|^2$. In other words, $p_k = |u_k|^2$ is the probability of finding the quantum system in the state $|\Psi\rangle$ after the unitary evolution from the initial state $|\Psi_0\rangle$ over the time t_k. Since the evolution is defined by the Hamiltonian H, we consider its eigenstates $\{|E_j\rangle\}_{j=1,...,N}$ and its corresponding eigenvalues $\{E_j\}_{j=1,...,N}$ and write the amplitude in spectral representation as

$$u_k = \sum_{j=1}^{N}\langle\Psi|E_j\rangle\langle E_j|\Psi_0\rangle e^{-iE_jt_k} \equiv \sum_{j=1}^{N} q_j e^{-iE_jt_k} \, . \tag{2}$$

Although the phases are not directly experimentally observable, their effects can be detected through the interference of different quantum states. For instance, the product of amplitudes at different times with probabilities of interfering amplitudes reads

$$\begin{aligned} u_k^* u_{k'} + u_{k'}^* u_k &= \tfrac{1}{2}\left(|u_k + u_{k'}|^2 - |u_k - u_{k'}|^2\right), \\ i(u_k^* u_{k'} - u_{k'}^* u_k) &= \tfrac{1}{2}\left(|u_k + iu_{k'}|^2 - |u_k - iu_{k'}|^2\right), \end{aligned} \tag{3}$$

where the probabilities can be detected in interferometric measurements. These relations suggest considering the correlation of the amplitudes u_k and $u_{k'}$ at different times through the TCM $\langle u_k^* u_{k'} \rangle_\tau$, where the average $\langle ... \rangle_\tau$ is taken from the distribution of times $\{t_k\}$ as a result of inaccurate clocks: The time is measured by a clock in each laboratory by counting time steps $\{\tau_n\}$. These clocks have a limited accuracy, such that the time steps vary randomly. This implies a sequence of measurements in each laboratory, where the clocks indicate k time steps and a total evolution time $t_k = \tau_1 + \cdots + \tau_k$ for different values of k. Now we can compare the measured sequences of different laboratories. This provides a distribution of results for $u_k^* u_{k'}$ due to different inaccurate clocks, where we assume that the fluctuations of the time steps $\{\tau_n\}$ are independently and equally distributed. Then the TCM is, as a spectral representation,

$$\langle u_k^* u_{k'} \rangle_\tau = \sum_{j,j'} q_j^* q_{j'} \langle e^{iE_j(\tau_1+\cdots+\tau_k)} e^{-iE_{j'}(\tau_1+\cdots+\tau_{k'})} \rangle_\tau \tag{4}$$

$$= \sum_{j,j'} q_j^* q_{j'} \begin{cases} \langle e^{i(E_j - E_{j'})(\tau_1+\cdots+\tau_k)} \rangle_\tau \langle e^{-iE_{j'}(\tau_{k+1}+\cdots+\tau_{k'})} \rangle_\tau & k' > k \\ \langle e^{i(E_j - E_{j'})(\tau_1+\cdots+\tau_{k'})} \rangle_\tau \langle e^{iE_{j}(\tau_{k'+1}+\cdots+\tau_{k})} \rangle_\tau & k' < k \\ \langle e^{i(E_j - E_{j'})(\tau_1+\cdots+\tau_k)} \rangle_\tau & k' = k \end{cases}. \tag{5}$$

When $\lambda_j = \langle e^{iE_j\tau} \rangle_\tau$ and $\lambda_{jj'} = \langle e^{i(E_j - E_{j'})\tau} \rangle_\tau$ the TCM elements become

$$\langle u_k^* u_{k'} \rangle_\tau = \sum_{j,j'} q_j^* q_{j'} \begin{cases} \lambda_{jj'}^k \lambda_{j'}^{k'-k} & k' \geq k \\ \lambda_{jj'}^{k'} \lambda_j^{*k-k'} & k' < k \end{cases}. \tag{6}$$

The TCM decays exponentially with $|k - k'|$, provided that $|\lambda_j| < 1$. Moreover, for fixed $|k - k'|$ the TCM is constant for the diagonal elements $\lambda_{jj} = 1$. This reflects the fact that a unitary evolution between the same energy eigenstates gives just a phase factor $e^{-iE_j\tau}$ (cf. Equation (2)). For different energy states, on the other hand, these phase factors lead to a decay due to interference effects after the time average.

2.1. Diagonal Elements of the TCM

The diagonal TCM element $\langle |u_k|^2 \rangle_\tau$ is the probability of measuring the state $|\Psi\rangle$ at time t_k. Before time averaging, the expression $|u_k|^2$ is a diagonal element of the density matrix $\rho(t_k)$ with respect to the state $|\Psi\rangle$. The trace of $|u_k|^2$ with respect to all states $|\Psi\rangle$ of the underlying Hilbert space is the spectral form factor, often used for the characterization of many-body quantum chaos [18–20]. We only mention this, but do not study it here. In comparison with the spectral form factor, the diagonal elements of the TCM refer to a specific measured state and to the time index k of the measurement. Thus, $\langle |u_k|^2 \rangle_\tau$ might be useful when we analyze a large set of experimental data for a specific state at different times.

According to Equation (6) the average transition probability $\langle |u_k|^2 \rangle_\tau$ reads

$$\langle |u_k|^2 \rangle_\tau = P_N + \sum_{j,j'=1;j'\neq j}^{N} q_j^* q_{j'} \lambda_{jj'}^k, \quad P_N = \sum_{j=1}^{N} |q_j|^2. \tag{7}$$

The term P_N describes the recurrent behavior, which does not depend on time. It is the asymptotic transition probability for $k \to \infty$

$$P_N = \lim_{k\to\infty} \langle |u_k|^2 \rangle_\tau , \tag{8}$$

provided that the energy levels are not degenerated. The case of degenerate energy levels is discussed in the next section. The second term in Equation (7) decays exponentially with time due to $|\lambda_{jj'}| < 1$, and only this term describes a change of the transition probability during the evolution of the quantum system. This result provides a separation of the diagonal elements of the TCM into a static recurrent term P_N and a dynamic term that decays quickly.

The recurrent term P_N provides important information regarding the properties of the quantum system. Since $|q_j|^2 = |\langle\Psi|E_j\rangle|^2 |\langle\Psi_0|E_j\rangle|^2$ is a product of the overlaps between the energy eigenstate $|E_j\rangle$ with the initial state and with the measured state, it is a measure of how much this energy eigenstate contributes to the transition $|\Psi_0\rangle \to |\Psi\rangle$ during the unitary evolution. This can be used, for instance, to describe localization with the asymptotic behavior of the return probability to the initial state $|\Psi_0\rangle \to |\Psi_0\rangle$: With the dimensionality N of the underlying Hilbert space, we get localization when $\lim_{N\to\infty} P_N > 0$, whereas the absence of localization is characterized by $\lim_{N\to\infty} P_N = 0$ [16]. This can be understood by noting that the normalization of quantum states implies $\sum_{j=1}^{N} |\langle\Psi_0|E_j\rangle|^2 = 1$, and that for a localized state, only a few energy eigenstates have nonzero overlaps with $|\Psi_0\rangle$. For a delocalized state, on the other hand, the overlaps are nonzero for a number of order N of energy eigenstates. A special case is when all these overlaps are equal. In that case we have $|\langle\Psi_0|E_j\rangle|^2 = 1/N$ due to the normalization, which implies $P_N = 1/N$. Anderson localization is associated with a random Hamiltonian [21]. According to the above described picture, we can also consider Hilbert-space localization for a deterministic Hamiltonian, which depends strongly on the initial state. For an energy eigenstate, the system will always remain in the latter under unitary evolution. More generally, if the initial state is a superposition of m energy eigenstates, the system will always remain inside the m–dimensional Hilbert space, spanned by these energy eigenstates. In the case, where the initial state is the eigenstate of H_0 of the Hamiltonian $H = H_0 + \eta H_1$. η is a small parameter. ηH_1 is a small perturbation. In that case, it is possible that this perturbation provides an exponentially decaying evolution away the initial state. This would be considered as exponential Hilbert-space localization.

In the subsequent discussion we focus on the diagonal elements of the TCM, since the off-diagonal TCM elements decay exponentially with $|k - k'|$ according to Equation (6).

2.2. *Effect of Spectral Degeneracies*

Assuming that there is a spectral degeneracy $E_1 = E_2$, we have $\lambda_{12} = \lambda_{21} = 1$, and the diagonal TCM elements in Equation (7) read in this case

$$\langle |u_k|^2 \rangle_\tau = \sum_{j,j'=1}^{N} q_j^* q_{j'} \lambda_{jj'}^k = P_N + q_1^* q_2 + q_2^* q_1 + \sum_{j,j'=1;j'\neq j;(j,j')\neq(1,2),(2,1)}^{N} q_j^* q_{j'} \lambda_{jj'}^k , \tag{9}$$

such that the recurrent part of the transition probability becomes

$$\lim_{k\to\infty} \langle |u_k|^2 \rangle_\tau = P_N + q_1^* q_2 + q_2^* q_1 = |q_1 + q_2|^2 + \sum_{j=3}^{N} |q_j|^2 . \tag{10}$$

Thus, the effect of a spectral degeneracy is a change of the recurrent and the decaying behavior, where the recurrent term changes by $|q_1 + q_2|^2 - |q_1|^2 - |q_2|^2$. This means that the diagonal elements of the TCM are very sensitive in terms of spectral degeneracies.

After applying a discrete Fourier transformation to the decaying part of $\langle |u_k|^2 \rangle_T$, we obtain the function

$$\begin{aligned} \tilde{U}_d(e^{i\omega}) &= \sum_{k\geq 1} e^{i\omega k} \sum_{j,j'=1;k'\neq k;(j,j')\neq(1,2),(2,1)}^{N} q_j^* q_{j'} \lambda_{jj'}^k \\ &= \sum_{j,j'=1;j'\neq j;(j,j')\neq(1,2),(2,1)}^{N} q_j^* q_{j'} \frac{\lambda_{jj'}}{e^{-i\omega} - \lambda_{jj'}}, \end{aligned} \tag{11}$$

which is a function of ω on the interval $[0, 2\pi)$. In other words, $\tilde{U}_d(z)$ is a sum of poles inside the unit circle due to $|\lambda_{jj'}| < 1$. The poles $\lambda_{nn'}$ and $\lambda_{n'n}$ approach the unit circle when we get closer to a degeneracy of E_n and $E_{n'}$. This should be visible in $\tilde{U}_d(e^{i\omega})$. The corresponding decay time $T_d = -1/\log|\lambda_{nn'}|$ diverges due to $|\lambda_{nn'}| \sim 1$. Therefore, the decay time T_d is a measure of the distance from a spectral degeneracy; it diverges when we approach the degeneracy. In general, we can define

$$T_m = \max_{j,j'=1,\ldots,N} -\frac{1}{\log|\lambda_{jj'}|} \tag{12}$$

as the largest decay time as a measure of level degeneracy.

3. Example: Bosonic Josephson Junction

In this section, we study the diagonal TCM elements of a bosonic Josephson junction (BJJ) with N bosons as a closed quantum system. The motivation for choosing this example is at least threefold: The model is (i) simple enough but not trivial, with interesting features based on tunneling and boson–boson interaction; (ii) it can be solved exactly; and (iii) it has been realized experimentally [22,23] with applications for commercial quantum computers [24]. The BJJ consists of two identical wells filled with interacting bosons and a tunneling junction between them. More formally, it is defined by the Bose–Hubbard Hamiltonian [25]:

$$H = -\frac{J}{2}(a_l^\dagger a_r + a_r^\dagger a_l) + U(n_l^2 + n_r^2), \quad n_{l,r} = a_{l,r}^\dagger a_{l,r}, \tag{13}$$

where $a_{l,r}^\dagger$ ($a_{l,r}$) are the creation (annihilation) operators in the left and right wells, respectively. The first term of H describes tunneling of atoms between the wells, and for $U > 0$ the second term represents a repulsive particle–particle interaction that favors energetically a symmetric distribution of bosons in the double well. Without tunneling (i.e., for $J = 0$) there are two-fold degenerate energy levels $E_k = U[(N-k)^2 + k^2]/2$ with eigenstates that are superpositions of the product Fock state $|k, N-k\rangle$ ($\equiv |k\rangle \otimes |N-k\rangle$) and its mirror image $|N-k, k\rangle$. This two-fold degeneracy is similar to the two-fold degeneracy of the Ising model, mentioned in the Introduction. In contrast to the Z_2 symmetry of the Ising model, the BJJ Hamiltonian has mirror-type symmetry due to the double well structure. Individual tunneling of bosons between the wells plays the role of the symmetry-breaking term, similarly to a local spin flip in the Ising model. Thus, an arbitrarily small tunneling parameter J will lift the two-fold degeneracy of the Fock states. A difference between the Ising model and the BJJ is that without tunneling in the latter, all energy levels are two-fold degenerate. Therefore, the initial state can be prepared in any of these energy levels to follow the evolution due to tunneling in the vicinity of degenerate levels. This is important because the degenerate ground state may not be reached due to the energy conservation in the closed quantum system.

For the following we use $|\Psi_0\rangle = |0, N\rangle$ as the initial state and $|\Psi\rangle = |N, 0\rangle$ as the measured state. Then we define the return probability (RP) and the transition probability (TP) as

$$|u_{r,k}|^2 = |\langle 0, N|e^{-iHt_k}|0, N\rangle|^2, \quad |u_{t,k}|^2 = |\langle N, 0|e^{-iHt_k}|0, N\rangle|^2. \tag{14}$$

Since both states $|\Psi_0\rangle$, $|\Psi\rangle$ are eigenstates of H in the absence of tunneling ($J = 0$), we get

$$|u_{r,k}|^2 = 1\,, \quad |u_{t,k}|^2 = 0 \tag{15}$$

for any k due to orthogonality. This reflects the fact that this pair of states breaks the mirror symmetry of the double well. The opposite extreme is the BJJ without boson–boson interaction ($U = 0$), which is more complex and will be discussed in the next subsection. As we will see, this case can be described by simple functions for $|u_{r,k}|^2$, $|u_{t,k}|^2$. For the interplay of tunneling and boson–boson interaction ($J, U \neq 0$), the behavior is more complex and we rely on the time averaged expressions $\langle|u_{r,k}|^2\rangle_\tau$, $\langle|u_{t,k}|^2\rangle_\tau$ with $\tau_k = \bar{\tau} + \tau'_k$ and an exponential distribution for τ'_k:

$$\langle ... \rangle_\tau = \int_0^\infty ... \prod_{n\geq 1} e^{-\tau'_n} d\tau'_n\,, \tag{16}$$

where the time is measured in units of $\hbar/J$.

3.1. Non-Interacting Bosons

For $U = 0$ the spectrum of H consists of equidistant energy levels $E_j = -J(N/2 - j)$ ($j = 0, 1, ..., N$) and eigenstates

$$|E_j\rangle = \frac{2^{-N/2}}{\sqrt{j!(N-j)!}}(a_l^\dagger + a_r^\dagger)^j (a_l^\dagger - a_r^\dagger)^{N-j}|0,0\rangle\,.$$

Then the RP and the TP at time t_k for N bosons read

$$\begin{aligned} |u_{r,k}|^2 &= |\langle 0,N|e^{-iHt_k}|0,N\rangle|^2 = |\cos^N(Jt_k/2)|^2\,, \\ |u_{t,k}|^2 &= |\langle N,0|e^{-iHt_k}|0,N\rangle|^2 = |\sin^N(Jt_k/2)|^2\,. \end{aligned} \tag{17}$$

For the average TP we get $t_k = k\bar{\tau} + \tau'_1 + \cdots + \tau'_k$ and the exponential distribution of Equation (16):

$$\langle|u_{t,k}|^2\rangle_\tau = 2^{-2N}\sum_{l,l'=0}^{N}\binom{N}{l}\binom{N}{l'}(-1)^{l+l'}\left[\frac{e^{-iJ\bar{\tau}(l-l')}}{1-iJ(l-l')}\right]^k$$

$$= 2^{-2N}\sum_{l=0}^{N}\binom{N}{l}^2 + 2^{-2N}\sum_{l,l'=0;l'\neq l}^{N}\binom{N}{l}\binom{N}{l'}(-1)^{l+l'}\left[\frac{e^{-iJ\bar{\tau}(l-l')}}{1-iJ(l-l')}\right]^k\,, \tag{18}$$

whereas the average RP is the same without the factor $(-1)^{l+l'}$. From these results we get for the asymptotic value at $k \sim \infty$ for both probabilities

$$P_N \sim \langle|u_{r,k}|^2\rangle_\tau \sim \langle|u_{t,k}|^2\rangle_\tau \sim 2^{-2N}\sum_{l=0}^{N}\binom{N}{l}^2\,, \tag{19}$$

which reflects the mirror symmetry of the BJJ.

3.2. Interacting Bosons

Exact solutions exist for this model also for $J, U \neq 0$, but in contrast to the non-interacting case they are complex and difficult to present in general. For instance, the resolvent is a meromorphic function with polynomials of orders N and $N+1$ [16,26]. Therefore, we only plot the results for the RP and the TP and their averaged counterparts here. For transparency, we chose for all subsequent plots $N = 20$ bosons.

First, the evolutions of the RP and the TP are presented in Figure 1 for two values ($u = 1, 2$) of the interaction parameter $u = NU/J$. This clearly reveals that the RP dominates over the TP for increasing u, as we expected from the results of the two limits

$J = 0$ and $U = 0$. It is interesting to note that in a mean-field (classical) approximation of the BJJ there is a sharp phase transition in terms of the interaction parameter, where the mean-field TP is completely suppressed when $u \geq u_c = 2$ [27]. The strong interaction phase is also called the self-trapping phase. The analogue of the latter in the quantum BJJ is Hilbert-space localization, reflected by the scaling behavior of the inverse participation ratio [16]. This also indicates the existence of a critical u_c.

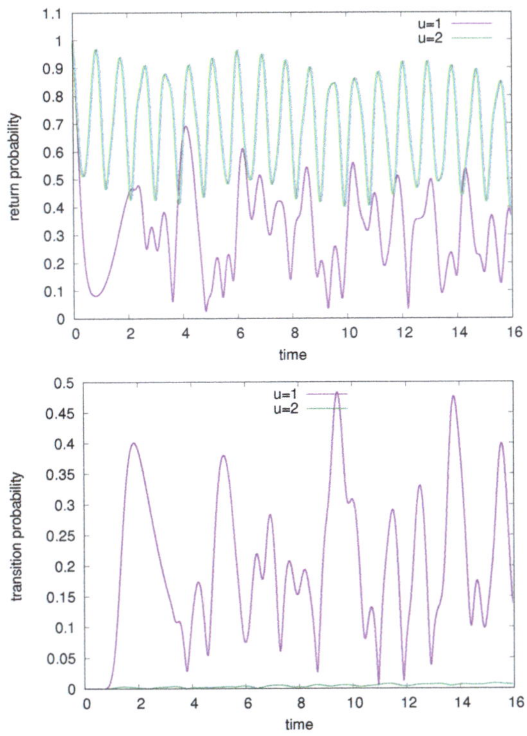

Figure 1. Signatures of a qualitative change in the evolution of a bosonic Josephson junction upon increasing interaction strength u. The plots represent the dynamics of 20 bosons for $u = 1, 2$, where the top panel gives the return probability $|u_{r,k}|^2$ and the bottom panel the transition probability $|u_{t,k}|^2$.

In Figure 2, the effect of time averaging on $|u_{r,k}|^2$ and $|u_{t,k}|^2$ for $u = 1$ is visualized. It reflects the smoothing of the strongly fluctuating dynamics with a recurrent and a decaying contribution according to Equations (7) and (9). It is obvious that the separation of the recurrent and the decaying behavior is not feasible without time averaging.

The existence of a critical interaction strength $u_c \approx 1.89...$ is demonstrated in Figure 3, where the $\langle |u_{r,k}|^2 \rangle_T$ jumps upon increasing u at u_c. Moreover, $\langle |u_{t,k}|^2 \rangle_T$ develops a characteristic peak at u_c. This behavior reflects the appearance of nearly degenerate energy levels, as described in Section 2.2.

Finally, in Figure 4 the change of the time scales for the decay of the average TP $\langle |u_{t,k}|^2 \rangle_T$ is visualized for $u = 1.7, ..., 2.2$. The decay is reduced by increasing interaction strength u. This reflects the fact that the splitting of the energy levels is reduced by the interaction, as we expected.

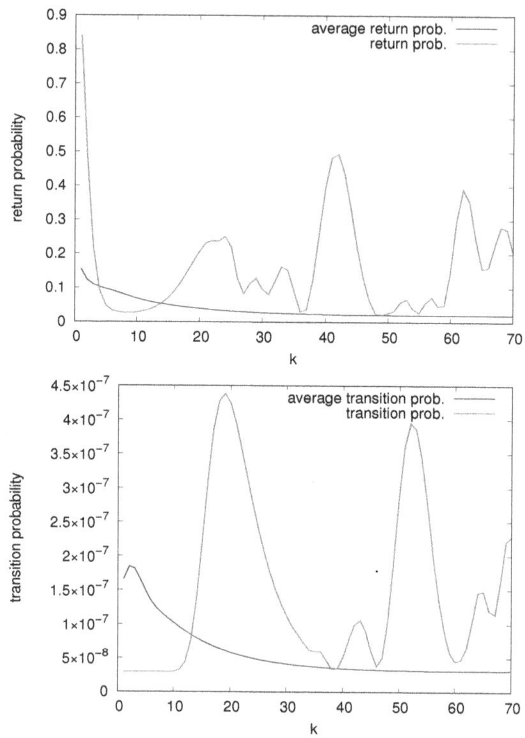

Figure 2. A comparison of the return probability $|u_{r,k}|^2$ and the average return probability $\langle |u_{r,k}|^2 \rangle_\tau$ (top panel) and of the corresponding transition probabilities (bottom panel). The average was taken with respect to the exponential distribution of Equation (16). The interaction parameter is $u = 1$, and $\bar{\tau} = 1/10$.

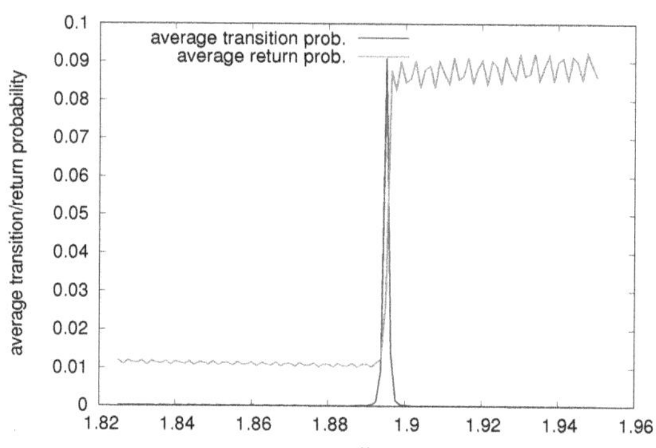

Figure 3. The critical regime of the Hilbert-space localization with $u_c \approx 1.89$ is visualized with $\langle |u_{r,k}|^2 \rangle_\tau$ and $\langle |u_{t,k}|^2 \rangle_\tau$ at $k = 70$.

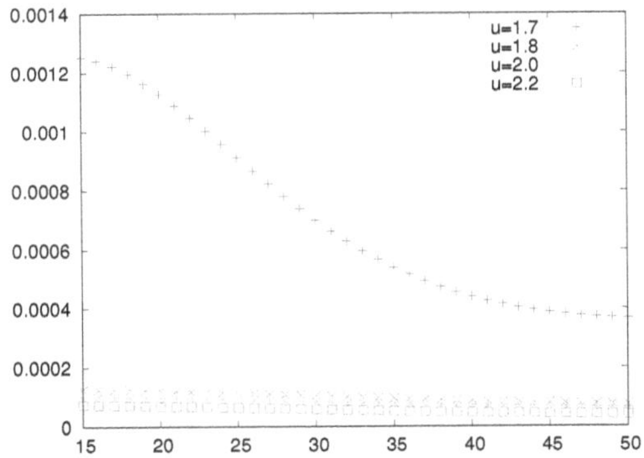

Figure 4. Decay of $\langle|u_{t,k}|^2\rangle_T$ for different interaction parameters $u = 1.7, \ldots, 2.2$.

4. Discussion and Conclusions

Our analysis of the quantum unitary evolution was strictly focused on the result of a single measurement in each of many identical experiments, which were subject of a unitary evolution. Averaging with respect to the statistical outcome due to measurements at randomly distributed times led to the TCM. We focused on the diagonal TCM elements to study the evolution of the quantum system. The analysis of the off-diagonal TCM elements was the subject of a previous work [15]. Similarly to the off-diagonal TCM elements, the diagonal TCM elements revealed separation of the evolution into a static recurrent part and a dynamic decaying part. We found that the decay rate of the latter is related to the spacing between energy levels, which diverges when the spacing vanishes near a degeneracy. Thus, the decay rate is a quantity that can be used to detect symmetry changes or the appearance of spontaneous symmetry breaking. This was observed in the example of the BJJ: In this model the energy levels were two-fold degenerate in the limit $u \to \infty$. This is reflected in Figure 4, where the decay decreases with increasing u.

Another interesting aspect of the BJJ is the transition to Hilbert-space localization [16]. This transition was also detected with the help of the average RP and average TP in Figure 3, where the average RP experiences a jump to a higher value for $u > u_c$. On the other hand, the average TP has only a sharp peak near u_c but has the same value away from u_c.

We can conclude that time averaging over an ensemble of measurements is crucial for extracting the (static) recurrent behavior and the (dynamic) decaying behavior. This can be formulated in terms of the TCM. The separation of static and dynamic behavior is generic for the unitary quantum evolution. The TCM provides a tool to analyze properties of the quantum system, which are associated with spectral degeneracy in the case of phase transitions. It can be applied to theoretical calculations and to experimental data, collected from many experiments in identical quantum systems. We have demonstrated in the case of the BJJ that the TCM approach delivers interesting generic information. The BJJ can be considered as a building block of the Bose–Hubbard model on a lattice. Therefore, the TCM approach should be applicable to more complex quantum models, including bosonic and fermionic Hubbard models and quantum spin systems.

Funding: This work was supported by the Julian Schwinger Foundation.

Acknowledgments: I am grateful to Eli Barkai for interesting discussions regarding the Hilbert-space fragmentation.

Conflicts of Interest: The author declares no conflict of interest.

References

1. Wigner, E.P. Characteristic Vectors of Bordered Matrices With Infinite Dimensions. *Ann. Math.* **1955**, *62*, 548–564. [CrossRef]
2. Porter, C.E.; Thomas, R.G. Fluctuations of Nuclear Reaction Widths. *Phys. Rev.* **1956**, *104*, 483–491. [CrossRef]

3. Dyson, F.J. Statistical Theory of the Energy Levels of Complex Systems. I. *J. Math. Phys.* **1962**, *3*, 140–156. [CrossRef]
4. Porter, C.E. Book Review: Statistical theories of spectra: fluctuations. C.E. PORTER (Academic Press, New York, 1965. xv-576 p. 5.95 paper, 9.50 cloth). *Nucl. Phys.* **1966**, *78*, 696–696. [CrossRef]
5. Wigner, E.P. Random Matrices in Physics. *SIAM Rev.* **1967**, *9*, 1–23. [CrossRef]
6. Mehta, M. *Random Matrices*; Number v. 142 in Pure and Applied Mathematics; Academic Press: Cambridge, MA, USA, 2004.
7. Beenakker, C.W.J. Random-matrix theory of quantum transport. *Rev. Mod. Phys.* **1997**, *69*, 731–808. [CrossRef]
8. Dyson, F.J. Statistical Theory of the Energy Levels of Complex Systems. II. *J. Math. Phys.* **1962**, *3*, 157–165. doi: 10.1063/ 1.1703774. [CrossRef]
9. Dyson, F.J. Statistical Theory of the Energy Levels of Complex Systems. III. *J. Math. Phys.* **1962**, *3*, 166–175. . [CrossRef]
10. van Enk, S.J.; Beenakker, C.W.J. Measuring Trρ^n on Single Copies of ρ Using Random Measurements. *Phys. Rev. Lett.* **2012**, *108*, 110503. [CrossRef]
11. Elben, A.; Vermersch, B.; Dalmonte, M.; Cirac, J.I.; Zoller, P. Rényi Entropies from Random Quenches in Atomic Hubbard and Spin Models. *Phys. Rev. Lett.* **2018**, *120*, 050406. [CrossRef]
12. Vermersch, B.; Elben, A.; Dalmonte, M.; Cirac, J.I.; Zoller, P. Unitary n-designs via random quenches in atomic Hubbard and spin models: Application to the measurement of Rényi entropies. *Phys. Rev. A* **2018**, *97*, 023604. [CrossRef]
13. Li, Y.; Chen, X.; Fisher, M.P.A. Measurement-driven entanglement transition in hybrid quantum circuits. *Phys. Rev. B* **2019**, *100*, 134306. [CrossRef]
14. Skinner, B.; Ruhman, J.; Nahum, A. Measurement-Induced Phase Transitions in the Dynamics of Entanglement. *Phys. Rev. X* **2019**, *9*, 031009. [CrossRef]
15. Ziegler, K. Quantized dynamics in closed quantum systems. *J. Phys. Math. Theor.* **2021**, *54*, 205303. [CrossRef]
16. Cohen, D.; Yukalov, V.I.; Ziegler, K. Hilbert-space localization in closed quantum systems. *Phys. Rev. A* **2016**, *93*, 042101. [CrossRef]
17. Moudgalya, S.; Motrunich, O.I. Hilbert Space Fragmentation and Commutant Algebras. *arXiv* **2021**, arXiv:2108.10324.
18. Cotler, J.S.; Gur-Ari, G.; Hanada, M.; Polchinski, J.; Saad, P.; Shenker, S.H.; Stanford, D.; Streicher, A.; Tezuka, M. Black holes and random matrices. *J. High Energy Phys.* **2017**, *2017*, 118. [CrossRef]
19. Kos, P.; Ljubotina, M.; Prosen, T.C.V. Many-Body Quantum Chaos: Analytic Connection to Random Matrix Theory. *Phys. Rev. X* **2018**, *8*, 021062. [CrossRef]
20. Chan, A.; De Luca, A.; Chalker, J.T. Solution of a Minimal Model for Many-Body Quantum Chaos. *Phys. Rev. X* **2018**, *8*, 041019. [CrossRef]
21. Anderson, P.W. Absence of Diffusion in Certain Random Lattices. *Phys. Rev.* **1958**, *109*, 1492–1505. [CrossRef]
22. Bar-Gill, N.; Kurizki, G.; Oberthaler, M.; Davidson, N. Dynamic control and probing of many-body decoherence in double-well Bose–Einstein condensates. *Phys. Rev. A* **2009**, *80*, 053613. [CrossRef]
23. Juliá-Díaz, B.; Zibold, T.; Oberthaler, M.K.; Melé-Messeguer, M.; Martorell, J.; Polls, A. Dynamic generation of spin-squeezed states in bosonic Josephson junctions. *Phys. Rev. A* **2012**, *86*, 023615. [CrossRef]
24. Barends, R.; Kelly, J.; Megrant, A.; Sank, D.; Jeffrey, E.; Chen, Y.; Yin, Y.; Chiaro, B.; Mutus, J.; Neill, C.; et al. Coherent Josephson Qubit Suitable for Scalable Quantum Integrated Circuits. *Phys. Rev. Lett.* **2013**, *111*, 080502. [CrossRef] [PubMed]
25. Gati, R.; Oberthaler, M.K. A bosonic Josephson junction. *J. Phys. At. Mol. Opt. Phys.* **2007**, *40*, R61–R89. [CrossRef]
26. Ziegler, K. Dynamical creation of entangled bosonic states in a double well. *J. Phys. At. Mol. Opt. Phys.* **2011**, *44*, 145302. [CrossRef]
27. Milburn, G.J.; Corney, J.; Wright, E.M.; Walls, D.F. Quantum dynamics of an atomic Bose–Einstein condensate in a double-well potential. *Phys. Rev. A* **1997**, *55*, 4318–4324. [CrossRef]

symmetry

MDPI

Article

Morphology of an Interacting Three-Dimensional Trapped Bose–Einstein Condensate from Many-Particle Variance Anisotropy

Ofir E. Alon [1,2]

[1] Department of Mathematics, University of Haifa, Haifa 3498838, Israel; ofir@research.haifa.ac.il
[2] Haifa Research Center for Theoretical Physics and Astrophysics, University of Haifa, Haifa 3498838, Israel

Abstract: The variance of the position operator is associated with how wide or narrow a wave-packet is, the momentum variance is similarly correlated with the size of a wave-packet in momentum space, and the angular-momentum variance quantifies to what extent a wave-packet is non-spherically symmetric. We examine an interacting three-dimensional trapped Bose–Einstein condensate at the limit of an infinite number of particles, and investigate its position, momentum, and angular-momentum anisotropies. Computing the variances of the three Cartesian components of the position, momentum, and angular-momentum operators we present simple scenarios where the anisotropy of a Bose–Einstein condensate is different at the many-body and mean-field levels of theory, despite having the same many-body and mean-field densities per particle. This suggests a way to classify correlations via the morphology of 100% condensed bosons in a three-dimensional trap at the limit of an infinite number of particles. Implications are briefly discussed.

Keywords: Bose-Einstein condensates; infinite-particle-number limit; many-body theory; mean-field theory; position variance; momentum variance; angular-momentum variance; solvable models; harmonic-interaction model; anisotropy

Citation: Alon, O.E. Morphology of an Interacting Three-Dimensional Trapped Bose–Einstein Condensate from Many-Particle Variance Anisotropy. *Symmetry* **2021**, *13*, 1237. https://doi.org/10.3390/sym13071237

Academic Editors: V. I. Yukalov, V. S. Bagnato and Rashid G. Nazmitdinov

Received: 13 June 2021
Accepted: 5 July 2021
Published: 9 July 2021

1. Introduction

There has been an increasing interest in the theory and properties of trapped Bose–Einstein condensates at the limit of an infinite number of particles [1–12]. Here, one may divide the research questions into two, inter-connected groups. The first group of research questions deals with rigorous results, mainly proving when many-body and mean-field, Gross–Pitaevskii theories coincide at this limit, whereas the second group of questions deals with characterizing correlations in a trapped Bose–Einstein condensate based on the difference between many-body and mean-field properties at the infinite-particle-number limit. In [2], it has been shown that the ground-state energy per particle and density per particle computed at the many-body level of theory coincide with the respective mean-field results. The infinite-particle-number limit is defined such that the interaction parameter, i.e., the product of the number of bosons times the interaction strength (proportional to the scattering length), is held constant. Similarly, in [3] it has been shown that the reduced one-body and any reduced finite-n-body density matrix [13,14] per particle are 100% condensed, and that the leading natural orbital boils down to the Gross–Pitaevskii single-particle function. Analogous results connecting time-dependent many-body and mean-field theories are given in [4,5] and developments for mixtures in [9–11]. In [12], the many-boson wave-function at the infinite-particle-number limit has been constructed explicitly.

The difference between many-body and mean-field theories at the limit of an infinite number of particles, which as stated above coincide at the level of the energy, densities, and reduced density matrices per particle, starts to show up in variances of many-particle observables [6,7]. Of course, the wave-functions themselves differ and their overlap is smaller than one [8]. In evaluating the variances of many-particle observables two-body operators emerge whose combination with the elements of the reduced two-body density matrix can pick up even the tiniest depletion, which always exist due to the inter-particle interaction [6,7]. Here, the quantitative difference between the many-body and mean-field variances is a useful tool to benchmark many-body numerical approaches [15,16], whereas the qualitative differences serve to define and characterize the nature of correlations in 100% condensed bosons at the infinite-particle-number limit.

Qualitative differences between the many-body and mean-field variances per particle depend on both the system and observable under investigation and emerge because the 100% condensed bosons are interacting. In the ground-state of a one-dimensional double-well potential, the mean-field position variance per particle increases monotonously with the interaction parameter whereas, once about a single particle is excited outside the condensed mode, the many-body position variance per particle starts to decrease [6]. In the analogous time-dependent setup of a bosonic Josephson junction, the mean-field variance is oscillating and bound by the size of the junction, whereas the many-body variance increases to 'sizes' several times larger [7]. In two spatial dimensions additional features come out. The position variance per particle in a thin annulus can exhibit a different dimensionality [17] and both the position and momentum variances can exhibit opposite anisotropies [18] when computed at the many-body and mean-field levels of theory in an out-of-equilibrium quench dynamics. In two spatial dimensions the many-particle variance of the $\hat{L}_Z$ component of the angular-momentum operator becomes available, and used to analyze the lack of conservation of symmetries in the mean-field dynamics [19].

In the present work we analyze the many-particle position, momentum, and angular-momentum variances of a three-dimensional anisotropic trapped Bose–Einstein condensate at the limit of an infinite number of particles, focusing on three-dimensional scenarios that do not have (one-dimensional and) two-dimensional analogs. Mainly, the available permutations between the three Cartesian components of a many-particle operator, such as the position and momentum operators, allow one for various different anisotropies of the respective mean-field and many-body variances than in two spatial dimensions [18]. Furthermore, anisotropy of the angular-momentum variance can only be investigated when there is more than one component, and this occurs with the three Cartesian components of the angular-momentum operator in three spatial dimensions.

The structure of the paper is as follows. In Section 2 theory and definitions are developed. In Section 3 we present two applications where a common methodological line of investigation is that the variances at the many-body level of theory can be computed analytically. In Section 3.1, the anisotropy of the position and momentum variances in the out-of-equilibrium breathing dynamics of a Bose–Einstein condensate in a three-dimensional anisotropic harmonic potential are analyzed, and in Section 3.2, a solvable model is devised which allows one to analyze the anisotropy of the angular-momentum variances in the ground state of interacting bosons in a three-dimensional anisotropic harmonic potential. All quantities are computed at the infinite-particle-number limit. Finally, we summarize in Section 4.

2. Theory

The variances per particle of a many-particle observable $\hat{O} = \sum_{j=1}^{N} \hat{o}_j$ computed at the many-body (MB) and mean-field, Gross–Pitaevskii (GP) levels of theory are connected at the limit of an infinite number of particles by the following relation [6]:

$$\text{Var}|_{\text{MB}}(\hat{O}) = \text{Var}|_{\text{GP}}(\hat{O}) + \text{Var}|_{\text{correlations}}(\hat{O}). \quad (1)$$

Here, $\text{Var}|_{\text{MB}}(\hat{O}) = \lim_{N\to\infty}\left[\frac{1}{N}(\langle\Psi|\hat{O}^2|\Psi\rangle - \langle\Psi|\hat{O}|\Psi\rangle^2)\right]$, where Ψ is the solution of the many-particle Schrödinger equation, and $\text{Var}|_{\text{GP}}(\hat{O}) = \langle\psi_{GP}|\hat{o}^2|\psi_{GP}\rangle - \langle\psi_{GP}|\hat{o}|\psi_{GP}\rangle^2$, where ψ_{GP} is the solution of the corresponding Gross–Pitaevskii equation. Recall that the infinite-particle-number limit is defined such that the interaction parameter, i.e., the product of the number of bosons times the interaction strength, is kept fixed. Furthermore, only in the infinite-particle-number limit the density per particle is identical at the many-body and mean-field levels of theory, and thus (1) compares the variances per particle of the same density per particle. The correlations term, $\text{Var}|_{\text{correlations}}(\hat{O})$, quantifying the difference between the mean-field and many-body variances, depends on the elements of the reduced two-body density matrix where at least one of the indexes corresponds to a natural orbital higher than the condensed mode [6]. For non-interacting bosons the correlations term obviously vanishes. As stated above, one is interested in qualitative differences between $\text{Var}|_{\text{GP}}(\hat{O})$ and $\text{Var}|_{\text{MB}}(\hat{O})$ and their origin.

Consider a Bose–Einstein condensate for which the many-body variances of the three Cartesian components of, say, the position operator are different and satisfy, without loss of generality, the anisotropy

$$\text{Var}|_{\text{MB}}(\hat{X}) > \text{Var}|_{\text{MB}}(\hat{Y}) > \text{Var}|_{\text{MB}}(\hat{Z}). \tag{2a}$$

We define the following classification with respect to the possible different anisotropies of the respective mean-field position variances:

$$\begin{aligned}
\{1\}: \quad & \Big\{\text{Var}|_{\text{GP}}(\hat{X}) > \text{Var}|_{\text{GP}}(\hat{Y}) > \text{Var}|_{\text{GP}}(\hat{Z})\Big\}, \\
\{1,2\}: \quad & \Big\{\text{Var}|_{\text{GP}}(\hat{Y}) > \text{Var}|_{\text{GP}}(\hat{X}) > \text{Var}|_{\text{GP}}(\hat{Z}); \\
& \text{Var}|_{\text{GP}}(\hat{X}) > \text{Var}|_{\text{GP}}(\hat{Z}) > \text{Var}|_{\text{GP}}(\hat{Y}); \\
& \text{Var}|_{\text{GP}}(\hat{Z}) > \text{Var}|_{\text{GP}}(\hat{Y}) > \text{Var}|_{\text{GP}}(\hat{X})\Big\}, \\
\{1,2,3\}: \quad & \Big\{\text{Var}|_{\text{GP}}(\hat{Y}) > \text{Var}|_{\text{GP}}(\hat{Z}) > \text{Var}|_{\text{GP}}(\hat{X}); \\
& \text{Var}|_{\text{GP}}(\hat{Z}) > \text{Var}|_{\text{GP}}(\hat{X}) > \text{Var}|_{\text{GP}}(\hat{Y})\Big\}.
\end{aligned} \tag{2b}$$

Naturally, the classification (2b) follows the classes of the S_3 permutation group denoted by $\{1\}$, $\{1,2\}$, and $\{1,2,3\}$. If the mean-field variances exhibit anisotropy other than the anisotropy of the respective many-body variances, i.e., the ordering of the former does not belong to the class $\{1\}$, we may interpret that the mean-field and many-body morphologies of the Bose–Einstein condensate with respect to the operators under investigation are distinct. This implies that the correlations term in (1) becomes dominate for the variances of these operators. In the present work we investigate manifestations of definition (2) utilizing the many-particle position $(\hat{X}, \hat{Y}, \hat{Z})$, momentum $(\hat{P}_X, \hat{P}_Y, \hat{P}_Z)$, and angular-momentum $(\hat{L}_X, \hat{L}_Y, \hat{L}_Z)$ operators for classifying the morphology of 100% condensed trapped bosons at the limit of an infinite number of particles.

3. Applications

3.1. Position and Momentum Variances in an Out-of-Equilibrium Dynamics of a Three-Dimensional Trapped Bose–Einstein Condensate

Consider N structureless bosons trapped in a three-dimensional anisotropic harmonic potential and interacting by a general two-body interaction $\hat{W}(\mathbf{r} - \mathbf{r}')$. The frequencies of the trap satisfy, without loss of generality, $\omega_x < \omega_y < \omega_z$. We work with dimensionless quantities, $\hbar = m = 1$. Using Jacobi coordinates, $\mathbf{Q}_k = \frac{1}{\sqrt{k(k+1)}} \sum_{j=1}^{k} (\mathbf{r}_{k+1} - \mathbf{r}_j)$, $k = 1, \ldots, N-1$, $\mathbf{Q}_N = \frac{1}{\sqrt{N}} \sum_{j=1}^{N} \mathbf{r}_j$, where $\mathbf{r}_1, \ldots, \mathbf{r}_N$ are the coordinates in the laboratory frame, the Hamiltonian can be written as:

$$\hat{H}(\mathbf{Q}_1, \ldots, \mathbf{Q}_N) = -\frac{1}{2}\frac{\partial^2}{\partial \mathbf{Q}_N^2} + \frac{1}{2}\left(\omega_x^2 Q_{N,x}^2 + \omega_y^2 Q_{N,y}^2 + \omega_z^2 Q_{N,z}^2\right) + \hat{H}_{rel}(\mathbf{Q}_1, \ldots, \mathbf{Q}_{N-1}). \tag{3}$$

The 'relative' Hamiltonian $\hat{H}_{rel}$ collects all terms depending on the relative coordinates $\mathbf{Q}_1, \ldots, \mathbf{Q}_{N-1}$, and $\mathbf{Q}_k = (Q_{k,x}, Q_{k,y}, Q_{k,z})$. Suppose now that the bosons are prepared in the ground state of the non-interacting system. The ground-state is separable in the Jacoby coordinates and reads $\Phi(\mathbf{Q}_1,\ldots,\mathbf{Q}_N) = \left(\frac{\omega_x}{\pi}\right)^{\frac{N}{4}}\left(\frac{\omega_y}{\pi}\right)^{\frac{N}{4}}\left(\frac{\omega_z}{\pi}\right)^{\frac{N}{4}} e^{-\frac{1}{2}\sum_{k=1}^{N}\left(\omega_x Q_{k,x}^2+\omega_y Q_{k,y}^2+\omega_z Q_{k,z}^2\right)}$, where the relation $\sum_{j=1}^{N}\mathbf{r}_j^2 = \sum_{k=1}^{N}\mathbf{Q}_k^2$ connecting the laboratory and Jacoby coordinates is used. The solution of the time-dependent many-boson Schrödinger equation, $\hat{H}\Psi(t) = i\frac{\partial\Psi(t)}{\partial t}$, where Φ is the initial condition, reads $\Psi(t) = e^{-i\hat{H}t}\Phi$. Consequently, because of the center-of-mass separability of the Hamiltonian $\hat{H}$ and of Φ, the position and momentum variances per particle of the time-dependent state $\Psi(t)$, for a general inter-particle interaction $\hat{W}(\mathbf{r}-\mathbf{r}')$, are those of the static, non-interacting system:

$$\begin{aligned}
&\text{Var}|_{\text{MB}}(\hat{X}) = \frac{1}{2\omega_x}, \quad \text{Var}|_{\text{MB}}(\hat{Y}) = \frac{1}{2\omega_y}, \quad \text{Var}|_{\text{MB}}(\hat{Z}) = \frac{1}{2\omega_z},\\
&\text{Var}|_{\text{MB}}(\hat{P}_X) = \frac{\omega_x}{2}, \quad \text{Var}|_{\text{MB}}(\hat{P}_Y) = \frac{\omega_y}{2}, \quad \text{Var}|_{\text{MB}}(\hat{P}_Z) = \frac{\omega_z}{2}.
\end{aligned} \tag{4}$$

In other words, the anisotropies $\text{Var}|_{\text{MB}}(\hat{X}) > \text{Var}|_{\text{MB}}(\hat{Y}) > \text{Var}|_{\text{MB}}(\hat{Z})$ of the position operator and $\text{Var}|_{\text{MB}}(\hat{P}_Z) > \text{Var}|_{\text{MB}}(\hat{P}_Y) > \text{Var}|_{\text{MB}}(\hat{P}_X)$ of the momentum operator, when computed at the many-body level of theory, hold for all times during the out-of-equilibrium dynamics, see the constant-value (dashed) curves in Figures 1 and 2. We note that the variances per particle (4) hold for any number of bosons due to the separability of the center-of-mass. However, only at the limit of an infinite number of particles the density per particle coincides within many-body and mean-field levels of theory and can thus be exactly computed from the Gross–Pitaevskii equation.

What happens at the Gross–Pitaevskii level of theory? Can the mean-field variances have different orderings than the many-body variances, i.e., belong to other anisotropy classes based on the S_3 permutation group, see (2b), than to $\{1\}$? If yes, then why and how? The Gross–Pitaevskii or non-linear Schrödinger equation is given by $\left[-\frac{1}{2}\frac{\partial^2}{\partial\mathbf{r}^2} + \frac{1}{2}\left(\omega_x^2 x^2 + \omega_y^2 y^2 + \omega_z^2 z^2\right) + g|\psi_{GP}(\mathbf{r},t)|^2\right]\psi_{GP}(\mathbf{r},t) = i\frac{\partial\psi_{GP}(\mathbf{r},t)}{\partial t}$, where $g = 4\pi N a_s$ is the coupling constant and a_s the s-wave scattering length of the above two-body interaction $\hat{W}(\mathbf{r}-\mathbf{r}')$. The initial condition, as above, is the ground state of the non-interacting system, $\psi_{GP}(\mathbf{r},0) = \left(\frac{\omega_x}{\pi}\right)^{\frac{1}{4}}\left(\frac{\omega_y}{\pi}\right)^{\frac{1}{4}}\left(\frac{\omega_z}{\pi}\right)^{\frac{1}{4}} e^{-\frac{1}{2}\left(\omega_x x^2+\omega_y y^2+\omega_z z^2\right)}$. The Gross–Pitaevskii equation does not maintain the center-of-mass separability of the initial condition because of its non-linear term, which, therefore, can lead to variations of the position and momentum variances when computed at the mean-field level of theory.

Figures 1 and 2 display the Gross–Pitaevskii dynamics of the position and momentum variances per particle, respectively, for four coupling constants, $g = 0.18, 9.0, 18.0$, and 27.0. To integrate the three-dimensional Gross–Pitaevskii equation we use a box of size $[-10,10) \times [-10,10) \times [-10,10)$, a Fourier-discrete-variable-representation with 128^3 grid points and periodic boundary conditions, and the numerical implementation embedded in [20]. The dynamics are computed for the four coupling constants g and depicted by the oscillating (solid) curves in Figures 1 and 2. The left columns are for a 10% anisotropy of the harmonic trap, i.e., $\omega_x = 0.9$, $\omega_y = 1.0$, and $\omega_z = 1.1$, and the right columns are for a 20% anisotropy of the harmonic trap, namely, $\omega_x = 0.8$, $\omega_y = 1.0$, and $\omega_z = 1.2$. We remark that the expectation values per particle of the position $(\hat{X},\hat{Y},\hat{Z})$ and momentum $(\hat{P}_X,\hat{P}_Y,\hat{P}_Z)$ operators computed at the mean-field and many-body levels of theory coincide at the limit of an infinite number of particles and are all equal to zero in the present scenario.

For the smallest coupling constant, $g = 0.18$, we see that the mean-field variances oscillate with very small amplitudes around the respective constant values of the many-body variances. This means that the mean-field anisotropy of the position variance, $\text{Var}|_{\text{GP}}(\hat{X}) > \text{Var}|_{\text{GP}}(\hat{Y}) > \text{Var}|_{\text{GP}}(\hat{Z})$, and its many-body anisotropy, $\text{Var}|_{\text{MB}}(\hat{X}) > \text{Var}|_{\text{MB}}(\hat{Y}) > \text{Var}|_{\text{MB}}(\hat{Z})$, are alike. A similar situation is found for the momentum variance, namely, that the mean-field momentum anisotropy, $\text{Var}|_{\text{GP}}(\hat{P}_Z) > \text{Var}|_{\text{GP}}(\hat{P}_Y) > \text{Var}|_{\text{GP}}(\hat{P}_X)$, and the many-body anisotropy, $\text{Var}|_{\text{MB}}(\hat{P}_Z) > \text{Var}|_{\text{MB}}(\hat{P}_Y) > \text{Var}|_{\text{MB}}(\hat{P}_X)$, are the same. Consequently, we may conclude that for small coupling constants the anisotropy class of the position operator is $\{1\}$ and, likewise, the anisotropy class of the momentum operator is $\{1\}$, see (2). In other words, the contribution of the correlations term in (1) is marginal.

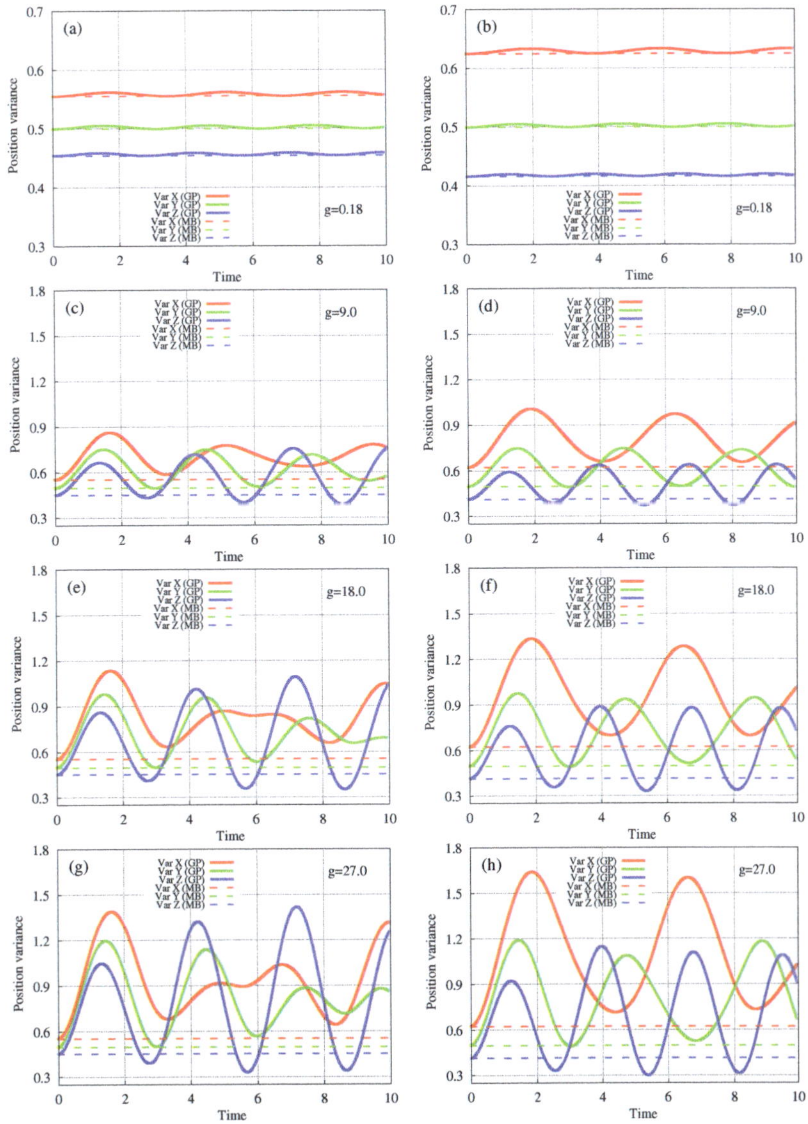

Figure 1. Many-particle position ($\hat{X}$, $\hat{Y}$, and $\hat{Z}$; in red, green, and blue) variance per particle as a function of time computed at the limit of an infinite number of particles within many-body (dashed lines) and mean-field (solid lines) levels of theory in an interaction-quench scenario. The harmonic trap is 10% anisotropic in panels (**a**,**c**,**e**,**g**) and 20% anisotropic in panels (**b**,**d**,**f**,**h**). The coupling constant g is indicated in each panel. Different anisotropy classes of the position variance emerge with time. See the text for more details. The quantities shown are dimensionless.

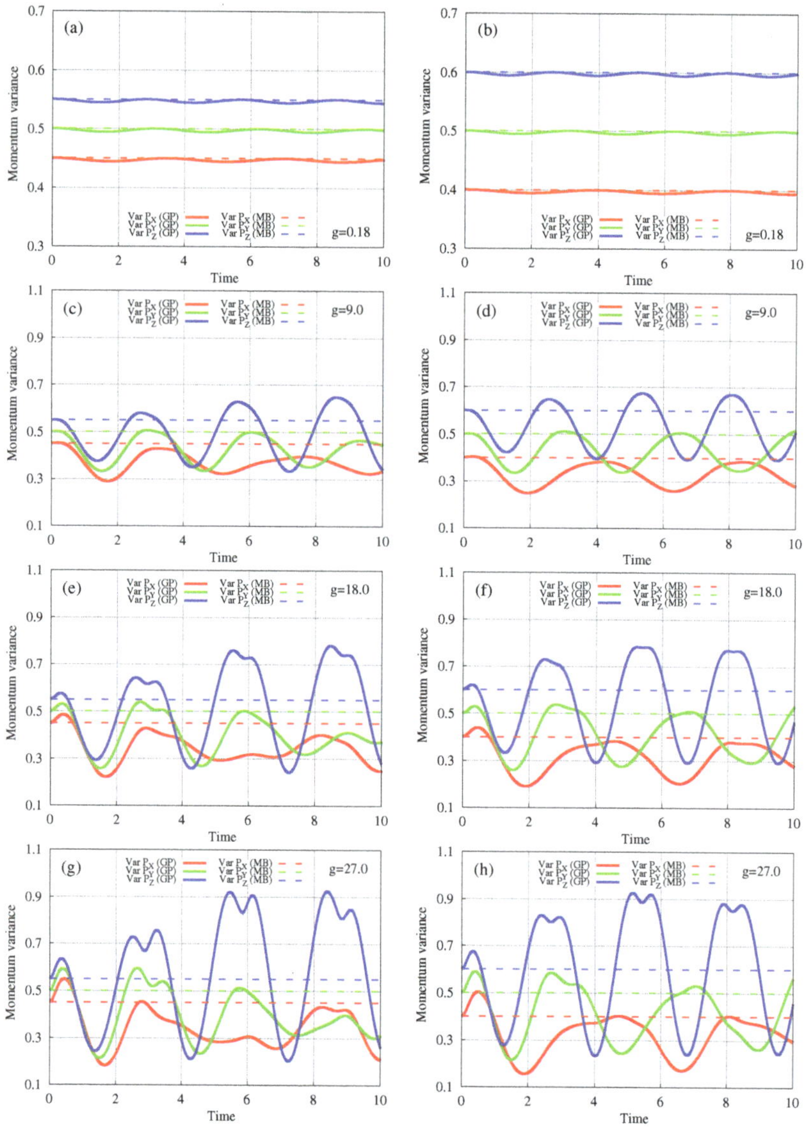

Figure 2. Many-particle momentum ($\hat{P}_X$, $\hat{P}_Y$, and $\hat{P}_Z$; in red, green, and blue) variance per particle as a function of time computed at the infinite-particle-number limit within many-body (dashed lines) and mean-field (solid lines) levels of theory in an interaction-quench scenario. The harmonic trap is 10% anisotropic in panels (**a**,**c**,**e**,**g**) and 20% anisotropic in panels (**b**,**d**,**f**,**h**). The coupling constant g is indicated in each panel. Different anisotropy classes of the momentum variance emerge with time. See the text for more details. The quantities shown are dimensionless.

The situation becomes more interesting for the larger coupling constants, $g = 9.0, 18.0$, and 27.0. We begin with the position variances, Figure 1. The variances are found to oscillate prominently, with much larger amplitudes than for $g = 0.18$, and, subsequently, to cross each other. There are three ingredients that enable and govern this crossing dynamics. The first, is that the amplitudes of oscillations of $\text{Var}|_{GP}(\hat{X})$, $\text{Var}|_{GP}(\hat{Y})$, and $\text{Var}|_{GP}(\hat{Z})$ are slightly different already at short times, with the former being the larger and the latter being the smaller (more prominent for 20% than for 10% trap anisotropy). The second, is that the respective frequencies of oscillations are also slightly different at short times, with the former being the smaller and the latter being the larger. Both features correlate with the ordering of the frequencies of the trap, $\omega_x < \omega_y < \omega_z$. The third ingredient is that the three Cartesian components are coupled to each other during the dynamics, what impacts the oscillatory pattern at intermediate and later times (more prominent for 10% than for 20% trap anisotropy, see Figure 1).

Combing the above, we find for the 10% trap anisotropy that around $t = 4.0$ $\text{Var}|_{GP}(\hat{Z}) > \text{Var}|_{GP}(\hat{Y}) > \text{Var}|_{GP}(\hat{X})$ takes place, around $t = 6.0$ $\text{Var}|_{GP}(\hat{X}) > \text{Var}|_{GP}(\hat{Y}) > \text{Var}|_{GP}(\hat{Z})$ holds (again), and around $t = 7.0$ $\text{Var}|_{GP}(\hat{Z}) > \text{Var}|_{GP}(\hat{Y}) > \text{Var}|_{GP}(\hat{X})$ occurs. In other words, the anisotropy class of the position variance starts as $\{1\}$ for $t = 0$, changes to $\{1,2\}$ around $t = 4$, is back to $\{1\}$ around $t = 6$, and becomes $\{1,2,3\}$ around $t = 7$. Furthermore, this pattern is found to be robust for different, increasing coupling constants, see Figure 1. For 20% trap anisotropy we find a different crossing patten of the position variances. The anisotropy class begins as $\{1\}$ for $t = 0$, changes at around $t = 3.0$ to $\{1,2\}$, and immediately after, at around $t = 3.75$, it is $\{1,2,3\}$. Now, around $t = 7.0$ there is a broad regime of anisotropy class $\{1,2\}$. Another difference of a geometrical origin between the dynamics in the 20% and 10% trap anisotropies can be seen for $g = 9$, see Figure 1c,d. Here, the coupling constant is sufficiently large to lead to crossing of all position variances for the 10% anisotropy trap, and, consequently, to the position anisotropy class $\{1,2,3\}$ (at around $t = 7$). On the other hand, for the 20% anisotropy trap the coupling constant is just short of allowing all position variances to cross each other and, clearly, the anisotropy class $\{1,2,3\}$ cannot occur (as it happens at around $t = 3.75$ for the further larger coupling constants, $g = 18$ and 27). All in all, we have demonstrated in a rather common (out-of-equilibrium quench) scenario the emergence of anisotropy classes other than $\{1\}$, i.e., $\{1,2\}$ and $\{1,2,3\}$, for the position operator of a Bose–Einstein condensate at the infinite-particle-number limit. Hence, the correlations term in (1) for the position variance becomes dominant in the dynamics.

The results for the momentum variances per particle, see Figure 2, follow similar and corresponding trends as those for the position operator, albeit the crossings of the respective momentum curves take place during slightly narrower time windows than for the position operator for the parameters used. Thus, for the 10% trap anisotropy we have $\text{Var}|_{GP}(\hat{P}_X) > \text{Var}|_{GP}(\hat{P}_Y) > \text{Var}|_{GP}(\hat{P}_Z)$ around $t = 4.25$, $\text{Var}|_{GP}(\hat{P}_Z) > \text{Var}|_{GP}(\hat{P}_Y) > \text{Var}|_{GP}(\hat{P}_X)$ in the vicinity of $t = 6.0$, and $\text{Var}|_{GP}(\hat{P}_Y) > \text{Var}|_{GP}(\hat{P}_X) > \text{Var}|_{GP}(\hat{P}_Z)$ around $t = 7.25$. Therefore, the anisotropy class of the momentum variance starts at $\{1\}$ for $t = 0$, turns to $\{1,2\}$ around $t = 4.25$, returns to $\{1\}$ for a wider time window around $t = 6.0$, and changes to $\{1,2,3\}$ around $t = 7.25$. For the 20% trap anisotropy we find, starting from the anisotropy class $\{1\}$ for $t = 0$, the class $\{1,2,3\}$ at around $t = 4.0$, the class $\{1,2\}$ at around $t = 4.75$, again the class $\{1\}$ at around $t = 6$, and once more the class $\{1,2,3\}$ at around $t = 9.5$. As for the position variance, we find the pattern to be robust for different, increasing coupling constants, see Figure 2. Furthermore, the above-discussed difference of a geometrical origin between the position-variance dynamics in the 20% and 10% trap anisotropies for $g = 9$ emerges also for the momentum variance, see Figure 2c,d. Here, the coupling constant is sufficiently large to lead to crossing of all momentum variances for the 10% trap anisotropy, but not for the 20% trap anisotropy. As a result, the former system exhibits also the anisotropy class $\{1,2,3\}$ for the momentum variance, whereas the latter only the $\{1,2\}$ anisotropy class. Summarizing, we have demonstrated in a simple scenario, of an out-of-equilibrium breathing dynamics, the emergence of anisotropy classes other than $\{1\}$, namely, $\{1,2\}$ and $\{1,2,3\}$, for the many-particle position as well as many-particle momentum operators of a trapped Bose–Einstein condensate at the limit of an infinite number of particles. When these latter anisotropy classes describe the morphology of the Bose–Einstein condensate, it implies that the correlations term in (1) governs the position and momentum variance dynamics at the infinite-particle-number limit.

3.2. Angular-Momentum Variance in the Ground State of a Three-Dimensional Trapped Bose–Einstein Condensate

The possibility to learn on the relations governing correlations and variance anisotropy between the different components of the angular-momentum operator opens up only in three spatial dimensions. Here, in the context of the present work, the challenge is to find a many-particle model where angular-momentum properties can be treated analytically at the many-body level of theory and in the limit of an infinite number of particles. Such a model is the three-dimensional anisotropic harmonic-interaction model, and the results presented below build on and clearly extends the investigation of the two-dimensional anisotropic harmonic-interaction model reported in [21]. The harmonic-interaction model has been used quite extensively including to model Bose–Einstein condensates [22–36]. Finally, and as a bonus, we mention that the three-dimensional anisotropic harmonic-interaction model can also be solved analytically at the mean-field level of theory, which is useful for the analysis.

In the laboratory frame the three-dimensional anisotropic harmonic-interaction model reads: $\hat{H}(\mathbf{r}_1,\ldots,\mathbf{r}_N) = \sum_{j=1}^N \left[-\frac{1}{2}\frac{\partial^2}{\partial \mathbf{r}_j^2} + \frac{1}{2}\left(\omega_x^2 x_j^2 + \omega_y^2 y_j^2 + \omega_z^2 z_j^2\right)\right] + \lambda_0 \sum_{1\le j<k}^N (\mathbf{r}_j - \mathbf{r}_k)^2$, i.e., it is obtained from the Hamiltonian (3) when the two-body interaction is $\hat{W}(\mathbf{r}-\mathbf{r}') = \lambda_0(\mathbf{r}-\mathbf{r}')^2$. Then, the 'relative' Hamiltonian is given explicitly by

$$\hat{H}_{rel}(\mathbf{Q}_1,\ldots,\mathbf{Q}_{N-1}) = \\ = \sum_{k=1}^{N-1}\left[-\frac{1}{2}\frac{\partial^2}{\partial \mathbf{Q}_k^2} + \frac{1}{2}\left(\sqrt{\omega_x^2 + 2N\lambda_0}Q_{k,x}^2 + \sqrt{\omega_y^2 + 2N\lambda_0}Q_{k,y}^2 + \sqrt{\omega_z^2 + 2N\lambda_0}Q_{k,z}^2\right)\right]. \tag{5}$$

The many-body ground state of $\hat{H}$ is readily obtained and given by

$$\Psi(\mathbf{Q}_1,\ldots,\mathbf{Q}_N) = \\ = \left(\frac{\omega_x}{\pi}\right)^{\frac{1}{4}}\left(\frac{\omega_y}{\pi}\right)^{\frac{1}{4}}\left(\frac{\omega_z}{\pi}\right)^{\frac{1}{4}}\left(\frac{\sqrt{\omega_x^2+2N\lambda_0}}{\pi}\right)^{\frac{N-1}{4}}\left(\frac{\sqrt{\omega_y^2+2N\lambda_0}}{\pi}\right)^{\frac{N-1}{4}}\left(\frac{\sqrt{\omega_z^2+2N\lambda_0}}{\pi}\right)^{\frac{N-1}{4}} \times \\ e^{-\frac{1}{2}\left(\sqrt{\omega_x^2+2N\lambda_0}\sum_{k=1}^{N-1}Q_{k,x}^2+\omega_x Q_{N,x}^2\right)} e^{-\frac{1}{2}\left(\sqrt{\omega_y^2+2N\lambda_0}\sum_{k=1}^{N-1}Q_{k,y}^2+\omega_y Q_{N,y}^2\right)} e^{-\frac{1}{2}\left(\sqrt{\omega_z^2+2N\lambda_0}\sum_{k=1}^{N-1}Q_{k,z}^2+\omega_z Q_{N,z}^2\right)}. \tag{6}$$

As states above, it is also possible to solve analytically the three-dimensional anisotropic harmonic-interaction model at the mean-field level of theory by generalizing [21,25]. The final result for the mean-field solution of the ground state reads

$$\psi_{GP}(\mathbf{r}) = \left(\frac{\sqrt{\omega_x^2+2\Lambda}}{\pi}\right)^{\frac{1}{4}}\left(\frac{\sqrt{\omega_y^2+2\Lambda}}{\pi}\right)^{\frac{1}{4}}\left(\frac{\sqrt{\omega_z^2+2\Lambda}}{\pi}\right)^{\frac{1}{4}} \times \\ \times e^{-\frac{1}{2}\sqrt{\omega_x^2+2\Lambda}\,x^2} e^{-\frac{1}{2}\sqrt{\omega_y^2+2\Lambda}\,y^2} e^{-\frac{1}{2}\sqrt{\omega_z^2+2\Lambda}\,z^2}, \tag{7}$$

where $\Lambda = \lambda_0(N-1)$ is the interaction parameter. For reference, $\psi_{GP}(\mathbf{r})$ solves the Gross–Pitaevskii equation $\left[-\frac{1}{2}\frac{\partial^2}{\partial \mathbf{r}^2} + \frac{1}{2}\left(\omega_x^2x^2 + \omega_y^2y^2 + \omega_z^2z^2\right) + \Lambda\int d\mathbf{r}'|\psi_{GP}(\mathbf{r}')|^2(\mathbf{r}-\mathbf{r}')^2\right]\psi_{GP}(\mathbf{r}) = \mu\psi_{GP}(\mathbf{r})$, where μ is the chemical potential. Note that both many-body and mean-field solutions can be written as products of the respective solutions in one dimension along the x, y, and z directions.

Before we arrive at the angular-momentum variances and for our needs, see below, we make a stopover and compute the position and momentum variances per particle in the model. At the many-body level we obviously have the result (4), since for the interacting ground-state the center-of-mass is separable and, hence, the position and momentum variances are independent of the two-body interaction. At the mean-field level we readily find from (7) the result

$$\mathrm{Var}|_{\mathrm{GP}}(\hat{X}) = \frac{1}{2\sqrt{\omega_x^2+2\Lambda}}, \quad \mathrm{Var}|_{\mathrm{GP}}(\hat{Y}) = \frac{1}{2\sqrt{\omega_y^2+2\Lambda}}, \quad \mathrm{Var}|_{\mathrm{GP}}(\hat{Z}) = \frac{1}{2\sqrt{\omega_z^2+2\Lambda}},$$
$$\mathrm{Var}|_{\mathrm{GP}}(\hat{P}_X) = \frac{\sqrt{\omega_x^2+2\Lambda}}{2}, \quad \mathrm{Var}|_{\mathrm{GP}}(\hat{P}_Y) = \frac{\sqrt{\omega_y^2+2\Lambda}}{2}, \quad \mathrm{Var}|_{\mathrm{GP}}(\hat{P}_Z) = \frac{\sqrt{\omega_z^2+2\Lambda}}{2}. \tag{8}$$

The mean-field variances (8) depend on the interaction parameter Λ, unlike the respective many-body variances (4). It turns out that this property would be instrumental when analyzing the anisotropy of the angular-momentum variance below. We briefly comment on the anisotropies of the position and momentum variances in the model. Comparing the mean-field (8) and many-body (4) variances per particle we find that the former belong to the anisotropy class $\{1\}$ independently of the interaction parameter Λ both for the position and momentum operators. For the mean-field variance of the ground state at the infinite-particle-number limit to belong to an anisotropy class other than $\{1\}$, one would have to go beyond the simple single-well geometry, see the anisotropy of the position variance in a double-well potential in two spatial dimensions [37].

We can now move to the expressions for the angular-momentum variances at the limit of an infinite number of particles, by generalizing results obtained in two spatial dimensions [21] to three spatial dimensions. The calculation at the mean-field level using (7) readily gives

$$\mathrm{Var}|_{\mathrm{GP}}(\hat{L}_X) = \frac{1}{4}\frac{\left(\sqrt{\omega_y^2+2\Lambda}-\sqrt{\omega_z^2+2\Lambda}\right)^2}{\sqrt{\omega_y^2+2\Lambda}\sqrt{\omega_z^2+2\Lambda}} = \frac{1}{4}\frac{\left(\frac{\sqrt{\omega_y^2+2\Lambda}}{\sqrt{\omega_z^2+2\Lambda}}-1\right)^2}{\frac{\sqrt{\omega_y^2+2\Lambda}}{\sqrt{\omega_z^2+2\Lambda}}},$$
$$\mathrm{Var}|_{\mathrm{GP}}(\hat{L}_Y) = \frac{1}{4}\frac{\left(\sqrt{\omega_z^2+2\Lambda}-\sqrt{\omega_x^2+2\Lambda}\right)^2}{\sqrt{\omega_z^2+2\Lambda}\sqrt{\omega_x^2+2\Lambda}} = \frac{1}{4}\frac{\left(\frac{\sqrt{\omega_z^2+2\Lambda}}{\sqrt{\omega_x^2+2\Lambda}}-1\right)^2}{\frac{\sqrt{\omega_z^2+2\Lambda}}{\sqrt{\omega_x^2+2\Lambda}}}, \tag{9}$$
$$\mathrm{Var}|_{\mathrm{GP}}(\hat{L}_Z) = \frac{1}{4}\frac{\left(\sqrt{\omega_x^2+2\Lambda}-\sqrt{\omega_y^2+2\Lambda}\right)^2}{\sqrt{\omega_x^2+2\Lambda}\sqrt{\omega_y^2+2\Lambda}} = \frac{1}{4}\frac{\left(\frac{\sqrt{\omega_x^2+2\Lambda}}{\sqrt{\omega_y^2+2\Lambda}}-1\right)^2}{\frac{\sqrt{\omega_x^2+2\Lambda}}{\sqrt{\omega_y^2+2\Lambda}}}.$$

In the absence of interaction these expressions boil down, respectively, to $\frac{1}{4}\frac{\left(\frac{\omega_y}{\omega_z}-1\right)^2}{\frac{\omega_y}{\omega_z}}$, $\frac{1}{4}\frac{\left(\frac{\omega_z}{\omega_x}-1\right)^2}{\frac{\omega_z}{\omega_x}}$, and $\frac{1}{4}\frac{\left(\frac{\omega_x}{\omega_y}-1\right)^2}{\frac{\omega_x}{\omega_y}}$, the angular-momentum variances of a single particle in a three-dimensional anisotropic harmonic potential. We see that for non-interacting particles and at the mean-field level the angular-momentum variances per particle depend on the ratios of frequencies, not on their absolute magnitudes. In the first case these are the bare frequencies of the harmonic trap whereas in the second case these are the interaction-dressed frequencies (7) resulting from the non-linear term.

The computation of the many-body variances is lengthier. It amounts to computing the angular-momentum variances for finite systems which exhibit an explicit dependence on the number of bosons N, and then performing the infinite-particle-number limit where several terms fall. Using [21] the final expressions for the correlations terms (1) are

$$\begin{aligned}
\text{Var}|_{\text{correlations}}(\hat{L}_X) &= \frac{1}{4}\frac{\left(\sqrt{\omega_y^2+2\Lambda}-\sqrt{\omega_z^2+2\Lambda}\right)^2}{\sqrt{\omega_y^2+2\Lambda}\sqrt{\omega_z^2+2\Lambda}}\left(\sqrt{1+\frac{2\Lambda}{\omega_y^2}}-1\right)\left(\sqrt{1+\frac{2\Lambda}{\omega_z^2}}-1\right), \\
\text{Var}|_{\text{correlations}}(\hat{L}_Y) &= \frac{1}{4}\frac{\left(\sqrt{\omega_z^2+2\Lambda}-\sqrt{\omega_x^2+2\Lambda}\right)^2}{\sqrt{\omega_z^2+2\Lambda}\sqrt{\omega_x^2+2\Lambda}}\left(\sqrt{1+\frac{2\Lambda}{\omega_z^2}}-1\right)\left(\sqrt{1+\frac{2\Lambda}{\omega_x^2}}-1\right), \\
\text{Var}|_{\text{correlations}}(\hat{L}_Z) &= \frac{1}{4}\frac{\left(\sqrt{\omega_x^2+2\Lambda}-\sqrt{\omega_y^2+2\Lambda}\right)^2}{\sqrt{\omega_x^2+2\Lambda}\sqrt{\omega_y^2+2\Lambda}}\left(\sqrt{1+\frac{2\Lambda}{\omega_x^2}}-1\right)\left(\sqrt{1+\frac{2\Lambda}{\omega_y^2}}-1\right).
\end{aligned} \tag{10}$$

Hence, adding (9) and (10) we readily have from (1) the many-body variances per particle at the infinite-particle-number limit, $\text{Var}|_{\text{MB}}(\hat{L}_X)$, $\text{Var}|_{\text{MB}}(\hat{L}_Y)$, and $\text{Var}|_{\text{MB}}(\hat{L}_Z)$.

We remark that the expectation values per particle of the angular-momentum operator $(\hat{L}_X, \hat{L}_Y, \hat{L}_Z)$, as well as the respective expectation values of the position and momentum operators, computed at the mean-field and many-body levels of theory coincide at the limit of an infinite number of particles and are all equal to zero in the ground state.

We investigate and discuss an example. Let the frequencies of the three-dimensional anisotropic harmonic trap be $\omega_x = 0.7$, $\omega_y = 5.0$, and $\omega_z = 10.5$. Their ratios from large to small are: $\frac{\omega_z}{\omega_x} = 15$, $\frac{\omega_y}{\omega_x} = 7\frac{1}{7}$, and $\frac{\omega_z}{\omega_y} = 2\frac{1}{10}$. Then, the values of the angular-momentum variances per particle at zero interaction parameter, $\Lambda = 0$, are given from large to small by $\text{Var}|_{\text{GP}}(\hat{L}_Y) = \text{Var}|_{\text{MB}}(\hat{L}_Y) = \frac{7^2}{15} \approx 3.267$, $\text{Var}|_{\text{GP}}(\hat{L}_Z) = \text{Var}|_{\text{MB}}(\hat{L}_Z) = \frac{43^2}{1400} \approx 1.321$, and $\text{Var}|_{\text{GP}}(\hat{L}_X) = \text{Var}|_{\text{MB}}(\hat{L}_X) = \frac{11^2}{840} \approx 0.144$. Indeed, as the ratio of frequencies with respect to two axes is bigger, the corresponding angular-momentum variance per particle with respect to the third axis is larger, and vise versa.

What happens as the interaction sets in? Figure 3a depicts the many-body and mean-field angular-momentum variances as a function of the interaction parameter Λ. We examine positive values of Λ which correspond to the attractive sector of the harmonic-interaction model, see, e.g., [21,25,29]. Let us analyze the observations. With increasing interaction parameter the density narrows, along the x, y, and z directions. This is clear because the interaction between particles is attractive, and is manifested by the monotonously decreasing values of the position variances per particle (8). Furthermore, the density becomes less anisotropic, because the ratios of the dressed frequencies $\frac{\sqrt{\omega_z^2+2\Lambda}}{\sqrt{\omega_x^2+2\Lambda}}$, $\frac{\sqrt{\omega_y^2+2\Lambda}}{\sqrt{\omega_x^2+2\Lambda}}$, and $\frac{\sqrt{\omega_z^2+2\Lambda}}{\sqrt{\omega_y^2+2\Lambda}}$ monotonously decrease with increasing Λ. Consequently, the angular-momentum variances per particle decrease with the interaction parameter as well, see Figure 3a. The mean-field angular-momentum variances (9) are monotonously decreasing because of the just-described decreasing ratios of the dressed frequencies. The many-body angular-momentum variances are decreasing, at least for the values of interaction parameters studied here, because the positive-value correlations terms (10) grow slower than the mean-field angular-momentum variances decrease with Λ.

All in all, the anisotropy of the angular-momentum variance can now be determined. We find the anisotropy $\text{Var}|_{\text{MB}}(\hat{L}_Y) > \text{Var}|_{\text{MB}}(\hat{L}_Z) > \text{Var}|_{\text{MB}}(\hat{L}_X)$ to hold for all interaction parameters at the many-body level of theory. At the mean-field level of theory we find the same anisotropy, namely, $\text{Var}|_{\text{GP}}(\hat{L}_Y) > \text{Var}|_{\text{GP}}(\hat{L}_Z) > \text{Var}|_{\text{GP}}(\hat{L}_X)$, to hold for small interaction parameters. However, then, at just about $\Lambda = 5.0$ the mean-field anisotropy changes to $\text{Var}|_{\text{GP}}(\hat{L}_Y) > \text{Var}|_{\text{GP}}(\hat{L}_X) > \text{Var}|_{\text{GP}}(\hat{L}_Z)$, and this anisotropy continues for larger interaction parameters. Hence, we have found that the anisotropy of the angular-momentum operator in the ground state of the three-dimensional anisotropic harmonic-interaction model at the infinite-particle-number limit changes as a function of the interaction parameter from the anisotropy class $\{1\}$ to $\{1,2\}$, see Figure 3a. In terms of the correlations term (1), the anisotropy of the variance is governed then by many-body effects.

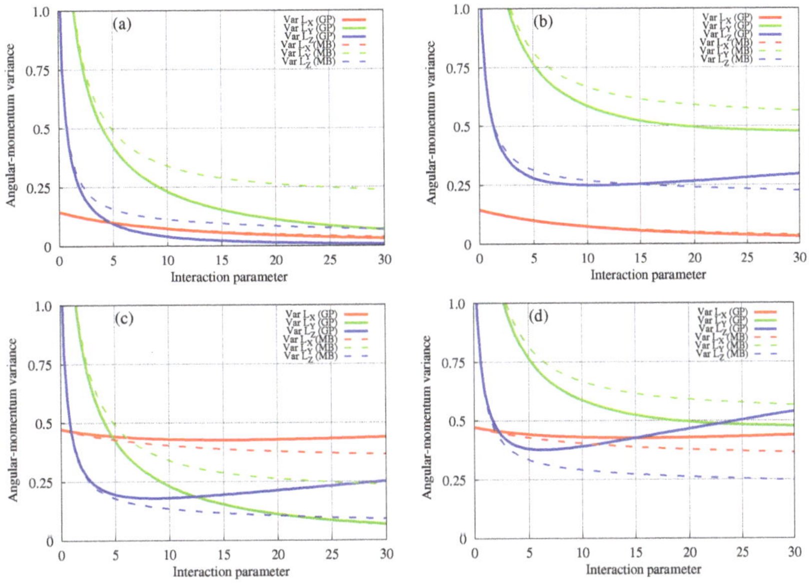

Figure 3. Many-particle angular-momentum ($\hat{L}_X$, $\hat{L}_Y$, and $\hat{L}_Z$; in red, green, and blue) variance per particle as a function of the interaction parameter Λ computed at the limit of an infinite number of particles within many-body (dashed lines) and mean-field (solid lines) levels of theory for the ground state of the three-dimensional anisotropic harmonic-interaction model. The frequencies of the trap are $\omega_x = 0.7$, $\omega_y = 5.0$, and $\omega_z = 10.5$. Results at several translations $\mathbf{r}_0$ of the center of the trap are shown in the panels: (**a**) $\mathbf{r}_0 = (0,0,0)$; (**b**) $\mathbf{r}_0 = (0.25,0,0)$; (**c**) $\mathbf{r}_0 = (0,0.25,0)$; (**d**) $\mathbf{r}_0 = (0.25,0.25,0)$. Different anisotropy classes of the angular-momentum variance emerge with the interaction parameter. See the text for more details. The quantities shown are dimensionless.

Still, can the above-found picture of angular-momentum variance anisotropy be made richer? The answer is positive and requires one to dive deeper into the properties of angular-momentum variances under translations. To this end, we employ and extend prior work in two spatial dimensions [21]. Suppose now that the harmonic trap is located not in the origin but translated to a general point $\mathbf{r}_0 = (x_0, y_0, z_0)$. The expectation values per particle of the momentum and angular-momentum operators are still zero, whereas the expectation value of the position operator is, of course, $\mathbf{r}_0$. Whereas the position and momentum variances are invariant to translations, the angular-momentum variances are not, which open up another degree of freedom to investigate the anisotropy class of the angular-momentum variances in three spatial dimensions. Mathematically, the transformation properties of the angular-momentum operator and its square combine in a non-trivial form those of the position and momentum operators. Physically, in a similar manner that angular momentum is defined with respect to a reference point, and it is different with respect to another reference point, so does the variance of the angular-momentum operator which changes with respect to distinct reference points.

Using the transformation properties of the angular-momentum operator $(\hat{L}_X, \hat{L}_Y, \hat{L}_Z)$ under translations, see Appendix A and [21], the final expressions for the translated angular-momentum variances per particle read explicitly

$$\begin{aligned}
\mathrm{Var}|_{\mathrm{MB}}(\hat{L}_X;\mathbf{r}_0) &= \frac{1}{4}\frac{\left(\sqrt{\omega_y^2+2\Lambda}-\sqrt{\omega_z^2+2\Lambda}\right)^2}{\sqrt{\omega_y^2+2\Lambda}\sqrt{\omega_z^2+2\Lambda}}\left[1+\left(\sqrt{1+\frac{2\Lambda}{\omega_y^2}}-1\right)\left(\sqrt{1+\frac{2\Lambda}{\omega_z^2}}-1\right)\right]+ \\
&+y_0^2\frac{\omega_z}{2}+z_0^2\frac{\omega_y}{2}, \\
\mathrm{Var}|_{\mathrm{MB}}(\hat{L}_Y;\mathbf{r}_0) &= \frac{1}{4}\frac{\left(\sqrt{\omega_z^2+2\Lambda}-\sqrt{\omega_x^2+2\Lambda}\right)^2}{\sqrt{\omega_z^2+2\Lambda}\sqrt{\omega_x^2+2\Lambda}}\left[1+\left(\sqrt{1+\frac{2\Lambda}{\omega_z^2}}-1\right)\left(\sqrt{1+\frac{2\Lambda}{\omega_x^2}}-1\right)\right]+ \\
&+z_0^2\frac{\omega_x}{2}+x_0^2\frac{\omega_z}{2}, \\
\mathrm{Var}|_{\mathrm{MB}}(\hat{L}_Z;\mathbf{r}_0) &= \frac{1}{4}\frac{\left(\sqrt{\omega_x^2+2\Lambda}-\sqrt{\omega_y^2+2\Lambda}\right)^2}{\sqrt{\omega_x^2+2\Lambda}\sqrt{\omega_y^2+2\Lambda}}\left[1+\left(\sqrt{1+\frac{2\Lambda}{\omega_x^2}}-1\right)\left(\sqrt{1+\frac{2\Lambda}{\omega_y^2}}-1\right)\right]+ \\
&+x_0^2\frac{\omega_y}{2}+y_0^2\frac{\omega_x}{2}
\end{aligned} \tag{11}$$

at the many-body level of theory and

$$\begin{aligned}
\mathrm{Var}|_{\mathrm{GP}}(\hat{L}_X;\mathbf{r}_0) &= \frac{1}{4}\frac{\left(\sqrt{\omega_y^2+2\Lambda}-\sqrt{\omega_z^2+2\Lambda}\right)^2}{\sqrt{\omega_y^2+2\Lambda}\sqrt{\omega_z^2+2\Lambda}}+y_0^2\frac{\sqrt{\omega_z^2+2\Lambda}}{2}+z_0^2\frac{\sqrt{\omega_y^2+2\Lambda}}{2}, \\
\mathrm{Var}|_{\mathrm{GP}}(\hat{L}_Y;\mathbf{r}_0) &= \frac{1}{4}\frac{\left(\sqrt{\omega_z^2+2\Lambda}-\sqrt{\omega_x^2+2\Lambda}\right)^2}{\sqrt{\omega_z^2+2\Lambda}\sqrt{\omega_x^2+2\Lambda}}+z_0^2\frac{\sqrt{\omega_x^2+2\Lambda}}{2}+x_0^2\frac{\sqrt{\omega_z^2+2\Lambda}}{2}, \\
\mathrm{Var}|_{\mathrm{GP}}(\hat{L}_Z;\mathbf{r}_0) &= \frac{1}{4}\frac{\left(\sqrt{\omega_x^2+2\Lambda}-\sqrt{\omega_y^2+2\Lambda}\right)^2}{\sqrt{\omega_x^2+2\Lambda}\sqrt{\omega_y^2+2\Lambda}}+x_0^2\frac{\sqrt{\omega_y^2+2\Lambda}}{2}+y_0^2\frac{\sqrt{\omega_x^2+2\Lambda}}{2}
\end{aligned} \tag{12}$$

at the mean-field level of theory. Let us examine expressions (11) and (12) more closely. The terms added to the translated angular-momentum variances at the many-body level depend on the corresponding components of the translation vector but not on the interaction parameter, whereas the added terms at the mean-field level of theory depend on and increase with Λ, see Appendix A for more details. The combined effects can be seen in Figure 3, compare panels Figure 3b–d with panel Figure 3a.

For the translation by $\mathbf{r}_0 = (0.25, 0, 0)$, the angular-momentum variances $\mathrm{Var}|_{\mathrm{MB}}(\hat{L}_X;\mathbf{r}_0)$ and $\mathrm{Var}|_{\mathrm{GP}}(\hat{L}_X;\mathbf{r}_0)$ are invariant quantities, $\mathrm{Var}|_{\mathrm{MB}}(\hat{L}_Y;\mathbf{r}_0)$ and $\mathrm{Var}|_{\mathrm{MB}}(\hat{L}_Z;\mathbf{r}_0)$ are shifted by interaction-independent values, and $\mathrm{Var}|_{\mathrm{GP}}(\hat{L}_Y;\mathbf{r}_0)$ and $\mathrm{Var}|_{\mathrm{GP}}(\hat{L}_Z;\mathbf{r}_0)$ increase by interaction-dependent values, see (11) and (12), and compare Figure 3a,b. The combined effect is that now both $\mathrm{Var}|_{\mathrm{MB}}(\hat{L}_Y;\mathbf{r}_0) > \mathrm{Var}|_{\mathrm{MB}}(\hat{L}_Z;\mathbf{r}_0) > \mathrm{Var}|_{\mathrm{MB}}(\hat{L}_X;\mathbf{r}_0)$ and $\mathrm{Var}|_{\mathrm{GP}}(\hat{L}_Y;\mathbf{r}_0) > \mathrm{Var}|_{\mathrm{GP}}(\hat{L}_Z;\mathbf{r}_0) > \mathrm{Var}|_{\mathrm{GP}}(\hat{L}_X;\mathbf{r}_0)$ hold for all interaction parameters Λ in the range studied. Consequently, the anisotropy class of the angular-momentum variance per particle for $\mathbf{r}_0 = (0.25, 0, 0)$ is $\{1\}$ only.

Next, for $\mathbf{r}_0 = (0, 0.25, 0)$, the angular-momentum variances $\mathrm{Var}|_{\mathrm{MB}}(\hat{L}_Y;\mathbf{r}_0)$ and $\mathrm{Var}|_{\mathrm{GP}}(\hat{L}_Y;\mathbf{r}_0)$ are invariant quantities, $\mathrm{Var}|_{\mathrm{MB}}(\hat{L}_X;\mathbf{r}_0)$ and $\mathrm{Var}|_{\mathrm{MB}}(\hat{L}_Z;\mathbf{r}_0)$ are shifted by interaction-independent values, and $\mathrm{Var}|_{\mathrm{GP}}(\hat{L}_X;\mathbf{r}_0)$ and $\mathrm{Var}|_{\mathrm{GP}}(\hat{L}_Z;\mathbf{r}_0)$ grow by interaction-dependent values, contrast Figure 3a,c. The combined effect is, of course, different than with $\mathbf{r}_0 = (0.25, 0, 0)$, and we discuss its main features, focusing on the regime of interaction parameters larger than about $\Lambda = 6$, for which the two many-body curves $\mathrm{Var}|_{\mathrm{MB}}(\hat{L}_X;\mathbf{r}_0)$ and $\mathrm{Var}|_{\mathrm{MB}}(\hat{L}_Y;\mathbf{r}_0)$ cross, see Figure 3c. The many-body angular-momentum variances satisfy $\mathrm{Var}|_{\mathrm{MB}}(\hat{L}_X;\mathbf{r}_0) > \mathrm{Var}|_{\mathrm{MB}}(\hat{L}_Y;\mathbf{r}_0) > \mathrm{Var}|_{\mathrm{MB}}(\hat{L}_Z;\mathbf{r}_0)$ for all studied interaction parameters. So, the effect of this translation is to alter the order of the many-body variances, i.e., to change the many-body anisotropy. Now, at the mean-field level, we find $\mathrm{Var}|_{\mathrm{GP}}(\hat{L}_X;\mathbf{r}_0) >$

$\text{Var}|_{\text{GP}}(\hat{L}_Y; \mathbf{r}_0) > \text{Var}|_{\text{GP}}(\hat{L}_Z; \mathbf{r}_0)$ up to about $\Lambda = 12$ and $\text{Var}|_{\text{GP}}(\hat{L}_X; \mathbf{r}_0) > \text{Var}|_{\text{GP}}(\hat{L}_Z; \mathbf{r}_0) > \text{Var}|_{\text{GP}}(\hat{L}_Y; \mathbf{r}_0)$ for the interaction parameters larger than about $\Lambda = 12$. All in all, for the translation by $\mathbf{r}_0 = (0.25, 0, 0)$ the above-described relations correspond, respectively, to the anisotropy classes $\{1\}$ and $\{1,2\}$ of the angular-momentum variances per particle.

Finally, for the translation by $\mathbf{r}_0 = (0.25, 0.25, 0)$ non of the angular-momentum variances is invariant, see (11) and (12). The many-body angular-momentum variances $\text{Var}|_{\text{MB}}(\hat{L}_X; \mathbf{r}_0)$, $\text{Var}|_{\text{MB}}(\hat{L}_Y; \mathbf{r}_0)$, and $\text{Var}|_{\text{MB}}(\hat{L}_Z; \mathbf{r}_0)$ are shifted by interaction-independent values and the mean-field angular-momentum quantities $\text{Var}|_{\text{GP}}(\hat{L}_X; \mathbf{r}_0)$, $\text{Var}|_{\text{GP}}(\hat{L}_Y; \mathbf{r}_0)$, and $\text{Var}|_{\text{GP}}(\hat{L}_Z; \mathbf{r}_0)$ increase by interaction-dependent values, see Figure 3d. We examine the overall effect, concentrating on the regime of interaction parameters larger than about $\Lambda = 2.5$, where the two many-body curves $\text{Var}|_{\text{MB}}(\hat{L}_X; \mathbf{r}_0)$ and $\text{Var}|_{\text{MB}}(\hat{L}_Z; \mathbf{r}_0)$ cross, see Figure 3d. The many-body angular-momentum variances satisfy $\text{Var}|_{\text{MB}}(\hat{L}_Y; \mathbf{r}_0) > \text{Var}|_{\text{MB}}(\hat{L}_X; \mathbf{r}_0) > \text{Var}|_{\text{MB}}(\hat{L}_Z; \mathbf{r}_0)$ for all interaction parameters in the range studied. Once again, the effect of the translation is to change the order of the many-body variances, thereby altering the many-body anisotropy, compare panels Figure 3a,c,d. At the mean-field level one finds $\text{Var}|_{\text{GP}}(\hat{L}_Y; \mathbf{r}_0) > \text{Var}|_{\text{GP}}(\hat{L}_X; \mathbf{r}_0) > \text{Var}|_{\text{GP}}(\hat{L}_Z; \mathbf{r}_0)$ up to about $\Lambda = 15$, then $\text{Var}|_{\text{GP}}(\hat{L}_Y; \mathbf{r}_0) > \text{Var}|_{\text{GP}}(\hat{L}_Z; \mathbf{r}_0) > \text{Var}|_{\text{GP}}(\hat{L}_X; \mathbf{r}_0)$ till about $\Lambda = 22.5$, and $\text{Var}|_{\text{GP}}(\hat{L}_Z; \mathbf{r}_0) > \text{Var}|_{\text{GP}}(\hat{L}_Y; \mathbf{r}_0) > \text{Var}|_{\text{GP}}(\hat{L}_X; \mathbf{r}_0)$ for all the interaction parameters larger than about $\Lambda = 22.5$ studied, see Figure 3d. Therefore, for $\mathbf{r}_0 = (0.25, 0.25, 0)$ the above-discussed findings imply that all anisotropy classes can be attained by the angular-momentum variances per particle in the three-dimensional harmonic-interaction model at the infinite-particle-number limit. These are, respectively, $\{1\}$, $\{1,2\}$, and $\{1,2,3\}$. In other words, we have shown that the correlations term (1) can dominate the angular-momentum properties of a trapped Bose–Einstein condensate at the limit of an infinite number of particles, which rounds off the present work.

4. Summary

The present work deals with a connection between anisotropy of and correlations in a three-dimensional trapped Bose–Einstein condensate. The merit of treating the limit of an infinite number of bosons is appealing, since the system is known to be 100% condensed in this limit, and some of its properties, notably the density per particle, are identical at the many-body and mean-field levels of theory.

We have analyzed the variances per particle of the three Cartesian components of the position $(\hat{X}, \hat{Y}, \hat{Z})$, momentum $(\hat{P}_X, \hat{P}_Y, \hat{P}_Z)$, and angular-momentum $(\hat{L}_X, \hat{L}_Y, \hat{L}_Z)$ operators at the many-body and mean-field levels of theory. In general, for small interaction parameters the differences between the many-body and mean-field quantities are quantitative whereas for larger interaction parameters qualitative differences emerge. We define the anisotropy class of the variance according to the different orderings from large to small, or permutations, of the three respective many-body and mean-field quantities. The anisotropy class $\{1\}$ implies the same ordering, the class $\{1,2\}$ implies that two of the components are permuted, and the anisotropy class $\{1,2,3\}$ means that the three components are permuted.

Two relatively transparent applications are presented, the first is the breathing of an anisotropic three-dimensional trapped Bose–Einstein condensate, and the second is the ground state of the anisotropic three-dimensional harmonic-interaction model. The former exhibits different anisotropy classes of the position and momentum variances per particle, because at the many-body level the variances are constant, while at the mean-field level the variance of each component oscillates with a different amplitude and frequency, and consequently different anisotropy classes occur in time. The latter application shows different anisotropy classes of the angular-momentum variance per particle, owing to the intricate transformation properties of the angular-momentum variance when the wave-packet is translated from the origin. The challenge, which obviously goes beyond the scope of the present work, would be the experimental observation of such anisotropies and correlations. To access many-particle variances one would need to measure explicitly the positions, momenta, or angular-momenta of, in principle, all particles in the Bose–Einstein condensate, rather than just the total density from which the expectation values of these observables can be deduced.

To sum up, the anisotropy or morphology of a three-dimensional trapped Bose–Einstein condensate can look quite different when examined through the 'glasses' of many-body and mean-field theories, even for 100% condensed bosons at the limit of an infinite number of particles. It would be interesting to conduct the investigation presented here and analyze the results in more complicated numerical many-particle scenarios. Last but not least, it is possible to envision classifying the morphology of a Bose–Einstein condensate beyond that emanating from the variances of the position, momentum, and angular-momentum operators. Furthermore, in four spatial dimensions one could expect an additional richness of the variance [38].

Funding: This research was funded by Israel Science Foundation (Grants No. 600/15 and 1516/19).

Institutional Review Board Statement: Not applicable.

Informed Consent Statement: Not applicable.

Data Availability Statement: Not applicable.

Acknowledgments: This research was supported by the Israel Science Foundation (Grants No. 600/15 and 1516/19). We thank Anal Bhowmik and Lorenz Cederbaum for discussions. Computation time on the BwForCluster, the High Performance Computing system Hive of the Faculty of Natural Sciences at University of Haifa, and the Hawk at the High Performance Computing Center Stuttgart (HLRS) is gratefully acknowledged.

Conflicts of Interest: The author declares no conflict of interest.

Appendix A. Translated Angular-Momentum Variances in Three Spatial Dimensions at the Limit of an Infinite Number of Particles

Consider interacting bosons at the limit of an infinite number of particles trapped in the ground state of the three-dimensional anisotropic harmonic potential (or any other potential which is reflection symmetric in the x, y, and z directions) centered at the origin. The expectation values per particle of the position, momentum, and angular-momentum operators vanish. Suppose now that the harmonic trap is translated to the location $\mathbf{r}_0 = (x_0, y_0, z_0)$. The translated and untranslated angular-momentum variances at the many-body and mean-field levels of theory can, respectively, be related as follows:

$$\begin{aligned}
\text{Var}|_{\text{MB}}(\hat{L}_X;\mathbf{r}_0) &= \text{Var}|_{\text{MB}}(\hat{L}_X) + y_0^2\text{Var}|_{\text{MB}}(\hat{P}_Z) + z_0^2\text{Var}|_{\text{MB}}(\hat{P}_Y), \\
\text{Var}|_{\text{MB}}(\hat{L}_Y;\mathbf{r}_0) &= \text{Var}|_{\text{MB}}(\hat{L}_Y) + z_0^2\text{Var}|_{\text{MB}}(\hat{P}_X) + x_0^2\text{Var}|_{\text{MB}}(\hat{P}_Z), \\
\text{Var}|_{\text{MB}}(\hat{L}_Z;\mathbf{r}_0) &= \text{Var}|_{\text{MB}}(\hat{L}_Z) + x_0^2\text{Var}|_{\text{MB}}(\hat{P}_Y) + y_0^2\text{Var}|_{\text{MB}}(\hat{P}_X)
\end{aligned} \tag{A1a}$$

and

$$\begin{aligned}
\text{Var}|_{\text{GP}}(\hat{L}_X;\mathbf{r}_0) &= \text{Var}|_{\text{GP}}(\hat{L}_X) + y_0^2\text{Var}|_{\text{GP}}(\hat{P}_Z) + z_0^2\text{Var}|_{\text{GP}}(\hat{P}_Y), \\
\text{Var}|_{\text{GP}}(\hat{L}_Y;\mathbf{r}_0) &= \text{Var}|_{\text{GP}}(\hat{L}_Y) + z_0^2\text{Var}|_{\text{GP}}(\hat{P}_X) + x_0^2\text{Var}|_{\text{GP}}(\hat{P}_Z), \\
\text{Var}|_{\text{GP}}(\hat{L}_Z;\mathbf{r}_0) &= \text{Var}|_{\text{GP}}(\hat{L}_Z) + x_0^2\text{Var}|_{\text{GP}}(\hat{P}_Y) + y_0^2\text{Var}|_{\text{GP}}(\hat{P}_X).
\end{aligned} \tag{A1b}$$

The explicit expressions at the many-body and mean-field levels of theory are given in the main text, see (11) and (12), respectively. We remind for reference that the position and momentum variances are invariant to translations. Consequently, by subtracting the Gross-Piteavskii (A1b) from the many-body (A1a) results, see (1), we readily find for the translated correlations terms:

$$\begin{aligned}
\text{Var}|_{\text{correlations}}(\hat{L}_X;\mathbf{r}_0) &= \text{Var}|_{\text{correlations}}(\hat{L}_X) + y_0^2\text{Var}|_{\text{correlations}}(\hat{P}_Z) + z_0^2\text{Var}|_{\text{correlations}}(\hat{P}_Y), \\
\text{Var}|_{\text{correlations}}(\hat{L}_Y;\mathbf{r}_0) &= \text{Var}|_{\text{correlations}}(\hat{L}_Y) + z_0^2\text{Var}|_{\text{correlations}}(\hat{P}_X) + x_0^2\text{Var}|_{\text{correlations}}(\hat{P}_Z), \\
\text{Var}|_{\text{correlations}}(\hat{L}_Z;\mathbf{r}_0) &= \text{Var}|_{\text{correlations}}(\hat{L}_Z) + x_0^2\text{Var}|_{\text{correlations}}(\hat{P}_Y) + y_0^2\text{Var}|_{\text{correlations}}(\hat{P}_X).
\end{aligned} \tag{A2}$$

The meaning of result (A2) is that the correlations terms of the translated angular-momentum variances depend on the respective correlations terms of the momentum variances and components of the translation vector $\mathbf{r}_0$. Consequently, the translated angular-momentum correlations terms (A2) generally depend more strongly on the interaction parameter than the untranslated ones. Indeed, Figure 3 plots some examples of angular-momentum variances for different $\mathbf{r}_0$, and shows that, once translations of the trap are included, the angular-momentum variances of a trapped Bose–Einstein condensate at the limit of an infinite number of particles can belong to any of the anisotropy classes, $\{1\}$, $\{1,2\}$, or $\{1,2,3\}$.

References

1. Castin, Y.; Dum, R. Low-temperature Bose–Einstein condensates in time-dependent traps: Beyond the U(1) symmetry breaking approach. *Phys. Rev. A* **1998**, *57*, 3008. [CrossRef]
2. Lieb, E.H.; Seiringer, R.; Yngvason, J. Bosons in a trap: A rigorous derivation of the Gross–Pitaevskii energy functional. *Phys. Rev. A* **2000**, *61*, 043602. [CrossRef]
3. Lieb, E.H.; Seiringer, R. Proof of Bose–Einstein Condensation for Dilute Trapped Gases. *Phys. Rev. Lett.* **2002**, *88*, 170409. [CrossRef] [PubMed]
4. Erdős, L.; Schlein, B.; Yau, H.-T. Rigorous Derivation of the Gross–Pitaevskii Equation. *Phys. Rev. Lett.* **2007**, *98*, 040404. [CrossRef]
5. Erdős, L.; Schlein, B.; Yau, H.-T. Derivation of the cubic non-linear Schrödinger equation from quantum dynamics of many-body systems. *Invent. Math.* **2007**, *167*, 515. [CrossRef]
6. Klaiman, S.; Alon, O.E. Variance as a sensitive probe of correlations. *Phys. Rev. A* **2015**, *91*, 063613. [CrossRef]
7. Klaiman, S.; Streltsov, A.I.; Alon, O.E. Uncertainty product of an out-of-equilibrium many-particle system. *Phys. Rev. A* **2016**, *93*, 023605. [CrossRef]
8. Klaiman, S.; Cederbaum, L.S. Overlap of exact and Gross–Pitaevskii wave functions in Bose–Einstein condensates of dilute gases. *Phys. Rev. A* **2016**, *94*, 063648. [CrossRef]
9. Anapolitanos, I.; Hott, M.; Hundertmark, D. Derivation of the Hartree equation for compound Bose gases in the mean field limit. *Rev. Math. Phys.* **2017**, *29*, 1750022. [CrossRef]
10. Michelangeli, A.; Olgiati, A. Mean-field quantum dynamics for a mixture of Bose–Einstein condensates. *Anal. Math. Phys.* **2017**, *7*, 377. [CrossRef]
11. Alon, O.E. Solvable model of a generic trapped mixture of interacting bosons: Reduced density matrices and proof of Bose–Einstein condensation. *J. Phys. A* **2017**, *50*, 295002. [CrossRef]
12. Cederbaum, L.S. Exact many-body wave function and properties of trapped bosons in the infinite-particle limit. *Phys. Rev. A* **2017**, *96*, 013615. [CrossRef]
13. Coleman, A.J.; Yukalov, V.I. *Reduced Density Matrices: Coulson's Challenge*; Lectures Notes in Chemistry; Springer: Berlin, Germany, 2000; Volume 72.
14. Mazziotti, D.A. (Ed.) *Reduced-Density-Matrix Mechanics: With Application to Many-Electron Atoms and Molecules*; Advances in Chemical Physics; Wiley: New York, NY, USA, 2007; Volume 134.
15. Lode, A.U.J.; Lévêque, C.; Madsen, L.B.; Streltsov, A.I.; Alon, O.E. *Colloquium*: Multiconfigurational time-dependent Hartree approaches for indistinguishable particles. *Rev. Mod. Phys.* **2020**, *92*, 011001. [CrossRef]
16. Bolsinger, V.J.; Krönke, S.; Schmelcher, P. Ultracold bosonic scattering dynamics off a repulsive barrier: Coherence loss at the dimensional crossover. *Phys. Rev. A* **2017**, *96*, 013618. [CrossRef]
17. Alon, O.E. Condensates in annuli: Dimensionality of the variance. *Mol. Phys.* **2019**, *117*, 2108. [CrossRef]
18. Klaiman, S.; Beinke, R.; Cederbaum, L.S.; Streltsov, A.I.; Alon, O.E. Variance of an anisotropic Bose–Einstein condensate. *Chem. Phys.* **2018**, *509*, 45. [CrossRef]
19. Sakmann, K.; Schmiedmayer, J. Conserving symmetries in Bose–Einstein condensate dynamics requires many-body theory. *arXiv* **2018**, arXiv:1802.03746v2
20. Streltsov, A.I.; Streltsova, O.I. MCTDHB-LAB, Version 1.5. 2015. Available online: http://www.mctdhb-lab.com (accessed on 10 May 2021)
21. Alon, O.E. Analysis of a Trapped Bose–Einstein Condensate in Terms of Position, Momentum, and Angular-Momentum Variance. *Symmetry* **2019**, *11*, 1344. [CrossRef]
22. Robinson, P.D. Coupled oscillator natural orbitals. *J. Chem. Phys.* **1977**, *66*, 3307. [CrossRef]
23. Hall, R.L. Some exact solutions to the translation-invariant *N*-body problem. *J. Phys. A* **1978**, *11*, 1227. [CrossRef]
24. Hall, R.L. Exact solutions of Schrödinger's equation for translation-invariant harmonic matter. *J. Phys. A* **1978**, *11*, 1235. [CrossRef]
25. Cohen, L.; Lee, C. Exact reduced density matrices for a model problem. *J. Math. Phys.* **1985**, *26*, 3105. [CrossRef]
26. Osadchii, M.S.; Muraktanov, V.V. The System of Harmonically Interacting Particles: An Exact Solution of the Quantum-Mechanical Problem. *Int. J. Quant. Chem.* **1991**, *39*, 173. [CrossRef]
27. Załuska-Kotur, M.A.; Gajda, M.; Orłowski, A.; Mostowski, J. Soluble model of many interacting quantum particles in a trap. *Phys. Rev. A* **2000**, *61*, 033613. [CrossRef]

28. Yan, J. Harmonic Interaction Model and Its Applications in Bose–Einstein Condensation. *J. Stat. Phys.* **2003**, *113*, 623. [CrossRef]
29. Gajda, M. Criterion for Bose–Einstein condensation in a harmonic trap in the case with attractive interactions. *Phys. Rev. A* **2006**, *73*, 023603. [CrossRef]
30. Armstrong, J.R.; Zinner, N.T.; Fedorov, D.V.; Jensen, A.S. Analytic harmonic approach to the *N*-body problem. *J. Phys. B* **2011**, *44*, 055303. [CrossRef]
31. Armstrong, J.R.; Zinner, N.T.; Fedorov, D.V.; Jensen, A.S. Virial expansion coefficients in the harmonic approximation. *Phys. Rev. E* **2012**, *86*, 021115. [CrossRef]
32. Schilling, C. Natural orbitals and occupation numbers for harmonium: Fermions versus bosons. *Phys. Rev. A* **2013**, *88*, 042105. [CrossRef]
33. Benavides-Riveros, C.L.; Toranzo, I.V.; Dehesa, J.S. Entanglement in N-harmonium: bosons and fermions. *J. Phys. B* **2014**, *47*, 195503. [CrossRef]
34. Bouvrie, P.A.; Majtey, A.P.; Tichy, M.C.; Dehesa, J.S.; Plastino, A.R. Entanglement and the Born-Oppenheimer approximation in an exactly solvable quantum many-body system. *Eur. Phys. J. D* **2014**, *68*, 346. [CrossRef]
35. Armstrong, J.R.; Volosniev, A.G.; Fedorov, D.V.; Jensen, A.S.; Zinner, N.T. Analytic solutions of topologically disjoint systems. *J. Phys. A* **2015**, *48*, 085301. [CrossRef]
36. Schilling, C.; Schilling, R. Number-parity effect for confined fermions in one dimension. *Phys. Rev. A* **2016**, *93*, 021601(R). [CrossRef]
37. Alon, O.E. Variance of a Trapped Bose–Einstein Condensate. *J. Phys. Conf. Ser.* **2019**, *1206*, 012009. [CrossRef]
38. Yukalov, V.I. Particle Fluctuations in Mesoscopic Bose Systems. *Symmetry* **2019**, *11*, 603. [CrossRef]

symmetry

MDPI

Article

Acoustic Plasmons in Graphene Sandwiched between Two Metallic Slabs

Luca Salasnich [1,2,3]

1 Dipartimento di Fisica e Astronomia "Galileo Galilei", Università di Padova, CNISM and INFN, Via Marzolo 8, 35131 Padova, Italy; luca.salasnich@unipd.it

2 Padua Quantum Technologies Research Center, Università di Padova, Via Gradenigo 6/b, 35131 Padova, Italy

3 Istituto Nazionale di Ottica (INO), Consiglio Nazionale delle Ricerche (CNR), Via Nello Carrara 1, 50019 Sesto Fiorentino, Italy

Abstract: We study the effect of two metallic slabs on the collective dynamics of electrons in graphene positioned between the two slabs. We show that if the slabs are perfect conductors, the plasmons of graphene display a linear dispersion relation. The velocity of these acoustic plasmons crucially depends on the distance between the two metal gates and the graphene sheet. In the case of generic slabs, the dispersion relation of graphene plasmons is much more complicated, but we find that acoustic plasmons can still be obtained under specific conditions.

Keywords: graphene; quantum many-body theory; acoustic plasmons

Citation: Salasnich, L. Acoustic Plasmons in Graphene Sandwiched between Two Metallic Slabs. *Symmetry* **2021**, *13*, 684. https://doi.org/10.3390/sym13040684

Academic Editor: V.I. Yukalov

Received: 14 March 2021
Accepted: 13 April 2021
Published: 15 April 2021

1. Introduction

In 2004, Graphene, a single layer of carbon atoms arranged in a two-dimensional honeycomb lattice, was isolated and characterized [1]. Since then, many electrical, thermal, chemical, optical, and mechanical properties of graphene have been studied, both experimentally and theoretically [2–5]. Quite remarkably, under appropriate conditions, the electrons in graphene behave as viscous fluids, exhibiting peculiar hydrodynamic effects [6]. In particular, it has been shown that the plasmons of graphene display a linear dispersion relation when, in the proximity of the graphene, a metallic slab screens the Coulomb potential of electrons in graphene [7,8].

In this brief communication, we extend the predictions obtained in [7] by considering a graphene sheet sandwiched between two metallic slabs. We find that, in this case, the electrons of graphene are characterized by acustic modes whose dispersion relation is linear in the long-wavelength regime. We obtain a simple analytical formula for the speed of these acustic modes.

2. Graphene Sandwiched between Two Materials

The monolayer graphene is a honeycomb lattice of carbon atoms in two spatial dimensions. Quasiparticles in graphene have the dispersion relation

$$E_{\mathbf{k}} = \pm v_F \hbar |\mathbf{k}| - \mu \, , \tag{1}$$

where v_F is the Fermi velocity, $\mathbf{k}$ is the two-dimensional (2D) quasiparticle wavevector, and μ the chemical potential. The Fermi wavenumber k_F depends on the chemical potential μ through the relation $k_F = \mu/(\hbar v_F)$. Note that, in 2D, $k_F = \sqrt{4\pi n/g}$, with n as the electron number density and g as the degeneracy. In graphene, $g = 4$: 2 for spin and 2, for inequivalent valleys in the Brillouin zone, and the chemical potential $\mu = \hbar v_F k_F$ is usually $\mu \simeq 10^2$ meV, while the Fermi velocity is $v_F \simeq 10^6$ m/s [2,3,9].

We initially assume that the graphene is sandwiched between two slabs made of generic materials, where L is the distance between the two slabs and d the distance between the lower slab and the graphene sheet. The Coulomb potential of charges in graphene is influenced by the the two slabs. We choose the z axis perpendicular to the graphene sheet, such that $z = 0$ fixes the position of graphene sheet. It follows that the lower slab is located at $z = -d$ and the upper slab at $z = L - d$.

Within the Random Phase Approximation (RPA) [10], the relative dielectric function of graphene is given by

$$\epsilon_g(q,\omega) = 1 - \tilde{V}(q,\omega)\,\Pi_0(q,\omega)\,, \tag{2}$$

where $\tilde{V}(q,\omega)$ is the Fourier transform of the screened (by the presence of the two slabs) Coulomb potential between quasiparticles of graphene and $\Pi_0(q,\omega)$ is the first-order dynamical polarization of non-interacting quasiparticles in graphene. Note that, for a very small wavenumber, $q = \sqrt{q_x^2 + q_y^2}$ and a frequency ω, such that $v_F q \ll \omega \ll 2\mu/\hbar$ the dynamical polarization reads [11,12]

$$\Pi_0(q,\omega) = \frac{\mu}{\pi\hbar^2}\frac{q^2}{\omega^2}\,. \tag{3}$$

The collective mode of plasmons in graphene is then obtained from the resonance condition [10]

$$\epsilon_g(q,\omega) = 0\,. \tag{4}$$

3. Perfect Conductors

Let us suppose that the two slabs are perfect conductors. A straightforward application of the method of image charges [13] gives the screened Coulomb potential between two particles, with electric charge e located in the plane $z = 0$ at distance $x^2 + y^2$ as

$$V(x,y) = e^2 \sum_{j=-\infty}^{+\infty} \left[\frac{1}{\sqrt{(x^2+y^2)+(2jL)^2}} - \frac{1}{\sqrt{(x^2+y^2)+(2d-2jL)^2}} \right], \tag{5}$$

where x and y are Cartesian coordinates in the plane of graphene. Performing the Fourier transform, we obtain

$$\tilde{V}(q) = \frac{2\pi e^2}{q} \sum_{j=-\infty}^{+\infty} \left[\mathrm{e}^{-2q|j|L} - \mathrm{e}^{-2q|d-jL|} \right] \tag{6}$$

with $q = \sqrt{q_x^2 + q_y^2}$. The series can be calculated explicity because is the sum of geometric series. After straightforward calculations, we obtain

$$\tilde{V}(q) = \frac{2\pi e^2}{q}\left(1 - \frac{\mathrm{e}^{-2qd} - 2\mathrm{e}^{-2qL} + \mathrm{e}^{-2q(L-d)}}{1 - \mathrm{e}^{-2qL}}\right). \tag{7}$$

From Equations (2)–(4) the dispersion relation of plasmons in graphene can be written as

$$\omega^2 = \frac{\mu}{\pi\hbar^2} q^2 \tilde{V}(q)\,, \tag{8}$$

or explicitly

$$\omega = \sqrt{\frac{2\mu e^2}{\hbar^2}}\, q^{1/2} \left(1 - \frac{\mathrm{e}^{-2qd} - 2\mathrm{e}^{-2qL} + \mathrm{e}^{-2q(L-d)}}{1 - \mathrm{e}^{-2qL}}\right)^{1/2}. \tag{9}$$

Thus, we have found an analytical formula for the dispersion relation of plasmons in the graphene sheet.

It is important to observe that, for small q, one obtains

$$\tilde{V}(q) = 4\pi e^2 d\left(1 - \frac{d}{L}\right) - \frac{4\pi e^2}{3} d^2 L \left(1 - 2\frac{d}{L} + \frac{d^2}{L^2}\right) q^2 + \ldots \tag{10}$$

Consequently, taking into account Equations (8) and (10), we finally obtain, for small q, the linear dispersion relation

$$\omega = c_p \, q \, , \tag{11}$$

where

$$c_p = \sqrt{\frac{4\mu e^2 d}{\hbar^2}\left(1 - \frac{d}{L}\right)} \tag{12}$$

is the speed of sound of acustic plasmons in graphene sandwiched between two ideal metal gates. Equation (12) is the main result of this brief paper. The velocity c_p can be controlled by varying the chemical potential μ but also the two distances d and L. In the limit $L \to +\infty$ from Equation (12), one finds

$$c_p = \sqrt{\frac{4\mu e^2 d}{\hbar^2}} \tag{13}$$

which is the result of Reference [7], namely, the velocity of acustic plasmons in graphene coupled to a single ideal metal gate.

4. Real Materials

For a generic material, the relative dielectric function ϵ_m depends on frequency ω and wavevector q. We set $\epsilon_{m,1}(q,\omega)$ and $\epsilon_{m,2}(q,\omega)$ as the relative dielectric functions of lower and upper materials, respectively. In this case, the derivation of the screened Coulomb potential is slightly more complicated but still analytically possible [13,14]. We obtain

$$\begin{aligned} \tilde{V}(q,\omega) = \frac{2\pi e^2}{q}\Big(1 - & \frac{r_1(q,\omega)\, e^{-2qd} + r_2(q,\omega)\, e^{-2q(L-d)}}{1 - r_1(q,\omega)\, r_2(q,\omega)\, e^{-2qL}} \\ & + \frac{2\, r_1(q,\omega)\, r_2(q,\omega)\, e^{-2qL}}{1 - r_1(q,\omega)\, r_2(q,\omega)\, e^{-2qL}}\Big) \, , \end{aligned} \tag{14}$$

where

$$r_1(q,\omega) = \frac{\epsilon_{m,1}(q,\omega) - 1}{\epsilon_{m,1}(q,\omega) + 1} \qquad \text{and} \qquad r_2(q,\omega) = \frac{\epsilon_{m,2}(q\omega) - 1}{\epsilon_{m,2}(q,\omega) + 1} \, . \tag{15}$$

Note that for two perfect metal gates, where $r_1 = r_2 = 1$, Equation (14) exactly becomes Equation (7).

4.1. Materials Sticked to Graphene

Setting $L = 2d$, in the limit $d \to 0$ Equation (14) gives

$$\tilde{V}(q,\omega) = \frac{2\pi e^2}{\epsilon_m(q,\omega)\, q} \, , \tag{16}$$

where

$$\epsilon_m(q,\omega) = \frac{1}{2}\left(\epsilon_{m,1}(q,\omega) + \epsilon_{m,2}(q,\omega)\right) . \tag{17}$$

Equation (16) is the screened Coulomb potential in a graphene sheet between two materials stuck onto it.

We adopt Equation (3) again, which is valid for $q \to 0$ and $v_F q \ll \omega \ll 2\mu/\hbar$ [11,12], and Equation (4). Then, for small q and assuming that ϵ_m is constant, we obtain

$$\omega = \sqrt{\frac{2e^2\mu}{\hbar^2 \epsilon_m}}\, \sqrt{q} \, , \tag{18}$$

which is the typical dispersion relation of plasmons in graphene exposed to two polar substrates [8].

4.2. Single Material Slab

In the absence of the upper slab, i.e., setting $r_2 = 0$, from Equation (14), we find

$$\tilde{V}(q,\omega) = \frac{2\pi e^2}{q}\left(1 - r(q,\omega)\, \mathrm{e}^{-2qd}\right) \tag{19}$$

removing the subindex 1 from $r_1(q,\omega)$. Then, for small q and assuming that r is constant, we obtain

$$\tilde{V}(q,\omega) = \frac{2\pi e^2(1-r)}{q} + 4\pi e^2 rd - 8\pi e^2 rd^2\, q + \dots \tag{20}$$

Clearly, only if $r = 1$ (perfect conductor), the $1/q$ term drops out and one again finds acoustic plasmons with the speed of sound given by Equation (13). More generally, the small-q dispersion relation of plasmons reads

$$\omega = \sqrt{\frac{2\mu e^2(1-r)}{\hbar^2} q + \frac{4\mu e^2 rd}{\hbar^2} q^2}\,, \tag{21}$$

which becomes acoustic-like under the condition

$$q \gg \frac{(1-r)}{2dr}\,. \tag{22}$$

For a real metal gate, the functional dependence of $r(q,\omega)$ is crucial. In this case, the relative dielectric function ϵ_m can be approximated as [15]

$$\epsilon_m(q,\omega) = 1 + \frac{q_{TF}^2}{q^2} - \frac{\omega_p^2}{\omega^2 + i\Gamma\,\omega}\,, \tag{23}$$

where q_{TF} is the Thomas–Fermi wavenumber, ω_p is the plasma frequency, and Γ the damping constant. Notice that the relative dielectric constant of a perfect conductor is $\epsilon_m = -\infty$.

5. Conclusions

We have derived a simple formula for the speed of sound of the acustic modes of electrons in a graphene sheet. The existence of these hydrodynamic effects is due to the presence of metallic slabs, which induce a screening the Coulomb potential of electrons in graphene. Our formula for the graphene sandwiched between two metallic slabs generalizes the one obtained in Reference [7] in the case of graphene coupled to a single metallic slab. In conclusion, it is important to stress that, very recently, acoustic plasmons have been observed, with a real-space imaging, in single graphene sheet over a dielectric-metal slab [14]. This graphene-dielectric-metal configuration is quite different with respect to the one considered in the present paper. However, for the sake of completeness, in the last section of our paper, we have also considered the effect of two generic slabs of the screened Coulomb potential of two electrons in graphene.

The author acknowledges Marco Polini, Alessandro Principi, and Flavio Toigo for useful discussions.

Funding: This research received no external funding.

Institutional Review Board Statement: Not applicable.

Informed Consent Statement: Not applicable.

Conflicts of Interest: The author declares no conflict of interest.

References

1. Novoselov, K.S.; Geim, A.K.; Morozov, S.V.; Jiang, D.; Zhang, Y.; Dubonos, S.V.; Grigorieva, I.V.; Firsov, A.A. Electric Field Effect in Atomically Thin Carbon Films. *Science* **2004**, *306*, 666. [CrossRef] [PubMed]
2. Geim, A.K.; Novoselov, K.S. The rise of graphene. *Nat. Mater.* **2007**, *6*, 183. [CrossRef] [PubMed]
3. Neto, A.H.C.; Guinea, F.; Peres, N.M.R.; Novoselov, K.S.; Geim, A.K. The electronic properties of graphene. *Rev. Mod. Phys.* **2009**, *81*, 109. [CrossRef]
4. Rozhkov, A.V.; Sboychakov, A.O.; Rakhmanov, A.L.; Nori, F. Electronic properties of graphene-based bilayer systems. *Phys. Rep.* **2016**, *648*, 1. [CrossRef]
5. Avsar, A.; Ochoa, H.; Guinea, F.; Ozyulmaz, B.; van Wees, B.J.; Vera-Maru, I.J. Spintronics in graphene and other two-dimensional materials. *Rev. Mod. Phys.* **2020**, *92*, 021003. [CrossRef]
6. Polini, M.; Geim, A.K. Viscous electron fluids. *Phys. Today* **2020**, *73*, 28. [CrossRef]
7. Principi, A.; Asgari, R.; Polini, M. Acoustic plasmons and composite hole-acoustic plasmon satellite bands in graphene on a metal gate. *Solid State Commun.* **2011**, *151*, 1627. [CrossRef]
8. Grigorenko, A.N.; Polini, M.; Novoselov, K.S. Graphene Plasmonics. *Nat. Photonics* **2012**, *6*, 749. [CrossRef]
9. Peres, N.M.R.; Guinea, F.; Neto, A.H.C. Electronic properties of disordered two-dimensional carbon. *Phys. Rev. B* **2006**, *73*, 125411. [CrossRef]
10. Fetter, A.L.; Walecka, J.F. *Quantum Theory of Many-Particle Systems*; Dover: New York, NY, USA, 2003.
11. Wunsch, B.; Stauber, T.; Sols, F.; Guinea, F. Dynamical polarization of graphene at finite doping. *New J. Phys.* **2006**, *8*, 318. [CrossRef]
12. Hwang, E.H.; Sarma, S.D. Acoustic phonon scattering limited carrier mobility in two-dimensional extrinsic graphene. *Phys. Rev. B* **2007**, *75*, 205418. [CrossRef]
13. Jakson, J.D. *Classical Electrodynamics*; Wiley: Hoboken, NJ, USA, 1998.
14. Menabde, S.G.; Lee, I.-H.; Lee, S.; Ha, H.; Heiden, J.T.; Yoo, D.; Kim, T.-T.; Low, T.; Lee, Y.H.; Oh, S.-H.; et al. Real-space imaging of acoustic plasmons in large-area graphene grown by chemical vapor deposition. *Nat. Commun.* **2021**, *12*, 938. [CrossRef] [PubMed]
15. Achcroft, N.W.; Mermin, D.N. *Solid State Physics*; Thomson Press: Delhi, India, 2003. [CrossRef] [PubMed]

symmetry

MDPI

Review

A Review of Many-Body Interactions in Linear and Nonlinear Plasmonic Nanohybrids

Mahi R. Singh

Department of Physics and Astronomy, The University of Western Ontario, London, ON N6A 3K7, Canada; msingh@uwo.ca

Abstract: In this review article, we discuss the many-body interactions in plasmonic nanohybrids made of an ensemble of quantum emitters and metallic nanoparticles. A theory of the linear and nonlinear optical emission intensity was developed by using the many-body quantum mechanical density matrix method. The ensemble of quantum emitters and metallic nanoparticles interact with each other via the dipole-dipole interaction. Surfaces plasmon polaritons are located near to the surface of the metallic nanoparticles. We showed that the nonlinear Kerr intensity enhances due to the weak dipole-dipole coupling limits. On the other hand, in the strong dipole-dipole coupling limit, the single peak in the Kerr intensity splits into two peaks. The splitting of the Kerr spectrum is due to the creation of dressed states in the plasmonic nanohybrids within the strong dipole-dipole interaction. Further, we found that the Kerr nonlinearity is also enhanced due to the interaction between the surface plasmon polaritons and excitons of the quantum emitters. Next, we predicted the spontaneous decay rates are enhanced due to the dipole-dipole coupling. The enhancement of the Kerr intensity due to the surface plasmon polaritons can be used to fabricate nanosensors. The splitting of one peak (ON) two peaks (OFF) can be used to fabricate the nanoswitches for nanotechnology and nanomedical applications.

Keywords: plasmonics; nano-optics; dipole-dipole interaction; density matrix method; dressed states; Kerr effect; quantum dots; metallic nanoparticles

Citation: Singh, M.R. A Review of Many-Body Interactions in Linear and Nonlinear Plasmonic Nanohybrids. *Symmetry* **2021**, *13*, 445. https://doi.org/10.3390/sym13030445

Academic Editor: V. I. Yukalov

Received: 19 January 2021
Accepted: 3 March 2021
Published: 9 March 2021

1. Introduction

Recently there has been considerable interest in studying the nonlinear properties of plasmonic nanohybrids (PNHs) made of metallic nanoparticles and quantum emitters (i.e., quantum dots (QDs)) [1–15].

The dipole-dipole interaction in the linear properties of PNHs and photonic crystals has been investigated [10–14]. The Kerr nonlinearity has been studied widely in quantum optics using three-level and four-level atoms [15–49]. For example, giant Kerr nonlinearities in atoms have been studied by using electromagnetically induced transparency. It was shown that the Kerr nonlinear refractive index of a three-level Λ atomic type is greatly enhanced inside an optical ring cavity. On the other hand, the enhancement of the Kerr nonlinearity was also investigated in a four-level atomic system. Experimentally, the enhancement of the Kerr-nonlinear coefficient was also observed in a three-level atomic system such as Rb atom.

Singh [40] studied the nonlinear second harmonic generation in nanohybrids made of a metallic nanoparticle and a QD. Some efforts have also been devoted to studying the Kerr nonlinearity in metallic nanohybrids. Recently, some effort has been devoted to studying the Kerr nonlinearity in PNHs [35–43]. For example, Terzis et al. [37] have fabricated optical systems from QD and gold-metallic nanoparticles. Singh experimentally and theoretically [42,43] have studied the nonlinear properties of metallic nanohybrids. The nonlinear effects can also be used for electro-optical device applications such as light modulators, optical switches, optical logic, and optical limiters. The nonlinear research in PNHs can create a revolutionary change in electronic and photonic nontechnology and nanomedicine.

In this review article, we outline a theory we have developed of the linear and nonlinear light emissions for PNHs by using the many-body quantum mechanical density matrix method. A schematic diagram of a nanohybrid containing interacting metallic nanoparticles (MNPs) and quantum emitters (QEs) is shown in Figure 1. In Section 1, we surveyed the literature for the Kerr nonlinearity. In Section 2, a theory of the surface plasmon polaritons due to the photon-plasmon interaction is discussed. In Section 3, the dipole-dipole interaction in the ensemble of QEs and MNPs is explained. In Section 4, a theory for the linear and nonlinear optical absorption intensity using the density matrix method is outlined. In Section 5, we derive the expressions of the spontaneous decay rates due to the dipole-dipole interaction. In Section 6, we show how we derived the expressions of dressed states in the strong exciton-DDI coupling limit. In Section 7, we discuss the simulation we performed on the nonlinear optical absorption intensity. Finally, in the last Section 8, we have summarized the findings of the review article.

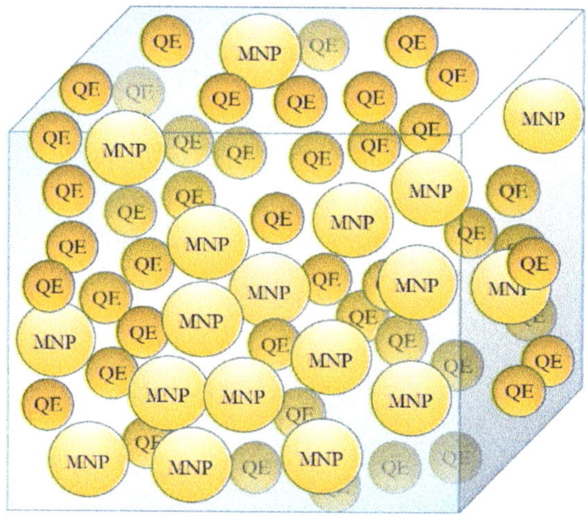

Figure 1. Schematic diagram of a nanohybrid which consists of a noninteracting metallic nanoparticles (MNPs) and interacting quantum emitters (QEs).

2. Surface Plasmon Polaritons

Let us consider a situation where a probe with a frequency ω_k and wave wavevector k is applied to study the PL in the PNH. We know that the surface plasmons are present on the surface of the MNP. The surface plasmons oscillate with a frequency ω_p. The probe photons interact with surface plasmons and this interaction creates new quasi-particles called surface plasmon polaritons. Let us calculate the surface plasmon polaritons due to the interaction of photon and surface plasmons interaction as follows.

The Hamiltonian of the surface plasmon and photons, and interactions between them can be written as

$$H_E = \sum_k \hbar\omega_k c_k^+ c_k + \sum_k \omega_p B_k^+ B_k + - \sum_k G_k \left(B_k^+ c_k + B_k c_k^+\right) \quad (1)$$

where c_k^+ and c_k are the photon creation and annihilation operators, respectively. Here B_k^+ and B_k are the surface plasmon creation and annihilation operators, respectively. The constant G_k is the coupling constant between the photons and plasmons. The first and second terms are for noninteracting photon and plasmon Hamilton. The last term is photon-plasmon interaction Hamiltonian.

Plasmons and photons operators satisfy the following commutation relation.

$$\begin{aligned} \left[B_k, B_k^+\right] &= \delta_{k,k'} \\ \left[c_k, c_k^+\right] &= \delta_{k,k'} \end{aligned} \tag{2}$$

Note that in the interaction, the Hamiltonian has a mixture of photon and plasmon operators. We aimed to diagonalize the Hamiltonian. The diagonalization of an interacting Hamiltonian is a standard method in the many-body theory. We defined new quasi-particle creation and annihilation operators p_{ik}^+ and p_{ik}^- with $i = 1,2$ We express the plasmon and photon operators in terms of these quasi-particle operators as

$$\begin{aligned} B_k &= u_{11}p_{1k} + u_{12}p_{2k} \\ B_k^+ &= u_{11}^* p_{1k}^+ + u_{12}^* p_{2k}^+ \\ c_k &= u_{21}p_{1k} + u_{22}p_{2k} \\ c_k^+ &= u_{21}^* p_{1k}^+ + u_{22}^* p_{2k}^+ \end{aligned} \tag{3}$$

New quasi-particle operators p_{ik}^+ and p_{ik}^- with i = 1,2 are called surface plasmon polariton operators, which have the combined effect of plasmons and photons.

The Hamiltonian in Equation (1) can be written in the simpler form for diagonalization. It can be expressed as

$$\begin{aligned} H &= \sum_k H_k, \\ H_k &= \hbar\omega_k c_k^+ c_k + \omega_p B_k^+ B_k - G_k\left(B_k^+ c_k + B_k c_k^+\right) \end{aligned} \tag{4}$$

We aimed to write the H_k in terms of SPP operators p_{ik}^+ and p_{ik}^-. Replacing the plasmon and exciton operators in terms of the SPP operators from Equation (3) into Equation (4) and performing extensive mathematical manipulation, we obtained the following expression as

$$H_k = \hbar\omega_k^+ p_{1k}^+ p_{1k} + \hbar\omega_k^- p_{2k}^+ p_{2k} \tag{5}$$

where ω_k^+ and ω_k^- are the energy of SPPs for the upper and lower branches, respectively. Note that coupling between photons and plasmon splits the SPP spectrum into two branches. The SPP spectrum is given by

$$\omega_k^{\pm} = \frac{1}{2}\left[(ck + \omega_p) \pm \sqrt{(ck - \omega_p)^2 + 4g_k^2}\right] \tag{6}$$

where we replaced $\omega_k = ck$. Finally, we obtained the expression for the expansion coefficients as

$$\begin{aligned} u_{11} &= \sqrt{\frac{g_k^2}{(\omega_p - \omega_k^+)^2 + g_k^2}}, \\ u_{21} &= \sqrt{\frac{(\omega_p - \omega_k^+)^2}{(\omega_p - \omega_k^+)^2 + g_k^2}} \\ u_{22} &= \sqrt{\frac{g_k^2}{(\omega_p - \omega_k^-)^2 + g_k^2}}, \\ u_{12} &= \sqrt{\frac{(\omega_p - \omega_k^-)^2}{\omega_p - \omega_k^-)^2 + g_k^2}} \end{aligned} \tag{7}$$

We then calculated the density of states (DOS) for SPPs. Later in the paper, we discuss how the DOS was used for the calculation of the SPP decay rates. When summation over wavevectors was converted into the integration, we needed to use the idea of the DOS

$$\begin{aligned} \sum_k &= \int D_k(\varepsilon_k) d\varepsilon_k, \\ D_k(\varepsilon_k) &= \frac{\Omega}{3\pi^2} k^2 \frac{dk}{d\varepsilon_k} \end{aligned} \tag{8}$$

where Ω is the volume of the PNH. Here D_k is called the DOS. The expression of the DOS can be expressed in term of the form factor Z_k as

$$\begin{aligned} D(\varepsilon_k) &= D_0(\varepsilon_k) Z_k(\varepsilon_k) \ , \\ D_0(\varepsilon_k) &= \frac{\Omega}{3\pi^2} \frac{\varepsilon_k^2}{(\hbar c)^3}, \\ Z_k(\varepsilon_k) &= (\hbar c) \frac{dk}{d\varepsilon_k} \end{aligned} \tag{9}$$

where D_0 is the DOS of photons in free space.

Inserting Equation (6) into Equation (9) and after some mathematical manipulation, the expression of the form factor Z_k could be found as

$$Z_k(\omega_k^{\pm}) = \left(\frac{(\omega_p - \omega_k^{\pm}) + |g_k|^2}{(\omega_p - \omega_k^{\pm})^2} \right). \tag{10}$$

We found that the form factor has a very large value when the surface plasmon frequency is the resonance with the SPP frequency, which is an interesting finding.

3. Dipole-Dipole Interactions

When the probe field falls on the PNH, induced dipoles are created in the QEs and MNPs. The dipole of one QE interacts with the dipoles of other QEs. This is called dipole-dipole interaction (DDI). Following reference [10], the DDI Hamiltonian can be written as

$$H_{DDI}^{QE} = \frac{1}{2} \sum_{i \rangle j}^{N} J_{ij} p_i \cdot p_j \tag{11}$$

where J_{ij} is the DDI coupling constant and is found as

$$J_{ij} = \frac{1}{\epsilon_b \epsilon_0 \, r_{ij}^3} \tag{12}$$

where p_i and p_j are the induced dipole moments in the ith-QE and jth-QE, respectively. Here r_{ij} is the distance between the ith and the jth QEs. In the mean-field approximation [10,44–49], the DDI Hamiltonian can be rewritten as

$$H_{DDI}^{QE} = \sum_i p_i E_{DDI}^{QE} \tag{13}$$

where E_{DDI}^{QE} is called the DDI electric field and is written as

$$E_{DDI}^{QE} = \left\langle \frac{1}{2} \sum_{j \neq i} J_{ij} p_j \right\rangle. \tag{14}$$

Here, the DDI field E_{DDI}^{QE} is the average dipole electric field created by all QEs on the ith-QE and that E_{DDI} has the effect of the long-range DDI (i.e., r^{-3}). The average in E_{DDI} has been evaluated in references [30–36] by using the method of Lorentz [49], The expression of the DDI field is found as and is written as

$$\begin{aligned} E_{DDI}^{QE} &= \frac{\lambda_{QD} \langle p_i \rangle}{3 \times 4\pi \epsilon_0 \epsilon_b}, \\ \langle p_i \rangle &= \frac{P_{QE}}{4\pi R_{QE}^3} \end{aligned} \tag{15}$$

where λ_{QD} is the DDI constant and $\langle p_i \rangle$ is the average polarization of the ith-QE. Here P_{QE} is the polarization of the QE and R_{QE} is the radius of the quantum emitter.

The probe electric field includes polarizations in the QE and MNP. They are denoted as P_{QE} and P_{MNP}. Due to these polarizations, the MNP produces the SPP field E_{SPP} and the QE produces an electric field denoted as E_{QE}. Solving Maxwell's equations in the quasi-static approximation [50,51], one can find the following expressions for polarizations as

$$\begin{aligned} P_{QE} &= 4\pi \in_0 \in_b R_{QE}^3 \zeta_{QE}(E_P + E_{SPP}) \\ \varsigma_{QE} &= \left[\frac{\epsilon_q - \epsilon_b}{\epsilon_q + 2\epsilon_b}\right] \\ P_{MNP} &= 4\pi \in_0 \in_b R_{MNP}^3 \zeta_{MNP}\left(E_P + E_{QE}\right), \\ \varsigma_{MNP} &= \left[\frac{\epsilon_m - \epsilon_b}{\epsilon_m + 2\epsilon_b}\right] \end{aligned} \quad (16)$$

Here ϵ_m, ϵ_q, and ϵ_b are the dielectric constant of the MNS, QE and background material, respectively. The electric fields produced by these polarizations were found as

$$\begin{aligned} E_{SPP} &= \frac{P_{MNP}}{4\pi \in_0 \in_b r^3}, \\ E_{QE} &= \frac{P_{QE}}{4\pi \in_0 \in_b r^3} \end{aligned} \quad (17)$$

Putting Equation (16) into Equation (17), we obtained

$$\begin{aligned} E_{SPP} &= \Pi_{SPP}(E_P + E_{QE}) \\ \Pi_{SPP} &= \frac{R_{MNP}^3 g_l \zeta_{MNP}}{r^3} \\ E_{QE} &= \Pi_{QE}(E_P + E_{SPP}) \\ \Pi_{QE} &= \frac{R_{QE}^3 g_l \zeta_{QE}}{r^3} \end{aligned} \quad (18)$$

The constant g_l is called the polarization parameter and it has values $g_l = 1$ and $g_l = -2$ for $\boldsymbol{P}_{MNP} \parallel \boldsymbol{r}$ and $\boldsymbol{P}_{MNP} \perp \boldsymbol{r}$. Note that both electric fields depend on r^{-3}. Putting Equation (18) into Equation (14), we obtained the expression of the DDI as

$$\begin{aligned} E_{DDI}^{QE} &= \Phi_{DDI}^{QE} E_P, \\ \Phi_{CDDI}^{QE} &= \frac{\lambda_{QD} g_l \varsigma_{MNP}}{3} \\ \Phi_{DDI}^{QE} &= \Phi_{CDDI}^{QE} + \Phi_{ADDI}^{QE} \\ \Phi_{ADDI}^{QE} &= \Phi_{CDDI}^{QE} \Pi_{SPP} \end{aligned} \quad (19)$$

Similarly, following the above method, we could also calculate the dipole-dipole interaction between MNPs as

$$\begin{aligned} E_{DDI}^{MNP} &= \Phi_{DDI}^{MNP} E_P \\ \Phi_{CDDI}^{MNP} &= \frac{\lambda_{MNP} g_l \varsigma_{MNP}}{3} \\ \Phi_{DDI}^{MNP} &= \Phi_{CDDI}^{MNP} + \Phi_{ADDI}^{MNP} \\ \Phi_{ADDI}^{MNP} &= \Phi_{CDDI}^{MNP} \Pi_{QE} \end{aligned} \quad (20)$$

Note that the Π-term and the Φ-term depend on r^{-3}. The higher-order terms r^{-6} were neglected because they are weak.

4. Linear and Nonlinear Plasmonics and Density Matrix Method

For this section, we calculated the plasmonic Kerr intensity. We considered that QD has four energy levels. They are denoted as |1>, |2>, |3> and |4>. The energy difference between levels |i> and |j> is expressed as ε_{ij}. To study the Kerr nonlinearity, we applied the probe field between $|\mathbf{1}\rangle$ and $|\mathbf{2}\rangle$. A schematic diagram of the four-level quantum emitter is shown in Figure 2.

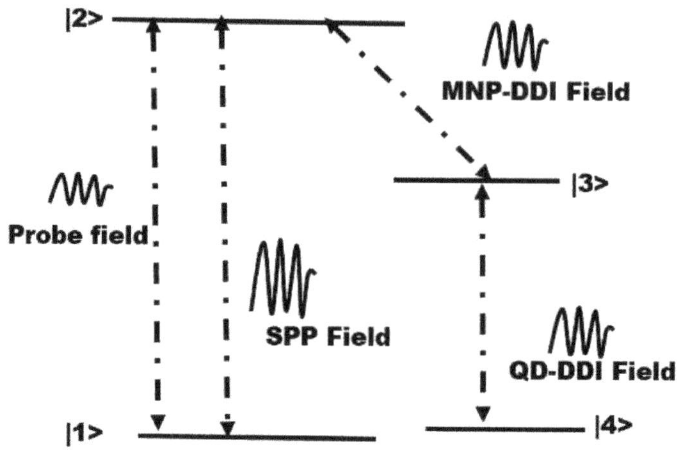

Figure 2. A schematic diagram of a four-level quantum dots (QDs) is plotted. Energy levels are denoted as |1>, |2>, |3> and |4>. The probe field and SPP field are applied in the transition $|1\rangle \leftrightarrow |2\rangle$. The DDI-MNP and DDI-QD fields are acting in the transitions $|2\rangle \leftrightarrow |3\rangle$ and $|3\rangle \leftrightarrow |4\rangle$, respectively.

Following the method of reference [52,53], the expression of the polarization was found as P_{QE}

$$P_{QE}(\omega_p) = P_{QE}^{(1)}(\omega_p) + P_{QE}^{(2)}(\omega_p, \omega_p) + P_{QE}^{(3)}(\omega_p, \omega_p, \omega_p) \tag{21}$$

$$\begin{aligned} P_{QE}^{(1)}(\omega_p) &= \epsilon_0\, \chi_{QE}^{(1)}(\omega_p) E_p(\omega_p) \\ P_{QE}^{(2)}(\omega_p, \omega_p) &= \epsilon_0\, \chi_{QE}^{(2)}(\omega_p, \omega_p) E_p(\omega_p) E_p(\omega_p) \\ P_{QE}^{(3)}(\omega_p, \omega_p, \omega_p) &= \epsilon_0\, \chi_{QE}^{(3)}(\omega_p, \omega_p, \omega_p) E_p(\omega_p) E_p(\omega_p) E_p(\omega_p). \end{aligned} \tag{22}$$

Here $\chi_{QE}^{(1)}$, $\chi_{QE}^{(2)}$, and $\chi_{QE}^{(3)}$ are the first, second the third-order expressions of the susceptibility. We know that the first-order susceptibility is responsible for the one-photon phenomena, whereas second-order susceptibility is responsible for the two-photon phenomena. Finally, the third-order susceptibility is responsible for the Kerr nonlinearity.

Fowling the method of references [52–55], the polarization of the QE can also be expressed in terms of the quantum density matrix operator (ρ) as follows

$$P_{QE}(\omega_p) = 2\mu_{21}[\rho_{21}(\omega_p) + \text{h.c.}] \tag{23}$$

where μ_{21} is the matrix elements of the dipole moment between transition $|1\rangle \leftrightarrow |2\rangle$ and ρ_{21} is the nonlinear density matrix operator (ρ) between transition $|1\rangle \leftrightarrow |2\rangle$. Expressing the nonlinear density matrix as follows

$$\rho_{21} = \rho_{21}^{(1)}(\omega_p) + \rho_{21}^{(2)}(\omega_p, \omega_p) + \rho_{21}^{(3)}(\omega_p, \omega_p, \omega_p). \tag{24}$$

Putting Equation (24) into Equation (23), we obtained

$$P_{QE}(\omega_p) = 2\mu_{21}\left[\rho_{21}^{(1)}(\omega_p) + \rho_{21}^{(2)}(\omega_p, \omega_p) + \rho_{21}^{(3)}(\omega_p, \omega_p, \omega_p) + h.c.\right]. \tag{25}$$

We compared Equations (21) and (25) and we found the relation between the susceptibility and the density matrix elements to be as follows

$$\begin{aligned} \chi^{(1)}(\omega_p) &= \frac{2\mu_{21}\rho_{21}^{(1)}(\omega_p)+hc}{\epsilon_0 E_p(\omega_p)} \\ \chi^{(2)}(\omega_p,\omega_p) &= \frac{2\mu_{21}\rho_{21}^{(2)}(\omega_p,\omega_p)+hc}{\epsilon_0 E_p(\omega_p)E_p(\omega_p)} \\ \chi^{(3)}(\omega_p,\omega_p,\omega_p) &= \frac{2\mu_{21}\rho_{21}^{(3)}(\omega_p,\omega_p,\omega_p)+hc}{\epsilon_0 E_p(\omega_p)E_p(\omega_p)E_p(\omega_p)} \end{aligned} \tag{26}$$

The above expression can be expressed in terms of the Rabi frequency as

$$\begin{aligned} \chi^{(1)}(\omega_p) &= \frac{2\mu_{21}^2\rho_{21}^{(1)}(\omega_p)+hc}{\epsilon_0\hbar\Omega_p}, \\ \chi^{(2)}(\omega_p,\omega_p) &= \frac{2\mu_{21}^3\rho_{21}^{(2)}(\omega_p,\omega_p)+hc}{\epsilon_0\hbar^2\Omega_p^2}, \\ \chi^{(3)}(\omega_p,\omega_p,\omega_p) &= \frac{2\mu_{21}^4\rho_{21}^{(3)}(\omega_p,\omega_p,\omega_p)+hc}{\epsilon_0\hbar^3\Omega_p^3} \end{aligned} \tag{27}$$

Following the method reference [52], the intensity of the fluorescence (light emission intensity) from the QE could be calculated in terms of the susceptibility as follows

$$\begin{aligned} I_{1ph} &= \left[\frac{\omega_p}{2c\epsilon_0\sqrt{\epsilon_b}}\left|\chi_{pp}^{(1)}(\omega_p)E_p(\omega_p)\right|^2\right] \\ I_{2ph} &= \left[\frac{\omega_p^2}{2c^2\epsilon_0\sqrt{\epsilon_b}}\left|\chi_{pp}^{(1)}(\omega_p,\omega_p)E_p(\omega_p)E_p(\omega_p)\right|^2\right] \\ I_{\text{kerr}} &= \left[\frac{\omega_p^3}{2c^3\epsilon_0\sqrt{\epsilon_b}}\left|\chi_{pp}^{(1)}(\omega_p,\omega_p,\omega_p)E_p(\omega_p)E_p(\omega_p)E_p(\omega_p)\right|^2\right]. \end{aligned} \tag{28}$$

The expression of the intensity of the fluorescence is given in Equation (28) can be expressed in terms of the density matrix elements by putting Equation (26) into Equation (28) as follows

$$\begin{aligned} I_{1ph} &= \frac{4\omega_p\omega_p\mu_{21}^2}{2c\epsilon_0\epsilon_0^2\sqrt{\epsilon_b}}\left|\rho_{21}^{(1)}(\omega_p)\right|^2 \\ I_{2ph} &= \frac{4\omega_p\omega_p\mu_{21}^2}{2c\epsilon_0\epsilon_0^2\sqrt{\epsilon_b}}\left|\rho_{21}^{(2)}(\omega_p,\omega_p)\right|^2 \\ I_{\text{kerr}} &= \frac{4\omega_p\omega_p\mu_{21}^2}{2c\epsilon_0\epsilon_0^2\sqrt{\epsilon_b}}\left|\rho_{21}^{(3)}(\omega_p,\omega_p,\omega_p)\right|^2 \end{aligned} \tag{29}$$

Note that the susceptibility (Equation (27)) and the fluorescence (Equation (29)) depend on the density matrix elements. Next using the quantum density matrix method, we evaluated the density matrix elements $\rho_{21}^{(1)}$, $\rho_{21}^{(2)}$, and $\rho_{21}^{(3)}$ for the first, second, and third-order in the probe field, respectively.

We know that the probe electric field can be denoted in terms of the Rabi frequency $\Omega_P = \mu_{21}E_P/h$ acting between transition $|\mathbf{1}\rangle \leftrightarrow |\mathbf{2}\rangle$. We consider that the SPP field represented in term of the SPP coupling constant Π_{SPP} is acting between transition $|\mathbf{1}\rangle \leftrightarrow |\mathbf{2}\rangle$. Similarly, the DDI-MNP field Φ_{DDI}^{MNP} is acting between transition $|\mathbf{2}\rangle \leftrightarrow |\mathbf{3}\rangle$ and the DDI-QD field Φ_{DDI}^{QE} is acting between transition $|\mathbf{3}\rangle \leftrightarrow |\mathbf{4}\rangle$. To make the calculation simple, we considered $\Phi_{DDI} = \Phi_{DDI}^{QE} = \Phi_{DDI}^{MNP}$. The physics of the problem do not change due to this approximation.

Three electric fields are falling on the QE: $E_T = E_P + E_{DDI}^{MNP} + E_{DDI}^{QE}$. These filed interact with the exciton of the QE. Using the dipole and rotating wave approximation [52–55], the interaction Hamiltonian between QE was found as

$$H_{in} = \hbar\Omega_P\sigma_{21} + \hbar\Omega_{SPP}\sigma_{21} + \hbar\Omega_{DDI}^{MNP}\sigma_{23} + \hbar\Omega_{DDI}^{QE}\sigma_{43} + h.c. \tag{30}$$

where

$$\begin{array}{c}\Omega_P = \frac{\mu_{21}E_P}{\hbar}, \Omega_{SPP} = \Omega_P\Pi_{SPP}, \\ \Omega_{DDI}^{MNP} = \Omega_P\Phi_{DDI}^{MNP}, \Omega_{DDI}^{QE} = \Omega_P\Phi_{DDI}^{QE}\end{array} \tag{31}$$

Here h.c. stands for the Hermitian conjugate. Here σ_{ij} = |*i*><*j*| is the exciton creation operators for $|i\rangle \leftrightarrow |j\rangle$. Parameter Ω_p is called the Rabi frequency. The first term in H_{int} is the exciton-probe field interaction. The second term is exciton-SPP field interaction. The third term is exciton-DDI-MNP interaction. The last term is exciton-DDI-QE field interaction.

With the help of Equation (30) and following the method of references [53–55] equations of motion for density matrix elements were found as follows

$$\begin{array}{c}\frac{d\rho_{11}}{dt} = 2\gamma_{23}\rho_{22} - i\Omega_P(\rho_{12} - \rho_{21}) \\ \frac{d\rho_{22}}{dt} = -2\gamma_{21}\rho_{22} + i\Omega_P(\rho_{12} - \rho_{21}) + i\Pi_{SPP}(\rho_{32} - \rho_{23}) \\ \frac{d\rho_{33}}{dt} = 2\gamma_{31}\rho_{22} + i\Pi_{SPP}(\rho_{23} - \rho_{32}) + i\Phi_{DDI}(\rho_{43} - \rho_{34})* \\ \frac{d\rho_{12}}{dt} = -d_{21}\rho_{12} - i\Pi_{SPP}\rho_{13} + i(\rho_{22} - \rho_{11})\Omega_P \\ \frac{d\rho_{13}}{dt} = -d_{13}\rho_{13} - i\rho_{12}\Pi_{SPP} + i\rho_{23}\Omega_P - i\rho_{14}\Phi_{DDI} \\ \frac{d\rho_{23}}{dt} = -d_{23}\rho_{23} + i(\rho_{33} - \rho_{22})\Pi_{SPP} + i\rho_{13}\Omega_P - i\rho_{24}\Phi_{DDI} \\ \frac{d\rho_{14}}{dt} = -d_{14}\rho_{14} - i\rho_{13}\Phi_{DDI} + i\rho_{24}\Omega_P \\ \frac{d\rho_{24}}{dt} = -d_{24}\rho_{24} + i\rho_{34}\Pi_{SPP} - i\rho_{23}\Phi_{DDI} + i\rho_{13}\Omega_P \\ \frac{d\rho_{34}}{dt} = -d_{34}\rho_{34} + i\rho_{24}\Pi_{SPP} + i\Phi_{DDI}(\rho_{44} - \rho_{33})\end{array} \tag{32}$$

where the parameters appearing in Equation (32) were found as

$$\begin{array}{c}d_{21} = \delta_{21} + i\gamma_{21}, d_{23} = \delta_{23} + i\gamma_{23}, d_{34} = \delta_{34} + i\gamma_{43}, \\ d_{41} = \delta_{41} + i\gamma_{41}, d_{42} = \delta_{42} + i\gamma_{42}, d_{31} = \delta_{31} + i\gamma_{31} \\ \delta_{21} = \omega_{21} - \omega_p, \delta_{34} = \omega_{34} - \omega_d, \delta_{23} = \omega_{23} - \omega_s \\ \delta_{42} = \delta_{23} + \delta_{34}, \delta_{14} = \delta_{21} - \delta_{23} - \delta_{34}, \delta_{13} = \delta_{21} - \delta_{23}\end{array} \tag{33}$$

Here δ_{ij} is called the field detuning. Physical quantity γ_{ij} is the exciton decay rates.

We solved Equation (32) in a steady state. We know that density matrix elements satisfy the condition $\rho_{11} + \rho_{22} + \rho_{33} + \rho_{44} = 0$. We tried to find the analytical expression of the density matrix elements in the steady-state. We considered that initially the ground state is filled and all excited states are empty: $\rho_{11}^{(0)} = 1$ and $\rho_{22}^{(0)} + \rho_{33}^{(0)} + \rho_{44}^{(0)} = 0$.

In the first order in the probe field, Ω_P, we solved Equation (32) using the above initial condition. After some mathematical manipulations, we obtained the following analytical expression of one-photon density matrix element $\rho_{12}^{(1)}$, as

$$\rho_{12}^{(1)} = -\frac{i\Omega_p(d_{13}d_{14} - \Phi_{DDI}^2)}{d_{21}(d_{13}d_{14} - \Phi_{DDI}^2) + d_{14}\Pi_{SPP}^2}. \tag{34}$$

The second-order two-photon density matrix element $\rho_{12}^{(2)}$ was calculated in the second order in the probe field, Ω_P^2 by solving Equation (32). After some mathematical manipulations, we obtained the following analytical expression of the element $\rho_{12}^{(2)}$,

$$\rho_{12}^{(2)} = +\frac{\rho_{23}^{(1)}\Pi_{SPP}\Omega_p d_{14} - \rho_{24}^{(1)}\Pi_{SPP}\Omega_p\Phi_{DDI}}{d_{21}\left(d_{13}d_{14} - d_{21}\Phi_{DDI}^2\right) + \Pi_{SPP}^2 d_{14}} + \frac{i\left(\rho_{22}^{(1)} - \rho_{11}^{(1)}\right)\Omega_p\left(d_{13}d_{14} - d_{21}\Phi_{DDI}^2\right)}{d_{21}\left(d_{13}d_{14} - d_{21}\Phi_{DDI}^2\right) + \Pi_{SPP}^2 d_{14}} \tag{35}$$

$$\begin{aligned}
\rho_{24}^{(1)} &= -\frac{i\rho_{23}^{(1)}\Phi_{DDI}d_{34}}{d_{24}d_{34}+\Pi_{SPP}^2}, \\
\rho_{24}^{(1)} &= -\frac{i\rho_{23}^{(1)}\Phi_{DDI}d_{34}}{d_{24}d_{34}+\Pi_{SPP}^2}, \\
\rho_{22}^{(1)} &= \frac{2\Pi_{SPP}\mathrm{Im}\alpha_{23}}{2\gamma_{21}+4\Pi_{SPP}\mathrm{Im}\alpha_{23}} \\
\rho_{23}^{(1)} &= \frac{+i\Pi_{SPP}\left(d_{24}d_{34}+\Pi_{SPP}^2\right)\left(\rho_{33}^{(1)}-\rho_{22}^{(1)}\right)}{d_{23}\left(d_{24}d_{34}+\Pi_{SPP}^2\right)+\Phi_{DDI}^2 d_{34}}, \\
\alpha_{23} &= \frac{+i\Pi_{SPP}\left(d_{24}d_{34}+\Pi_{SPP}^2\right)}{d_{23}\left(d_{24}d_{34}+\Pi_{SPP}^2\right)+\Phi_{DDI}^2 d_{34}}
\end{aligned} \tag{36}$$

The third order Kerr density matrix element $\rho_{12}^{(3)}$ was calculated in the third order in the probe field E_P^3 and E_{SPP}^3 by solving Equation (32). After some mathematical manipulations, we obtained the following analytical expression of the element $\rho_{12}^{(3)}$ as

$$\rho_{12}^{(3)} = \frac{\rho_{23}^{(2)}\Omega_p\Pi_{SPP}d_{14} + i\rho_{24}^{(2)}\Omega_p\Phi_{DDI}\Lambda_{SPP}}{d_{21}\left(d_{13}d_{14} - d_{21}\Phi_{DDI}^2\right) + \Pi_{SPP}^2 d_{14}} + \frac{i(\rho_{22}^{(2)} - \rho_{11}^{(2)})\Omega_p\left(d_{13}d_{14} - d_{21}\Pi_{SPP}^2\right)}{d_{21}\left(d_{13}d_{14} - d_{21}\Phi_{DDI}^2\right) + \Pi_{SPP}^2 d_{14}} \tag{37}$$

$$\begin{aligned}
\rho_{13}^{(1)} &= -\left(\frac{d_{14}\Phi_{DDI}\Omega_p}{d_{21}\left(d_{13}d_{14}-\Pi_{SPP}^2\right)+d_{14}\Phi_{DDI}^2}\right), \\
\rho_{24}^{(2)} &= -\frac{i\rho_{23}^{(2)}\Pi_{SPP}d_{34}}{d_{24}d_{34}+\Phi_{DDI}^2} \\
\rho_{23}^{(2)} &= \frac{\left(i\rho_{33}^{(2)}\Phi_{DDI}-i\rho_{22}^{(2)}\Phi_{DDI}+i\rho_{13}^{(1)}\Omega_P\right)\left(d_{24}d_{34}+\Phi_{DDI}^2\right)}{\left(d_{24}d_{34}+\Phi_{DDI}^2\right)+\Pi_{SPP}^2 d_{34}}, \\
\rho_{22}^{(2)} &= \frac{\Omega_p\mathrm{Im}\left(\rho_{12}^{(1)}\right)}{\gamma_{21}}
\end{aligned} \tag{38}$$

Note that the $\rho_{12}^{(1)}$, $\rho_{12}^{(2)}$, and $\rho_{12}^{(3)}$ depend on the SPP and DDI couplings.

We then calculated the linear fluorescence intensity due to the emission of one photon. Inserting Equation (34) into Equation (29) we obtained linear fluorescence intensity as

$$I_{1ph} = \frac{4\omega_p\omega_p\mu_{21}^2}{2c\,\epsilon_0\epsilon_0^2\sqrt{\epsilon_b}}\left|\frac{i\Omega_p\left(d_{13}d_{14} - \Phi_{DDI}^2\right)}{d_{21}\left(d_{13}d_{14} - \Phi_{DDI}^2\right) + d_{14}\Pi_{SPP}^2}\right|^2. \tag{39}$$

Similarly, we calculated the nonlinear fluorescence intensity due to the emission of two photons. Inserting Equation (35) into Equation (29) we obtained the two-photon fluorescence intensity as

$$\begin{aligned}
I_{2ph} &= \frac{4\omega_p\omega_p\mu_{21}^2}{2c\epsilon_0\epsilon_0^2\sqrt{\epsilon_b}}\left|\rho_A^{(2)}(\omega_p,\omega_p) + \rho_B^{(2)}(\omega_p,\omega_p)\right|^2 \\
\rho_A^{(2)}(\omega_p,\omega_p) &= +\frac{\rho_{23}^{(1)}\Pi_{SPP}\Omega_p d_{14}-\rho_{24}^{(1)}\Pi_{SPP}\Omega_p\Phi_{DDI}}{d_{21}\left(d_{13}d_{14}-d_{21}\Phi_{DDI}^2\right)+\Pi_{SPP}^2 d_{14}} \\
\rho_B^{(2)}(\omega_p,\omega_p) &= \frac{i\left(\rho_{22}^{(1)}-\rho_{11}^{(1)}\right)\Omega_p\left(d_{13}d_{14}-d_{21}\Phi_{DDI}^2\right)}{d_{21}\left(d_{13}d_{14}-d_{21}\Phi_{DDI}^2\right)+\Pi_{SPP}^2 d_{14}}.
\end{aligned} \tag{40}$$

Finally, we evaluated the nonlinear fluorescence intensity due to the emission of three photons. Inserting Equation (37) into Equation (29), we obtained three-photon fluorescence intensity (i.e., Kerr intensity) as

$$\begin{aligned} I_{\text{kerr}} &= I_0 \left| \frac{\left(\rho_A^{(3)} + \rho_A^{(3)}\right)}{d_{21}\left(d_{13}d_{14} - d_{21}\Lambda_{DDI}^2\right) + \Lambda_{SPP}^2 d_{14}} \right|^2 \\ \rho_A^{(3)} &= \rho_{23}^{(2)} \Omega_p \Pi_{SPP} d_{14} + i\rho_{24}^{(2)} \Omega_p \Phi_{DDI} \Pi_{SPP}, \\ I_0 &= \frac{4\omega_{21}\omega_{21}\mu_{21}^2}{2c\in_0\in_0^2\sqrt{\in_b}} \\ \rho_B^{(3)} &= i(\rho_{22}^{(2)} - \rho_{11}^{(2)})\Omega_P\left(d_{13}d_{14} - d_{21}\Pi_{SPP}^2\right). \end{aligned} \tag{41}$$

We found that the one-photon, two-photon, and three-photon fluorescence intensity depends on the SPP coupling and DDI couplings.

5. Exciton Decay Rates Due to Dipole-Dipole Interaction

The radiative decay rates γ_{ij} appear in Equations (27)–(35). The radiative decay rate γ_{21} is due to the spontaneous emission of exciton transition from $|2\rangle$ to $|1\rangle$. The decay rate γ_{23} is due to the exciton decay from $|2\rangle$ to $|3\rangle$ because of the exciton-DDI-MNP interaction. Similarly, the decay rate γ_{34} is also due to the exciton decay from $|3\rangle$ to $|4\rangle$ because of the exciton-DDI-QD interaction.

The SPP dispersion relation $\omega_k^{\pm}$ was evaluated in Equation (6) for the upper (+) and lower (−) SPP spectrum. The lower mode of the SPP is responsible for the enhancement of the SPP field. Hence, we consider the lower branch for the decay calculation. The spectrum of the lower branch was denoted as $\omega_k^s = \omega_k^-$. Henceforth, we take out superscript (−) for all terms.

The exciton interaction Hamiltonian with the probe, SPP, and DDI fields can be written in the dipole and rotating wave approximation as

$$H_{\text{int}} = H_0 + H_{ex-PP} + H_{ex-SPP} + H_{ex-DDI}^{MNP} + H_{ex-DDI}^{QD} \tag{42}$$

where the first term is the Hamiltonian for the noninteracting excitons and SPPs. The second term is due to the spontaneous emission coupling term due to the exciton-probe field interaction. The third term is due to the exciton-SPP field interaction. Finally, the fourth and fifth terms are due to the exciton-DDI field interaction. Their expressions were found as

$$\begin{aligned} H_0 &= \sum_{i=1,3} \hbar\omega_{i1}\sigma_{i1} + \sum_k \hbar\omega_k^s p_k^{\dagger} p_k \\ H_{ex-PP} &= \sum_k V_{PP}\left(p_k\sigma_{21}^{\dagger} + p_k^{\dagger}\sigma_{21}\right), \\ H_{ex-SPP} &= \sum_k V_{SPP}\left(p_k\sigma_{21}^{\dagger} + p_k^{\dagger}\sigma_{21}\right) \\ H_{ex-DDI}^{MNP} &= \sum_k V_{DDI}^{MNP}\left(p_k\sigma_{23}^{\dagger} + p_k^{\dagger}\sigma_{23}\right), \\ H_{ex-DDI}^{QE} &= \sum_k V_{DDI}^{QE}\left(p_k\sigma_{34}^{\dagger} + p_k^{\dagger}\sigma_{34}\right) \end{aligned} \tag{43}$$

where σ_{ij} = $|i\rangle\langle j|$ is the exciton creation operator for transition $|i\rangle \leftrightarrow |j\rangle$ where i and j stand for 1, 2, 3, and 4. Meanwhile, the operators $p_{nm,k_z}^{\dagger} p_{nm,k_z}$ are the photon creation and annihilation operators, respectively. The coupling constant appearing in Equation (43) was found as

$$\begin{aligned} V_{PP} &= i\left(\frac{\mu_{21}^2 \hbar\omega_k^s}{2\epsilon_0\epsilon_b V_{QE}}\right)^{1/2} \\ V_{SPP} &= i\left(\frac{\mu_{21}^2 \hbar\omega_k^s}{2\epsilon_0\epsilon_b V_{QE}}\right)^{1/2}[\Pi_{SPP}] \\ V_{DDI}^{MNP} &= i\left(\frac{\mu_{23}^2 \hbar\omega_k^s}{2\epsilon_0\epsilon_b V_{QD}}\right)^{1/2}[\Phi_{DDI}^{MNP}], \\ V_{DDI}^{QE} &= i\left(\frac{\mu_{34}^2 \hbar\omega_k^s}{2\epsilon_0\epsilon_b V_{QE}}\right)^{1/2}\left[\Pi_{DDI}^{QE}\right] \end{aligned} \tag{44}$$

where V_{QE} is the volume of the QE.

The Golden rule method of the quantum mechanical perturbation theory was used to calculate the decay rates. It is written as

$$\Gamma_{ij} = 2\pi\hbar\sum_k \int d\omega_k^s D_k(\omega_k^s)|V_{\text{int}}(\omega_k^s)|^2\delta(\omega_k^s - \varepsilon_{ij}) \tag{45}$$

where V_{int} is the interaction term given in Equation (44) and D_k is the DOS which has been calculated in Equation (8).

Putting Equations (44) and (8) into Equation (45) and doing extensive mathematical manipulations, we obtained the decay rates for exciton as

$$\Gamma_P = \gamma_0 Z_k(\omega_{21}) = \gamma_0\left(\frac{(\omega_p - \omega_{21}) + |g_k|^?}{(\omega_p - \omega_{21})^2}\right), \quad \gamma_0 = \left(\frac{\hbar^3\mu_{21}^2\omega_{21}^3}{\pi\ \epsilon_0\epsilon_b\ \hbar^4 c^3}\right) \tag{46}$$

$$\begin{aligned} \Gamma_{SPP} &= \gamma_0|\Pi_{SPP}|^2 Z_k(\omega_{21}) \\ \Gamma_{DDI}^{MNP} &= \gamma_0|\Lambda_{DDI}^{MNP}|Z_k(\omega_{23}) \\ \Gamma_{DDI}^{QD} &= \gamma_0\left|\Lambda_{DDI}^{QD}\right|^2 Z_k(\omega_{34}) \end{aligned} \tag{47}$$

Here γ_0 is the radiative decay rate when QD is in the vacuum. Please note the following relationship $\gamma_{21} = \Gamma_P + \Gamma_{SPP}$, $\gamma_{23} = \Gamma_{DDI}^{MNP}$ and $\gamma_{34} = \Gamma_{DDI}^{QE}$. Note that the radiative and nonradiative decay rates can have large values when exciton energies ε_{12}, ε_{23}, and ε_{34} are close to ε_{nm}. This is an interesting finding of the paper.

6. Dressed States: Exciton-DDI Coupling

We show in the next section that in the strong coupling between the exciton and DDI polaritons, the single peak in the Kerr absorption splits into two peaks. This phenomenon can be easily explained by the physics of the dressed state. We calculated the energies and eigenfunctions for exciton-DDI polariton Hamiltonian in the strong coupling. Note that the DDI field is acting between states |2> and |3>. For simplicity we denoted |3> and |2> as |a> and |b>, respectively. Here |a> and |b> are the ground state and the excited state of the exciton.

The polariton energy of the kth mode was denoted as $\varepsilon_k = \varepsilon_d$ and it was considered that the energy of the DDI polariton is very close to exciton energy i.e., $\varepsilon_k \sim \varepsilon_{ba}$. This means that only one mode of the polariton was acting with an exciton and we denoted $\varepsilon_k = \varepsilon_{DDI}$. The interaction Hamiltonian between exciton and the DDI polaritons was written as in the dipole and rotating wave approximation can be written as

$$H_T = \varepsilon_a\sigma_{aa} + \varepsilon_b\sigma_{bb} + \varepsilon_d p_k^+ p_k - \hbar\Omega_{DDI}(\sigma_{ba}^+ p_k + \sigma_{ba}^- p_k^+). \tag{48}$$

Here p_k^+ and p_k are the DDI polariton creation and annihilation operators, respectively. Here $|a><a|$ and $|b><b|$ are called the number operators for states $|a>$ and $|b>$, respectively. The first two terms are the noninteracting Hamilton for the exciton and kth mode DDI polariton. The last term is the exciton-polariton interaction Hamiltonian. The constant Ω_{DDI} is called the Rabi frequency for the DDI field.

We then calculated the eigenvalues and eigenfunctions of the total Hamiltonian as

$$H_T|\varphi_k\rangle = \varepsilon_T|\varphi_k\rangle \tag{49}$$

where are ε_T and $|\varphi_k\rangle$ are the eigenvalue and eigenket of the total Hamiltonian H_{int}. We considered two states which have the same number of particles, i.e., n_k. Therefore, we choose $|a,\ n_k\rangle$ and $|b,\ n_k-1\rangle$ states. The first has n_k polaritons and zero excitons and the second has one exciton and (n_k-1) polaritons. Note that both states form $|a,\ n_k\rangle$ and $|b,\ n_k-1\rangle$ form an orthonormal set and they satisfy the orthonormal conditions as follows

$$\langle a,n_k|a,n_k\rangle = 1,\ \langle b,n_k-1|b,n_k-1\rangle = 1,\ \langle a,n_k|b,n_k-1\rangle = 0. \tag{50}$$

The eigenket $|\varphi_k\rangle$ can be expressed as a linear combination of the orthonormal set $|a,\ n_k\rangle$ and $|b,\ n_k-1\rangle$ as

$$|\varphi_k\rangle = \alpha_a|a,n_k\rangle + \alpha_b|b,n_k-1\rangle \tag{51}$$

where α_a and α_b are expansion constants. Inserting Equation (51) into Equation (49) and using Equation (50), we obtained the following equations after some mathematical manipulations as

$$\begin{pmatrix} \{\varepsilon_T-(\varepsilon_a+\varepsilon_{n_k})\} & (\hbar\Omega_a\sqrt{n_k}) \\ -(\hbar\Omega_a\sqrt{n_k})\,\alpha_a & \{\varepsilon_T-(\varepsilon_b+\varepsilon_{n_k-1})\} \end{pmatrix}\begin{pmatrix} \alpha_a \\ \alpha_b \end{pmatrix} = 0. \tag{52}$$

The determinant of the above equation gives the eigenvalues of energy ε_T as

$$\begin{vmatrix} \{\varepsilon_T-(\varepsilon_a+\varepsilon_{n_k})\} & (\hbar\Omega_a\sqrt{n_k}) \\ -(\hbar\Omega_a\sqrt{n_k})\,\alpha_a & \{\varepsilon_T-(\varepsilon_b+\varepsilon_{n_k-1})\} \end{vmatrix} = 0. \tag{53}$$

The above expression reduces to the following equation as

$$\varepsilon_T^{\pm} = \left(\frac{\varepsilon_b+\varepsilon_{n_k-1}+\varepsilon_b+\varepsilon_{n_k}}{2}\right) \pm \frac{1}{2}\sqrt{(\varepsilon_d-\varepsilon_{ba})^2+4n_k\hbar^2|\Omega_{ddi}|^2}. \tag{54}$$

The ε_T^{+} and ε_T^{-} are eigenvalues for eigenkets $|\varphi_T^{+}\rangle$ and $|\varphi_T^{-}\rangle$ and they are called dressed states. Putting, $\varepsilon_{n_k} = n_k\varepsilon_d$ and $\varepsilon_{ba} = \varepsilon_b - \varepsilon_a$ into the above expression, we obtained

$$\varepsilon_T^{\pm} = \left(\frac{\varepsilon_{ba}}{2} + \left\{n_k - \frac{1}{2}\right\}\varepsilon_d\right) \pm \frac{1}{2}\sqrt{(\varepsilon_d-\varepsilon_{ba})^2+4n_k\hbar^2|\Omega_{ddi}|^2}. \tag{55}$$

We then calculated the eigenkets $|\varphi_T^{+}\rangle$ and $|\varphi_T^{-}\rangle$ from Equation (48) for eigenvalues ε_T^{+} and ε_T^{-}. The expansion coefficients $\alpha_a^{\pm}$ and $\alpha_b^{\pm}$ are provided in Equations (56) and (57). Using orthonormalization properties given in Equation (50) and after some mathematical manipulation, we obtained

$$\alpha_b^{\pm} = \left(\frac{\{\varepsilon_T^{\pm}-(\varepsilon_a+\varepsilon_{n_k})\}^2}{\{\varepsilon_T^{\pm}-(\varepsilon_a+\varepsilon_{n_k})\}^2+(\hbar\Omega_{ddi}\sqrt{n_k})^2}\right)^2 \tag{56}$$

$$(\alpha_a^{\pm})^2 = \left(\frac{\hbar\Omega_{ddi}\sqrt{n_k}}{\{\varepsilon_T^{\pm}-(\varepsilon_a+\varepsilon_{n_k})\}^2+(\hbar\Omega_{ddi}\sqrt{n_k})^2}\right)^{1/2}. \tag{57}$$

In most of the experiments, only one polariton was required to excite transition from ground state $|a\rangle$ to the excited state $|b\rangle$. Therefore, we added $n_k = 1$ in the above equations. For this case, we obtained the following expression for the eigenvalues

$$\varepsilon_T^{\pm} = \left(\frac{\delta_{ba}}{2} + \varepsilon_{ba}\right) \pm \frac{1}{2}\sqrt{(\delta_{ba})^2 + 4\hbar^2|\Omega_{ddi}|^2} \tag{58}$$

Here $\delta_{ba} = \varepsilon_{ba} - \varepsilon_d$ is called the detuning parameter in quantum optics. When the DDI polariton energy is in resonance with the exciton transition energy (i.e., $\delta_{ba} = \varepsilon_{ba} - \varepsilon_d = 0$), the above expression reduces to a simple form

$$\varepsilon_T^{\pm} = \varepsilon_{ab} \pm \hbar^2|\Omega_{ddi}|^2. \tag{59}$$

We found that when the polariton energy is in resonance with the exciton energy, the excited state splits into two dressed states $|\varphi_T^{+}\rangle$ and $|\varphi_T^{-}\rangle$. The energy difference between the two dressed state was proportional to the DDI coupling $\Omega_{ddi} = \Omega_P\Phi_{ddi}$.

7. Results and Discussion

In this section, we discuss the numerical simulations performed on the SPP dispersion relation, DOS of SPPs, and Kerr absorption intensity. The effect of the SPP and DDI couplings on the Kerr absorption intensity are also investigated. In our numerical simulations, we measured all physical qualities related to energy (frequency) are measured with respect to the decay rate γ_2. Some of the examples for the energy (frequency) physical quantities are Rabi frequency (Ω_P), exciton frequencies, probe detuning, and decay rates. In our numerical simulations, we used $\gamma_{23}/\gamma_{21} = \gamma_{34}/\gamma_{21}=1$ and $\Omega_P/\gamma_{21} = 1$. The probe detuning (δ_p) and DDI detuning (δ_d) were measured with respect to the decay rate γ_{21}.

We first calculated the dispersion relation for the SPPs and the density of states for the SPPs. Here the dispersion relation means the relation between the energy and wavevector of polaritons. The dispersion relation is calculated in Equation (6) and the DOS is calculated in Equation (9). The results are presented in Figure 3a,b. The normalized SPP energy ($\omega_k^{\pm}/\omega_p$) is plotted in Figure 3a as a function of a normalized wave vector (k/k_p). One can see that the SPP dispersion relation has a bandgap. The upper band does not participate in the enhancements of the Kerr intensity because its properties are similar to photons. On the other hand, the lower band plays an important role in plasmonics. The behavior of this band is a mixture of plasmons and photons (i.e., SPPs). These materials can be called the polaritonic bandgap materials, since they have a polaritonic bandgap in their band structure. They are similar to photonic bandgap materials that have a bandgap in their photonic dispersion relations.

The results for DOS are plotted in Figure 3b as a function of the normalized energy ($\omega_k^{\pm}/\omega_p$). We found that the DOS has large values near the band edges. This means that a large number of SPPs are located near the band edges. When a QE lies near the MNP, excitons of the QE interact with the SPPs of MNPs. However, if the energy of the excitons lies near the band edges, there will be huge exciton-SPP coupling since there are the huge number of polaritons are located near the band edges.

Next, we studied the effect of the SPP field on the Kerr absorption intensity (I_{kerr}/I_0). The results are plotted in Figure 4a as a function of normalized detuning ($\delta_p/\gamma_{21} = \delta_{21}/\gamma_{21}$) for different values of the SPP coupling (Π_{SPP}). For Figure 4a, we have considered that DDI coupling is absent. The solid, dash, and dash-dotted lines were plotted for the detuning parameter $\Pi_{SPP} = 0.1$, $\Pi_{SPP} = 0.7$, and $\Pi_{SPP} = 1.5$, respectively. Note that the SPP coupling is unitless. Other parameters are taken as $\Phi_{DDI} = 0.0$. Her the DDI coupling is also unitless. Note that at zero detuning (i.e., $\delta_p/\gamma_{21} = 0$), the Kerr intensity has a peak. The zero detuning means that the probe field frequency is in resonance with the exciton frequency ω_{21} (i.e., $\omega_p = \omega_{21}$).

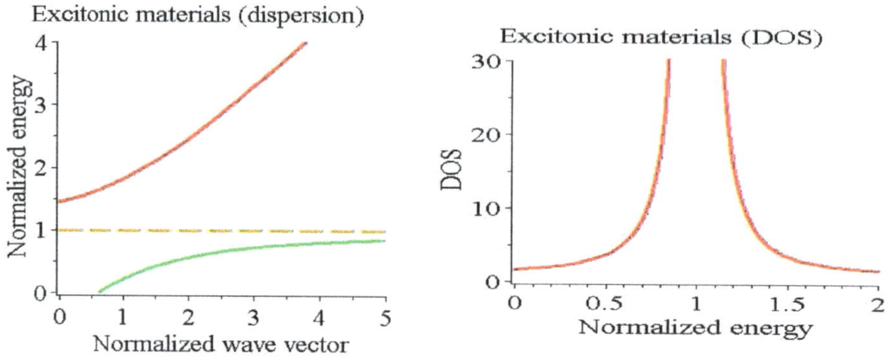

(**a**) (**b**)

Figure 3. (**a**) (left): The band structure of the SPPs is plotted as a function of the normalized energy and wavevectors. (**b**) (right): The DOS of the SPPs is plotted as a function of the normalized energy.

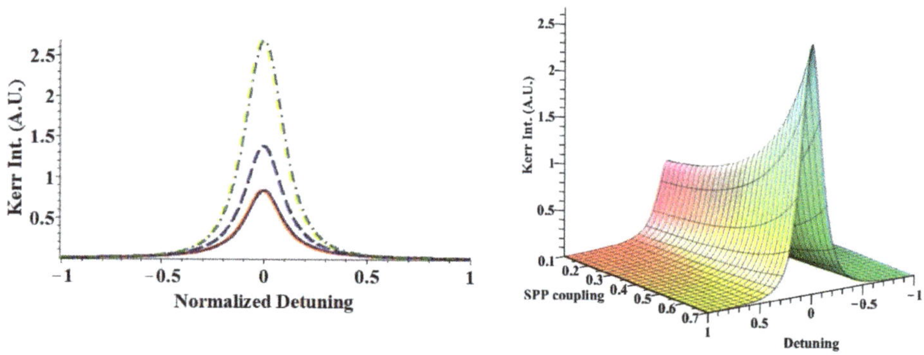

(**a**) (**b**)

Figure 4. (**a**) (left): The Kerr absorption intensity (I_{kerr}/I_0) is plotted as a function of the normalized probe detuning $\delta_p = \delta_{21}$. The solid, dash, and dash-dotted lines are plotted for $\Pi_{SPP} = 0.1$, $\Pi_{SPP} = 0.7$, and $\Pi_{SPP} = 0.7$, respectively. (**b**) (right): The Kerr absorption intensity (I_{kerr}/I_0) is plotted as a function of the SPP coupling Π_{SPP} (normalized unit) and the probe detuning (normalized unit).

We found in Figure 4a that as the SPP coupling increases, the Kerr intensity also increases. See the dotted and dash-dotted lines in the figure. This means that there is a large enhancement in the Kerr intensity due to the presence of SPPs. The enhancement occurred because the SPP coupling appears in the numerator of the Kerr intensity expression. See Equation (41). Further, we predicted that the enhancement has smaller values when the probe field is not in resonance with the exciton energy.

We also plotted a three-dimensional figure for the Kerr intensity (I_{kerr}/I_0) as a function of the SPP coupling (Π_{SPP}) and the probe detuning (δ_p/γ_{21}). The results are shown in Figure 4b. Note that as the SPP coupling increases, the Kerr effect also increases. There is a huge enhancement in the Kerr intensity due to the nonreality of the system.

The enhancement in both Figure 4a,b in the Kerr absorption intensity can be explained as follows. When the SPP field (i.e., MNS) is absent (i.e., $\Pi_{SPP} = 0$), the Kerr intensity is due to the contribution from three probe photons. However, the Kerr intensity has an extra contribution due to the SSP coupling. The extra contribution to the Kerr intensity is due to the three polaritons produced by the SPP field. The SPP contributions to the Kerr intensity are many times larger than the probe photons. In summary, we can say that the present finding can be used to fabricate nanosensor devices for nanotechnology and nanomedical applications by measuring the enhancement of the Kerr intensity.

Next, we investigated the effect of the DDI coupling (Φ_{DDI}) on the Kerr intensity (I_{kerr}/I_0). The results are plotted in Figure 5a as a function of the probe detuning (δ_p/γ_{21}). The solid, dash, and dash-dotted lines were plotted for the detuning parameter $\Phi_{DDI} = 0.1$, $\Phi_{DDI} = 0.5$, and $\Phi_{DDI} = 2.0$ respectively. The SPP coupling parameter was taken as $\Pi_{SPP} = 0.1$. The SPP and DDI coupling parameters were unitless. One can see from the figure that for the DDI coupling $\Phi_{DDI} = 0.5$, the Kerr intensity was enhanced. When the DDI coupling was further increased to $\Phi_{DDI} = 2.0$, the Kerr intensity decreased, and one peak split into two peaks.

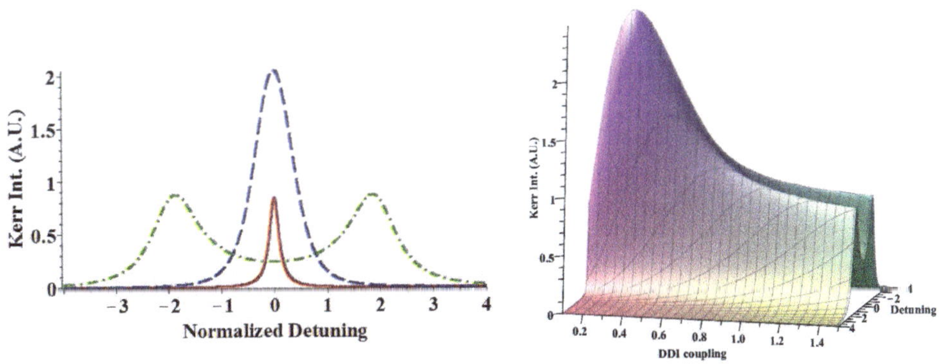

(**a**) (**b**)

Figure 5. (**a**) (left): The Kerr intensity (I_{kerr}/I_0) plotted as a function of the probe detuning $(\delta_p = \delta_{21})$. The solid, dash and dash-dotted lines are plotted for the detuning parameter $\Phi_{DDI} = 0.1$, $\Phi_{DDI} = 0.5$, and $\Phi_{DDI} = 2.0$, respectively. (**b**) (right): The Kerr intensity (I_{kerr}/I_0) plotted as a function of the probe detuning $(\delta_p = \delta_{21})$ and the DDI coupling Φ_{DDI}.

To make the effect of the DDI coupling on the Kerr intensity clearer, we plotted a three-dimensional figure of the Ker nonlinearity in Figure 5b. Note that when the DDI coupling is in the weak coupling limit, the Kerr intensity increases with the DDI coupling. It reaches a maximum value at a certain value of the DDI (i.e., $\Phi_{DDI} = 0.4$). When the DDI coupling increases further, the peak of the Kerr intensity splits into two-peaks. In summary, we can say that when $\Phi_{DDI} < 1$.

The Kerr intensity has one peak. On the other hand, when $\Phi_{DDI} > 1$, the Kerr intensity splits from one peak to two peaks. The condition $\Phi_{DDI} < 1$ is called the weak coupling limit, whereas the condition $\Phi_{DDI} > 1$ is called the strong coupling limit. This is an interesting finding of the paper.

We then defined the weak and strong DDI coupling limits. In the literature, the weak coupling limit is defined when the Rabi frequency of the DDI electric field (Ω_{DDI}) is smaller than the decay rate (i.e., $\Omega_{DDI} < \gamma_{21}$). Here the DDI Rabi frequency is defined as $\Omega_{DDI} = \Phi_{DDI}\Omega_P$. In this case, the DDI coupling Φ_{DDI} is in the weak coupling limit. On the other hand, the strong coupling limits are defined when the DDI Rabi frequency is larger than the decay rate ($\Omega_{DDI} < \gamma_{21}$). In this case, the DDI coupling is in the strong coupling limit. The above definitions are the approximate definition and are not applied in all problems. Note that in Figure 5b, the DDI coupling approximately satisfies the weak and strong coupling limit criteria for the enhancement and the splitting of the peak.

The splitting from one peak to two peaks due to the strong DDI coupling limit is explained by using the physics of dressed states, as discussed in Section 5. In the absence of the DDI coupling, the Kerr intensity is due to the three photons emission from the transition $|2\rangle \leftrightarrow |1\rangle$. On the other hand, in the strong DDI coupling limit the excited state $|2\rangle \leftrightarrow |3\rangle$ splits into two dressed states $|2_-\rangle \leftrightarrow |3\rangle$ and $|2_+\rangle \leftrightarrow |3\rangle$. Therefore, the Kerr intensity emissions have two peaks due to transitions $|2_-\rangle \leftrightarrow |1\rangle$ and $|2_+\rangle \leftrightarrow |1\rangle$. We have also found that the distance between the peaks increases as the DDI coupling increases. This is because the energy splitting is directly related to the DDI coupling. See Equation (59) for further details. In summary, we can say that the splitting of one peak (ON) into two peaks (OFF) can be used to fabricate nanoswitching devices for nanotechnology and nanomedical applications.

Finally, we provide comments on the giant nonlinearity found due to the Kerr effect in the present work. We have shown that there is a huge (giant) enhancement in Kerr nonlinearity due to the presence of SPPs. We have also found the due to the weak DDI coupling, there is also an enhancement in the Kerr nonlinearity. The giant Kerr nonlinearity found in plasmonic nanohybrids is of great importance in the context of quantum information theory and its applications. In particular, the photon/phonon blockade can appear in systems involving high Kerr-type nonlinearities [56,57]. Moreover, such systems can be applied in the maximally entangled state's generation [58] and as the source of quantum steering [59]. Quite recently, the model involving Kerr nonlinearities was considered in the context of the PT-symmetry breaking [60]. The present study can also be used for optical pumping [61] and photon transparency [62].

8. Conclusions

A theory of the Kerr nonlinear tensity was developed by using the many-body quantum mechanical density matrix method for plasmonic nanohybrids. We showed that the Kerr intensity enhances in the weak dipole-dipole coupling limits. On the other hand, in the strong dipole-dipole coupling limit, the single peak in the Kerr intensity splits into two peaks. Further, we found that the Kerr nonlinearity is also enhanced due to the SPP coupling. Next, we determined the spontaneous decay rates are enhanced due to the dipole-dipole coupling. The enhancement of the Kerr intensity due to the surface plasmon polaritons can be used to fabricate nanosensors. The splitting of one peak (ON) two peaks (OFF) can be used to fabricate the nanoswitches for nanotechnology and nanomedical applications.

Funding: This research was funded by the Natural Sciences and Engineering Research Council of Canada (NSERC) for the research grant (RGPIN-2018-05646).

Institutional Review Board Statement: Not applicable.

Informed Consent Statement: Not applicable.

Acknowledgments: The author is thankful to Grant Brassem for correcting English of the paper. The author is also thankful to the Discovery grant from the Natural Sciences and Engineering Research Council of Canada (NSERC) for the research grant (RGPIN-2018-05646).

Conflicts of Interest: Author does not have any conflict of interest.

References

1. Artuso, R.D.; Bryant, G.W. Strongly coupled quantum dot-metal nanoparticle systems: Exciton-induced transparency, discontinuous response, and suppression as driven quantum oscillator effects. *Phys. Rev. B* **2010**, *82*, 195419. [CrossRef]
2. Yannopapas, V.; Paspalakis, E. Giant enhancement of dipole-forbidden transitions via lattices of plasmonic nanoparticles. *J. Mod. Opt.* **2015**, *62*, 1435–1441. [CrossRef]
3. Tame, M.S.; McEnery, K.R.; Özdemir, Ş.K.; Lee, J.; Maier, S.A.; Kim, M.S. Quantum plasmonics. *Nat. Phys.* **2013**, *9*, 329–340. [CrossRef]
4. Törmä, P.; Barnes, W.L. Strong coupling between surface plasmon polaritons and emitters: A review. *Rep. Prog. Phys.* **2015**, *78*, 013901. [CrossRef] [PubMed]
5. Achermann, M. Exciton−Plasmon Interactions in Metal−Semiconductor Nanostructures. *J. Phys. Chem. Lett.* **2010**, *1*, 2837. [CrossRef]
6. Singh, M.R.; Guo, J.; Fanizza, E.; Dubey, M. Anomalous photoluminescence quenching in metallic nanohybrids. *J. Phys. Chem. C* **2019**, *123*, 10013–10020. [CrossRef]

7. Balakrishnan, S.; Najiminaini, M.; Singh, M.R.; Carson, J.J.L. A study of angle dependent surface plasmon polaritons in nano-hole array structures. *J. Appl. Phys.* **2016**, *120*, 034302. [CrossRef]
8. Antón, M.A.; Carreño, F.; Melle, S.; Calderon, O.G.; Granado, E.C.; Singh, M.R. Optical pumping of a single hole spin in ap-doped quantum dot coupled to a metallic nanoparticle. *Phys. Rev. B* **2013**, *87*, 195303. [CrossRef]
9. Racknor, C.; Singh, M.R.; Zhang, Y.; Birch, D.J.S.; Chen, Y. Energy transfer between a biological labelling dye and gold nanorods. *Methods Appl. Fluoresc.* **2013**, *2*, 015002. [CrossRef]
10. Singh, M.R.; Black, K. Anomalous dipole-dipole interaction between ensemble of quantum emitters in metallic nanoparticle hybrids. *J. Phys. Chem. C* **2018**, *122*, 26584–26591. [CrossRef]
11. Yudson, V.I.; Singh, M.R. Lattice-gas model for electron-hole coupling in disordered media. *Phys. Rev. B* **1998**, *58*, 16202. [CrossRef]
12. Singh, M.R.; Guo, J.; Chen, J. A Theoretical Study of Fluorescence Spectroscopy of Quantum Emitters Coupled with Plasmonic Dimer and Trimer. *J. Phys. Chem. C* **2019**, *123*, 17483–17490. [CrossRef]
13. Singh, M.R. The effect of the dipole–dipole interaction in electromagnetically induced transparency in polaritonic band gap materials. *J. Mod. Opt.* **2007**, *54*, 1739–1757. [CrossRef]
14. Singh, M.R. Dipole-Dipole Interaction in Photonic-Band-Gap Materials Doped with Nanoparticles. *Phys. Rev. A* **2007**, *75*, 043809. [CrossRef]
15. Schmidt, H.; Imamoglu, A. Giant Kerr nonlinearities obtained by electromagnetically induced transparency. *Opt. Lett.* **1996**, *21*, 1936–1938. [CrossRef] [PubMed]
16. Wang, H.; Goorskey, D.; Xiao, M. Enhanced Kerr Nonlinearity via Atomic Coherence in a Three-Level Atomic System. *Phys. Rev. Lett.* **2001**, *87*, 073601. [CrossRef] [PubMed]
17. Yan, X.-A.; Wang, L.-Q.; Yin, B.-Y.; Jiang, W.-J.; Zheng, H.-B.; Song, J.-P.; Zhang, Y.-P. Effect of spontaneously generated coherence on Kerr nonlinearity in a four-level atomic system. *Phys. Lett. A* **2008**, *372*, 6456–6460. [CrossRef]
18. Khoa, D.X.; Van Doai, L.; Son, D.H.; Bang, N.H. Enhancement of self-Kerr nonlinearity via electromagnetically induced transparency in a five-level cascade system: An analytical approach. *J. Opt. Soc. Am. B* **2014**, *31*, 1330–1334. [CrossRef]
19. Ren, J.; Chen, H.; Gu, Y.; Zhao, D.; Zhou, H.; Zhang, J.; Gong, Q. Plasmon-enhanced Kerr nonlinearity via subwavelength-confined anisotropic Purcell factors. *Nanotechnology* **2016**, *27*, 425205. [CrossRef] [PubMed]
20. Doai, L.V.; Khoa, D.X.; Bang, N.H. EIT enhanced self-Kerr nonlinearity in the three-level lambda system under Doppler broadening. *Phys. Scr.* **2015**, *90*, 45502. [CrossRef]
21. Sheng, J.; Yang, X.; Wu, H.; Xiao, M. Modified self-Kerr-nonlinearity in a four-level N-type atomic system. *Phys. Rev. A* **2011**, *84*, 053820. [CrossRef]
22. Singh, M.R. Two-photon absorption in photonic nanowires made from photonic crystals. *J. Opt. Soc. Am. B* **2009**, *26*, 1801–1807. [CrossRef]
23. Berland, K.; So, P.; Gratton, E. Two-photon fluorescence correlation spectroscopy: Method and application to the intracellular environment. *Biophys. J.* **1995**, *68*, 694–701. [CrossRef]
24. Jung, J.-M.; Yoo, H.-W.; Stellacci, F.; Jung, H.-T. Two-Photon Excited Fluorescence Enhancement for Ultrasensitive DNA Detection on Large-Area Gold Nanopatterns. *Adv. Mater.* **2010**, *22*, 2542–2546. [CrossRef]
25. Li, X.; Kao, F.-J.; Chuang, C.-C.; He, S. Enhancing fluorescence of quantum dots by silica-coated gold nanorods under one- and two-photon excitation. *Opt. Express* **2010**, *18*, 11335–11346. [CrossRef] [PubMed]
26. Gao, D.; Agayan, R.R.; Xu, H.; Philbert, M.A.; Kopelman, R. Nanoparticles for Two-Photon Photodynamic Therapy in Living Cells. *Nano Lett.* **2006**, *6*, 2383–2386. [CrossRef]
27. Yuan, H.; Khoury, C.G.; Hwang, H.; Wilson, C.M.; A Grant, G.; Vo-Dinh, T. Gold nanostars: Surfactant-free synthesis, 3D modelling, and two-photon photoluminescence imaging. *Nanotechnology* **2012**, *23*, 075102. [CrossRef]
28. Albota, M.; Beljonne, D.; Brédas, J.L.; Ehrlich, J.E.; Fu, J.Y.; Heikal, A.A.; Hess, S.E.; Kogej, T.; Levin, M.D.; Marder, S.R.; et al. Design of Organic Molecules with Large Two-Photon Absorption Cross Sections. *Science* **1998**, *281*, 1653–1656. [CrossRef]
29. Singh, M.R.; Persaud, P.D. Dipole—dipole Interaction in two-photon spectroscopy of metallic nanohybrids. *J. Phys. Chem. C.* **2020**, *124*, 6311–6320. [CrossRef]
30. Singh, M.R.; Persaud, P.D.; Yastrebov, S. A study of two-photon florescence in metallic nanoshells. *Nanotechnology* **2020**, *31*, 265203. [CrossRef] [PubMed]
31. Fejer, M.M. Nonlinear Optical Frequency Conversion. *Phys. Today* **1994**, *47*, 25–32. [CrossRef]
32. Cerullo, G.; De Silvestri, S. Ultrafast optical parametric amplifiers. *Rev. Sci. Instrum.* **2003**, *74*, 1–18. [CrossRef]
33. Sugioka, K. Progress in ultrafast laser processing and future prospects. *Nanophotonics* **2017**, *6*, 393–413. [CrossRef]
34. Potma, E.O.; De Boeij, W.P.; Wiersma, D.A. Nonlinear coherent four-wave mixing in optical microscopy. *J. Opt. Soc. Am. B* **2000**, *17*, 1678–1684. [CrossRef]
35. Yannopapas, V.; Paspalakis, E. Optical properties of hybrid spherical nanoclusters containing quantum emitters and metallic nanoparticles. *Phys. Rev. B* **2018**, *97*, 205433. [CrossRef]
36. Tohari, M.M.; Lyras, A.; AlSalhi, M.S. Giant Self-Kerr Nonlinearity in the Metal Nanoparticles-Graphene Nanodisks-Quantum Dots Hybrid Systems Under Low-Intensity Light Irradiance. *Nanomaterials* **2018**, *8*, 521. [CrossRef] [PubMed]
37. Terzis, A.; Kosionis, S.; Boviatsis, J.; Paspalakis, E. Nonlinear optical susceptibilities of semiconductor quantum dot—Metal nanoparticle hybrids. *J. Mod. Opt.* **2015**, *63*, 451–461. [CrossRef]

38. Liu, Q.; He, X.; Zhao, X.; Ren, F.; Xiao, X.; Jiang, C.; Zhou, X.; Lu, L.; Zhou, H.; Qian, S.; et al. Enhancement of third-order nonlinearity in Ag-nanoparticles-contained chalcohalide glasses. *J. Nanoparticle Res.* **2011**, *13*, 3693–3697. [CrossRef]
39. Kelly, K.L.; Coronado, E.; Zhao, L.L.; Schatz, G.C. The Optical Properties of Metal Nanoparticles: The Influence of Size, Shape, and Dielectric Environment. *J. Phys. Chem. B* **2003**, *107*, 668–677. [CrossRef]
40. Singh, M.R. Enhancement of the second-harmonic generation in a quantum dot–metallic nanoparticle hybrid system. *Nanotechnolgy* **2013**, *24*, 125701. [CrossRef] [PubMed]
41. Cox, J.D.; Singh, M.R.; Von Bilderling, C.; Bragas, A.V. A Nonlinear Switching Mechanism in Quantum Dot and Metallic Nanoparticle Hybrid Systems. *Adv. Opt. Mater.* **2013**, *1*, 460–467. [CrossRef]
42. Singh, M.R. Theory of all-optical switching based on the Kerr nonlinearity in metallic nanohybrids. *Phys. Rev. A* **2020**, *102*, 013708. [CrossRef]
43. Singh, M.R.; Yastrebov, S. Switching and Sensing Using Kerr Nonlinearity in Quantum Dots Doped in Metallic Nanoshells. *J. Phys. Chem. C* **2020**, *124*, 12065–12074. [CrossRef]
44. Mazenko, G.F. *Quantum Statisitcal Mechanics*; John Wiley and Sons Inc.: New York, NY, USA, 2000; Section 5.8.
45. Kittel, C. *Introduction to Solid State Physics*, 6th ed.; John Wiley and Sons Inc.: New York, NY, USA, 1996; Chapter 13.
46. Ali Omar, M. *Elementary Solid State Physics*; Addison-Wesley: New York, NY, USA, 1993; Section 8.11.
47. Gerstein, J.L.; Smith, F.W. *The Physics and Chemistry of Materials*; John Wiley and Sons Inc.: New York, NY, USA, 2001; Chapter 15.
48. Eyring, H. *Statistical Mechanics and Dynamics*; Dover: New York, NY, USA, 1952.
49. Lorentz, H. *The Theory of Electrons*; Dover: New York, NY, USA, 1952.
50. Novotny, L.; Hecht, B. *Principle of Nano-Optics*; Cambridge University Press: Cambridge, UK, 2006.
51. Sarid, D.; Challener, W.A. *Modern Introduction to Surface Plasmons: Theory, Mathematica Modeling, and Applications*; Cambridge University Press: Cambridge, UK, 2010.
52. Hanamura, E.; Kawabe, Y.; Yamanaka, A. *Quantum Nonlinear Optics*; Springer: Tokyo, Japan, 2007.
53. Boyd, R.W. *Nonlinear Optics*, 3rd ed.; Academic Press: New York, NY, USA, 2008.
54. Singh, M.R. *Electronic, Photonic, Polaritonic and Plasmonic Materials*; Wiley Custom: Toronto, ON, Canada, 2014.
55. Scully, M.O.; Zubairy, M.S. *Quantum Optics*; Cambridge University Press: London, UK, 1997.
56. Leoński, W.; Tanaś, R. Possibility of producing the one-photon state in a kicked cavity with a nonlinear Kerr medium. *Phys. Rev. A* **1994**, *49*, R20–R23. [CrossRef]
57. Imamoğlu, A.; Schmidt, H.; Woods, G.; Deutsch, M. Strongly Interacting Photons in a Nonlinear Cavity. *Phys. Rev. Lett.* **1997**, *79*, 1467–1470. [CrossRef]
58. Leoński, W.; Miranowicz, A. Kerr nonlinear coupler and entanglement. *J. Opt. B Quantum Semiclass. Opt.* **2004**, *6*, S37. [CrossRef]
59. Kalaga, J.K.; Leoński, W.; Szczęśniak, R. Quantum steering and entanglement in three-mode triangle Bose–Hubbard system. *Quantum Inf. Process.* **2017**, *16*, 265. [CrossRef]
60. Peřina, J., Jr.; Lukš, A.; Kalaga, J.K.; Leoński, W.; Miranowicz, A. Non-classical light at exceptional points of a quantum PT-symmetric two-mode system. *Phys. Rev. A* **2019**, *100*, 53820. [CrossRef]
61. Antón, M.; Carreño, F.; Melle, O.S.; Calderón, E.; Cabrera-Granado, M. Singh Transparency in semiconductor-metal nanoparticle hybrid system. *Phys. Rev. B* **2013**, *87*, 195303. [CrossRef]
62. Singh, M.R. Photon transparency in metallic photonic crystals doped with an ensemble of nanoparticles. *Phys. Rev. A* **2009**, *79*, 013826. [CrossRef]

MDPI
St. Alban-Anlage 66
4052 Basel
Switzerland
Tel. +41 61 683 77 34
Fax +41 61 302 89 18
www.mdpi.com

Symmetry Editorial Office
E-mail: symmetry@mdpi.com
www.mdpi.com/journal/symmetry

www.ingramcontent.com/pod-product-compliance
Lightning Source LLC
Chambersburg PA
CBHW061519210926
51850CB00008B/1969